Confining Electrochemistry to Nanopores

Detection Science Series

Titles in the Series:

1: Sensor Technology in Neuroscience
2: Detection Challenges in Clinical Diagnostics
3: Advanced Synthetic Materials in Detection Science
4: Principles and Practice of Analytical Techniques in Geosciences
5: Microfluidics in Detection Science: Lab-on-a-chip Technologies
6: Electrochemical Strategies in Detection Science
7: Peroxynitrite Detection in Biological Media: Challenges and Advances
8: Biological Fluid-Surface Interactions in Detection and Medical Devices
9: Advanced Environmental Analysis: Applications of Nanomaterials, Volume 1
10: Advanced Environmental Analysis: Applications of Nanomaterials, Volume 2
11: Quenched-phosphorescence Detection of Molecular Oxygen: Applications in Life Sciences
12: Carbon-based Nanomaterials in Analytical Chemistry
13: Forensic Analytical Methods
14: Immunosensors
15: Analytical Electrogenerated Chemiluminescence: From Fundamentals to Bioassays
16: *In Situ* Analysis of Cellular Functional Molecules
17: Confining Electrochemistry to Nanopores

How to obtain future titles on publication:
A standing order plan is available for this series. A standing order will bring delivery of each new volume immediately on publication.

For further information please contact:
Book Sales Department, Royal Society of Chemistry, Thomas Graham House, Science Park, Milton Road, Cambridge, CB4 0WF, UK
Telephone: +44 (0)1223 420066, Fax: +44 (0)1223 420247
Email: booksales@rsc.org
Visit our website at www.rsc.org/books

Confining Electrochemistry to Nanopores

Edited By

Yi-Tao Long
Nanjing University, China
Email: yitaolong@nju.edu.cn

ROYAL SOCIETY OF CHEMISTRY

Detection Science Series No. 17

Print ISBN: 978-1-78801-271-3
PDF ISBN: 978-1-78801-326-0
EPUB ISBN: 978-1-83916-090-5
Print ISSN: 2052-3068
Electronic ISSN: 2052-3076

A catalogue record for this book is available from the British Library

The Royal Society of Chemistry is a charity, registered in England and Wales, Number 207890, and a company incorporated in England by Royal Charter (Registered No. RC000524), registered office: Burlington House, Piccadilly, London W1J 0BA, UK, Telephone: +44 (0) 20 7437 8656.

For further information see our website at www.rsc.org

Printed in the United Kingdom by CPI Group (UK) Ltd, Croydon, CR0 4YY, UK

Preface

Nanopore electrochemistry refers to the promising measurement science based on elaborate pore structures that offer a well-defined geometric confined space to adopt and characterize single entities including single cells, single particles, and even single molecules, by electrochemical technology. The electrochemical confined effect within the nanopore displays the ability to achieve single entity discrimination by focusing energy (*e.g.*, electrochemical, light, and magnetic energies, *etc.*) into small areas, converting the intrinsic properties of single entities into electrochemical read-outs. Furthermore, the excellent resolution of confined nanopore technology also permits the possibility to resolve the transient signals for further revealing the information of single biomolecules dynamics. The chemical confinement inside nanopore provides the advanced electrochemically confined effects to convert transient information of chemical groups into the enhanced ion current signals. Recently, the nanopore electrochemistry has been further applied to biophysics, chemical biology, disease diagnostics, and other advanced disciplines by pushing the limitation of detection to micro and nano scale. In this book, we summarize the nanopore electrochemistry from the following three main areas. First, a nanopore-based single biomolecule sensing interface with chemical group precision is described, including the design of a single biomolecule interface, and instrumentation, with applications such as nucleotide detection, peptide discrimination, photo-responsive molecule analysis, and protein sequencing. Then, the inorganic material-based nanopore is summarized along with its fabrication process and application. Last, the glass nanopipette, which is a relatively feasible and cheap nanopore, is introduced to fill the gap of the complexity of the inorganic nanopore fabrication process. Such an electrochemical confinement enables the monitoring of biomolecule interactions and the electron-transfer process in single living cells with

Detection Science Series No. 17
Confining Electrochemistry to Nanopores
Edited by Yi-Tao Long

Published by the Royal Society of Chemistry, www.rsc.org

high resolution and negligible cell damage. With the advents of advanced measurement mechanism, instrumentation, and data algorithms, the electrochemically confined nanopore is undoubtedly an exciting and promising field. We expect the next avenue for the wide application of nanopore electrochemistry in a variety of disciplines, leading us to explore the new chemistry at a much smaller scale. Our hope is that this book will be useful to all interested in Nanopore Electrochemistry. We sincerely thank all authors who contributed to this book. We also thank our coworkers, colleagues, current and former group members for their contributions in Nanopores to advance this field.

Yi-Tao Long

Contents

Detection Science Series No. 17
Confining Electrochemistry to Nanopores
Edited by Yi-Tao Long

Published by the Royal Society of Chemistry, www.rsc.org

CHAPTER 1

Introduction

YI-TAO LONG

Nanjing University, Nanjing, China
Email: yitaolong@nju.edu.cn

This book aims to summarize the advancement of nanopore electrochemistry, from the fundamental sensing mechanism of pore–substrate interactions within a confined space to the exploration of rich single-molecule information from electrochemical signals. We present the single nanopore for single molecule sensing based on a comprehensive understanding of nanopore electrochemistry.[1,2] Two decades ago, researchers developed single peptide sensing using nanopore electrochemistry. The approach was simple but the scope "beyond sequencing of DNA" was foreseeable. Then the conceptual advances occurred, making it possible for single-molecule sensing in a label-free manner *via* "electrochemically confined space". Not only the biomolecule nanopore interface (*e.g.* protein channels) but also inorganic materials (*e.g.* graphene, Si, SiN_x) could build such a confined space. The nanopore sensing interface could be treated as a miniaturized electrode, it is feasible to manipulate every sensing element inside the biological nanopore sensing interface with chemical group precision. Therefore, nanopore electrochemistry achieved single molecule detection in a high-throughput manner with high amperometric resolution without labeling. Moreover, other electrochemical confined effects such as the confinement of biological interactions, the electron-transfer process, and the light inside nanopores, were also demonstrated. These confined effects provide the mutual information of a single molecule from the simultaneously recording of the electric and optical signals. The advancement of instrumentation with sufficient temporal-spatial resolution is also critical to sensitive electrochemical measurement. Many

Detection Science Series No. 17
Confining Electrochemistry to Nanopores
Edited by Yi-Tao Long

Published by the Royal Society of Chemistry, www.rsc.org

studies have demonstrated that the ions inside the nanopore confined space act as the smallest probes, since their vibrations demonstrate the intrinsic properties of analytes. Therefore, the application of time-frequency analysis of ionic current contributes to the appearance of a "single-molecule ionic spectrum", which enables the mapping of the single molecule weak interactions, such as hydrogen bonds, the Van der Waals force and the electrostatic effect. It is undeniable that nanopore electrochemistry, which shows the capability of addressing entirely new biological questions, is an exciting area for future research.

The understanding of physics, chemistry and the life sciences is determined by measurement science, deepening the precision and depth of observation. In order to process at ultra-fine resolution, for example at single-molecule level, a compatible sensing tool should be used to process the analytical methodologies. At this point, the single-molecule method has attracted increasing attention and has gradually become widespread in chemistry and biology. Single-molecule analysis, in terms of biology, has the capability to reveal brand new biological questions since it provides rich heterogeneities and stochastic information on dynamic biological systems.[3–5] Fundamentally, the sensing performance is mainly determined by the sensing interface. The sensing domain comprises reactive molecules for recognition and/or functional materials for signal amplification. Commonly, the sensing domain of the nanopore provides the strongest interaction with the bypass analytes; each substrate provides a unique interaction that turns out as signature readouts. Traditionally, the macroscale sensing domain has been applied in a wide range of versatile detecting cases. However, general sensing information can no longer support exploring fundamental chemical and biological reactions at the molecular level; as a result, single-molecule detection resolution is urgently needed. In order to push the limits of detection to the single molecule level, a comparable miniaturized sensor interface is a must. Adapted from traditional macroscale sensors, micro- or even nanoscopic sensors have been developed. We hereby think to turn over and roll the conventional sensing interface into a small structure for the sake of reducing the sensing domain of the sensor (Figure 1.1). Inspired by nature where membrane proteins on either eukaryotic or prokaryotic cells, acting as nanopores that are capable of probing the surrounding substrates and then selectively uptaking what is required, an artificial system has been created to accommodate such protein nanopores. Therefore, we propose that the nanopore is one of the most suitable structures for the sensing interface for single-molecule measurements as it accommodates a single molecule inside. The nanopore technique provides a well-defined three-dimensional (3D) structure that is comparable to the size of most target single molecules. The single entity is initially fitted into a proper spatial alignment, *i.e.* the nanopore-based sensing interface,[6] then an external stimulus is applied to the analyte, *i.e.* a bias voltage. Finally, the electrochemical readouts are acquired to be decoded that contain abundant single molecule information. In contrast to the conventional macroscopic strategy to dilute towards the single-molecule level, the nanopore provides a

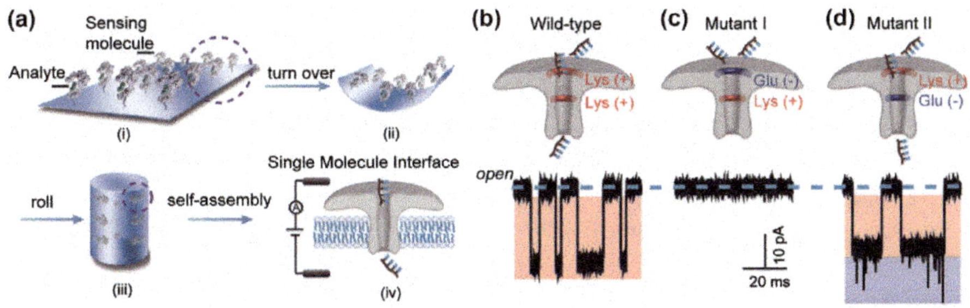

Figure 1.1 (a) A typical macro sensing interface undergoing turn-over, roll and finally self-assembly into a so-called single-molecule interface. Also, single-biomolecule interface manipulation of the aerolysin interaction with poly(dA)$_4$. (b) Wild-type aerolysin and (c) mutant I aerolysin with one amino acid at the nanopore opening modified with a negatively charged induced decreased capture rate. (d) Mutant II aerolysin with one amino acid at the nanopore lumen modified with a negative charged demonstrated stronger interaction.

confined space to accommodate only one single molecule from the bulk solution, which eliminates the assembly effect.[7,8] When a constant bias voltage is applied across the nanopore system, any accommodation in the substrate in single-molecule fashion offers unique electrochemical behavior that correlates with the spatial and temporal information. Once the sensing interface is compatible with the geometrical dimension of the bypass substrates, the biophysical properties of the analyzed substrates (*i.e.* physical dimension, polarity, charge state, dipole moment, *etc.*) can be revealed from the nanopore electrochemistry readouts. Herein, the single-molecule sensing interface provides a confinement for the analytes with different geometries and sizes.[1]

The pore-forming membrane protein, composed of several sub-units, is regarded as a single biomolecule interface (Figure 1.2), *i.e.* biological nanopore sensors. To achieve relatively high selectivity sensing of target analytes, it is necessary to modify the sensing domain to be specifically active to corresponding targets. Empowered by site-direct mutagenesis, one can manipulate the interface biophysical properties at each amino acid group. In general, under a constant bias voltage, each analyte is driven towards the nanopore opening, then it somehow interacts with each amino acid ladder and finally translocates/binds with the nanopore. One can manipulate a few molecules at the nanoscale interface to enhance the stochastic probabilities of analytes approaching the sensing hot spot so as to boost the electrochemical sensitivity. Each ion among the protein–molecule interaction can act as the smallest probe. The nanopore sensing mechanism involves reversible non-covalent interactions at the amino acid ladder to execute successive transport of ions and analyte into the channel, but at different rates. Such a dynamic process is dominated by ion redistribution over the sensing interface. Moreover, the handling of each amino acid inside the channel can

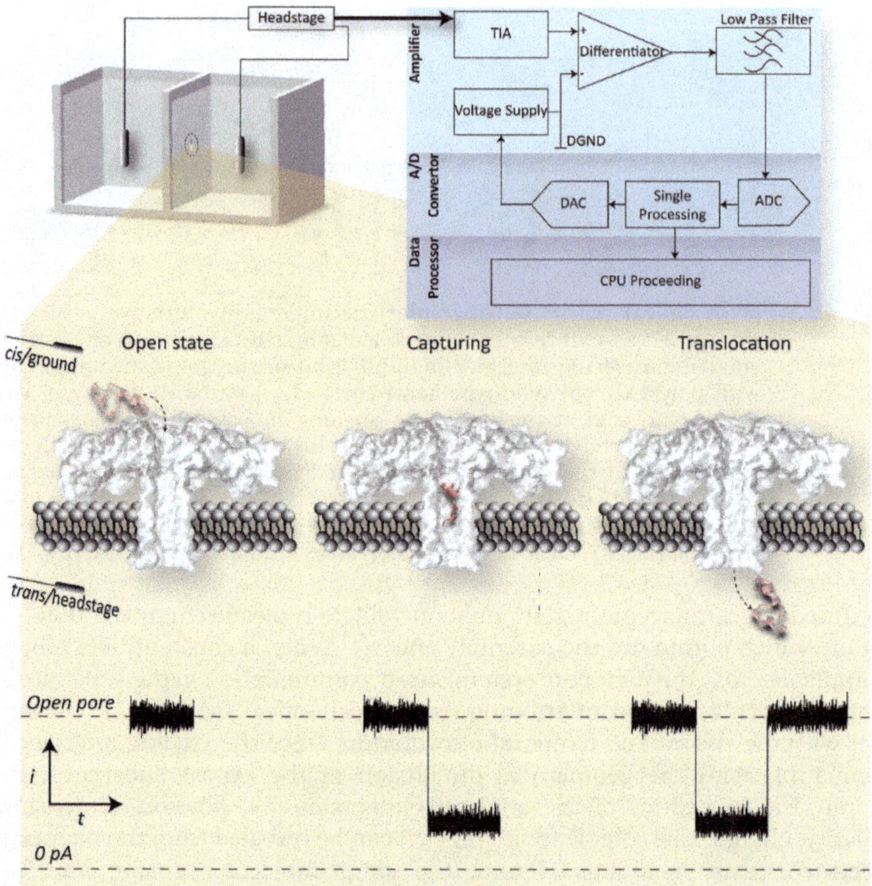

Figure 1.2 The sensing mechanism of a biomolecule nanopore interface. Top: the electrochemistry testing cell and the related data acquisition block diagram. Two compartments are separated by a Teflon septum with an aperture (*ca.* 50 µm) at the center. The aperture is further used to support an artificial lipid bilayer where a biological nanopore can be reconstituted and functionalized. Each compartment is connected *via* a pair of Ag/AgCl electrodes toward the headstage. The signal acquired by the headstage is first amplified and filtered at high signal-to-noise, then sampled and converted into a digital signal that can be recognized by the data processor. Middle: a cross-sectional view of the biological nanopore example illustrating that a polymer segment is brought near the nanopore "mouth", then captured by the nanopore lumen and finally released to the exit (an example of a translocation process). Bottom: the corresponding amperometric response of the nanopore electrochemical recording. A constant open current (pA) is obtained under a constant bias voltage (mV) when the substrate approaches the nanopore "mouth" before entering the lumen. A sudden drop in the current (conductance) is observed as soon as the substrate enters the nanopore lumen. Finally, the current immediately returns back to the open current once the substrate is released from the lumen region.

determine the pore–analyte interaction reaching the amplification of the non-covalent interactions.[9]

The nanopore-based single-molecule interface modulates the sensitivity and selectivity of single-molecule sensing. Further analysis allows us to understand that a unique electrochemical signal is generated not only due to spatial occupancy but also due to being dominated by the individual non-covalent bond-making and bond-breaking step. Such weak interactions occur due to nanopore-analyte-induced dynamic conformational changes and the electron/charge transfer process. In order to amplify and acquire the slightest electrochemical responses, several approaches such as high bandwidth electrochemical instrumentation, delicate sensing tools and an advanced data algorithm have been developed. In this book, we discuss the fabrication process of a single-molecule interface. There are pore forming materials adopted from nature that are synthesized into nanopores as well as pores fabricated in inorganic substrates. The single-biomolecule composed nanopore, also dubbed single-biomolecule interface, owes unique specificity to each residue inside the pore lumen being controllable. Solid-state nanopores fabricated with the aid of advanced fabrication processes can reach any desired pore opening dimension. One of the feasible fabrications processes based on a glass capillary provides an approach of high bandwidth recording of the electrochemical redox process for single-entity analysis.

We will also discuss the electrochemical confinement effects on nanopore sensing, especially on measuring the single-molecule non-covalent reactions and the conformational dynamics of individual molecules. Besides electrochemical information being read from each individual molecule's translocation signatures, electrochemically confined effects also provide correlation with other external driving forces, such as light scattering or bipolar electrochemistry, generating new sensing mechanisms along with the application.

Taking another biological nanopore, α-hemolysin, as an example, a dsDNA (double strand DNA) whose projection area is larger than the nanopore lumen diameter experiences an unzipping process providing a partial blockage. The unzipped dsDNA transverses through the narrowest constriction zone in its single-strand formation and generates a unique electrochemical signal in accordance with each nucleotide base. Apart from the direct readout information, the weak non-covalent interactions between substrates and nanopores may be hard to observe as these small fluctuations may be submerged in the noise level. Nonetheless, the hidden dynamic information containing such weak non-covalent interactions influences the ion movement in the nanopore lumen. Fundamentally, these weak interactions contain the continuous vibration of charge and electron density from the molecule or inter-molecule. In the entire interaction process, the charged ions or electrons from the bulk solution or those introduced along with the analyte, frequently collide and interact in the nanopore lumen, *i.e.* the nanoscale confined space. The ions inside such a nanopore confined space form a dynamic net that correlates with the electron density. The entrance of the analyte inevitably modifies this dynamic net, which can be regarded as each ion becoming the smallest sensor. These smallest sensor elements perceive and reflect non-covalent interactions into ionic vibration, which should be observed at the frequency domain. In short, all the frequency modification contains the information that should reflect the various (*i.e.* weak/strong non-covalent interactions typically) biophysical interactions among the analytes, ionic dynamic nets and each amino acid group that compose the nanopore. These reflections are commonly buried in the noise of the ionic current traces in the time domain acquired with the current state-of-art hardware.

Recently, we adopted the Hilbert–Huang transform (HHT), which has the capability to obtain high precision in the frequency and time domains, to analyze the nonlinear and non-stationary nanopore data.[10,11] Taking the example of the oligonucleotide poly(dA)$_4$ interaction with the wild type (WT) and the K238E aerolysin nanopores (Figure 1.3), the WT nanopore gives a larger dispersion degree, forwarding the energy axis in normalized energy (*z*-axis)–probability (*x*-axis) spectrum compared to the K238E analog. This observation demonstrates the fingerprint spectra, instead of the fingerprint

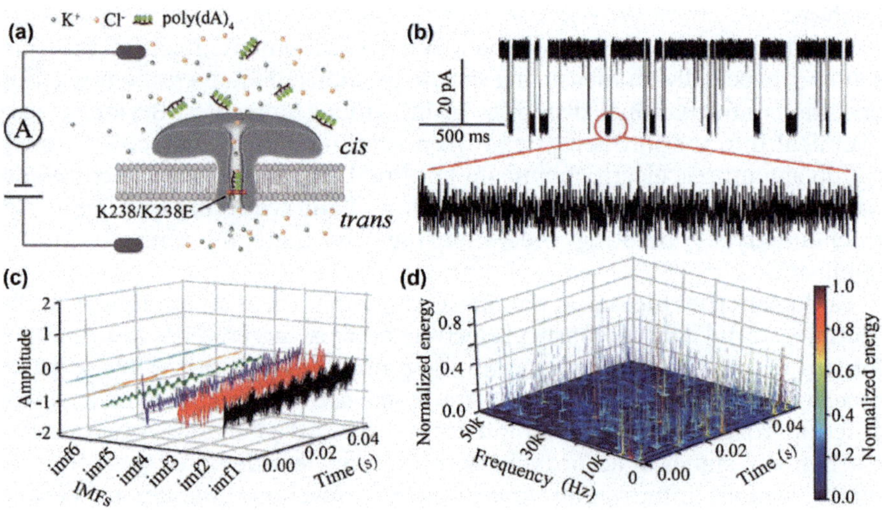

Figure 1.3 Schematic diagram of the frequency analysis of ionic current traces using the Hilbert–Huang transform method. (a) The illustration for sensing poly(dA)$_4$ using WT aerolysin and K238E analog. (b) Instance of the ionic current for the poly(dA)$_4$ interaction WT aerolysin nanopore, with a zoomed-in figure in the sub-diagram. (c) Representation of each intrinsic mode function (IMF) from the empirical mode decomposition (EMD) analyzes the enlarged part in (b). (d) The 3D spectrum energy–frequency–time distribution of the enlarged blockade shown in (b). Reproduced from ref. 11 with permission from the Royal Society of Chemistry.

Figure 1.4 Single-molecule music. Ideally, transferring the frequency–energy spectrum from the ionic current into the voice frequency could possibly let us hear the rhythm from membrane channels.
Reproduced from ref. 2 [https://doi.org/10.1093/nsr/nwy029] under the terms of a CC BY 4.0 license [https://creativecommons.org/licenses/by/4.0/].

current response, for the capability to analysis the non-covalent interactions.[12] By mimicking the frequency–energy spectrum of the typical ion current response into the hearable frequency, we illustrate that a nanopore sensor is composed of a single biomolecule interface akin to a tuba (Figure 1.4). When the molecule of interest flows into the "tuba" successively, the intrinsic properties of individual molecules can consequently be revealed by the dynamic modulation of the interactions ("button") between the "molecule flow" and sensing interface. Consequently, as the translocation of individual molecules through the confined pore, the beautiful molecule music of pore–analyte weak interactions will be "played".

References

1. Y.-L. Ying, J. Zhang, R. Gao and Y.-T. Long, *Angew. Chem., Int. Ed.*, 2013, **52**, 13154–13161.
2. Y.-L. Ying, C. Cao, Y.-X. Hu and Y.-T. Long, *Natl. Sci. Rev.*, 2018, **5**, 450–452.
3. J. J. Gooding and K. Gaus, *Angew. Chem., Int. Ed.*, 2016, **55**, 11354–11366.
4. T. Ha, *Nat. Methods*, 2014, **11**, 1015–1018.

5. Y. F. Dufrêne, E. Evans, A. Engel, J. Helenius, H. E. Gaub and D. J. Müller, *Nat. Methods*, 2011, **8**, 123–127.
6. E. J. G. Peterman and G. J. L. Wuite, *Single Molecule Analysis: Methods and Protocols*, Humana Press, 2011.
7. C. Cao, Y.-L. Ying, Z.-L. Hu, D.-F. Liao, H. Tian and Y.-T. Long, *Nat. Nanotechnol.*, 2016, **11**, 713–718.
8. J. Wang, J. Yang, Y.-L. Ying and Y.-T. Long, *Chem. – Asian J.*, 2019, **14**, 389–397.
9. C. Cao and Y. T. Long, *Acc. Chem. Res.*, 2018, **51**, 331–341.
10. H. Alzahrani, C. Antoine, L. Baker, S. Balme, G. Bhattacharya, P. W. Bohn, Q. Cai, C. Chikere, R. M. Crooks, N. Das, M. Edwards, C. Ehi-Eromosele, N. Ermann, L. Jiang, F. Kanoufi, C. Kranz, Y. Long, J. MacPherson, K. McKelvey, M. Mirkin, R. Nichols, W. Nogala, J. Pelta, H. Ren, J. Rudd, E. Schuhmann, Z. Siwy, Z. Tian, P. Unwin, L. Wen, H. White, K. Willets, Y. Wu and Y. Ying, *Faraday Discuss.*, 2018, **210**, 145–171.
11. S.-C. Liu, M.-X. Li, M.-Y. Li, Y.-Q. Wang, Y.-L. Ying, Y.-J. Wan and Y.-T. Long, *Faraday Discuss.*, 2018, **210**, 87–99.
12. M.-Y. Li, Y.-L. Ying, X.-X. Fu, J. Yu, S.-C. Liu, Y.-Q. Wang, S. Li, C. Cao, Y.-J. Wan and Y.-T. Long, *ChemRxiv. Preprint*, 2018, DOI: 10.26434/chemrxiv.7251683.v1.

CHAPTER 2

Fundamentals of Biological Nanopore Electrochemistry

YI-LUN YING,* JIAJUN WANG, XUE-YUAN WU AND YI-TAO LONG

Nanjing University, Nanjing, China
*Email: yilunying@nju.edu.cn

2.1 Basic Knowledge of Biological Electrochemistry in Nanopore Confined Spaces

As a reasonable model, the nanopore composed of a single biomolecule interface that has the comparable size of an analyte has attracted incredible attention.[1–6] Nature employs this single biomolecule interface to capture and identify each bypass entity at the single molecule level. To achieve single molecule sensing, nearly 30 years ago researchers developed nanopore electrochemistry by mimicking the translocation of targets through nanopores.[7–9]

When a single nanoscale pore-forming biomaterial (*e.g.* aerolysin or α-hemolysin) is self-assembled into an ion impermeable membrane, the pore lumen filled with water performs the only pathway for conducting two chambers. A pair of electrodes placed on each side of the nanopore drives the ions across the nanopore (Figure 2.1a).

Comparable to the molecule of interest, the ionic flux is confined by the nanopore and each entity from the bulk solution transits across the pore one-by-one under the bias voltage. The transit stay of a molecule inside the nanopore modulates the ions though the nanopore confined space, resulting

Detection Science Series No. 17
Confining Electrochemistry to Nanopores
Edited by Yi-Tao Long

Published by the Royal Society of Chemistry, www.rsc.org

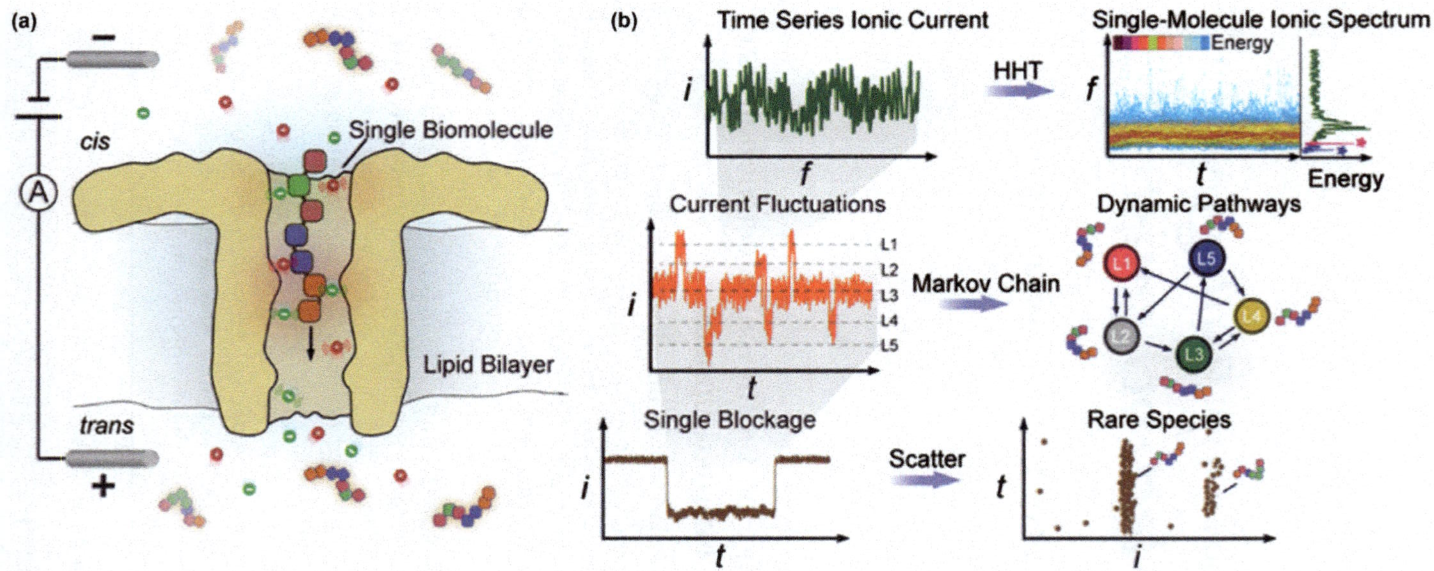

Figure 2.1 Nanopore-based single biomolecule interface for single molecule measurements. (a) A biological membrane protein molecule (*e.g.* aerolysin) inserts into the lipid bilayer. The analyte is electrochemically captured by the nanopore from the bulk solution, which regulates the ions flowing through the pore. Each ion acts as the smallest sensor to feedback the interaction between pore and analyte. (b) The obtained information from a single blockade event. Bottom: the statistical analysis of raw current blockades into scatter plots of blockage current (i) *versus* duration (t) can help to classify the species in the solution. Middle: tracing the multi-level current variations inside a blockade event can reveal the dynamic kinetics for the conversion of multi-intermediates. The data analysis methods for finding the relationship between multi-level currents can be, but are not limited to, the Markov chain. Top: analysing the hidden information in the noisy current traces using the Hilbert–Huang transform shows the "single molecule ionic spectrum", which could be used to quantify the non-covalent interaction between nanopore and analyte.
Reproduced from ref. 21 with permission from the American Chemical Society, Copyright 2019.

in transient electrochemical responses corresponding to each substrate according to the predicted model. Due to the stochastic behaviour of single molecules, each molecule owns a characteristic shape, secondary structure, conformation, dipole moment and diffuse coefficients. The characteristics of each molecule are explored from each blockade event, with higher temporal resolution, hidden states are resolved and a frequency spectrum is applied (Figure 2.1b). These features allow unique pore–analyte non-covalent bond interactions, regulating the dynamic re-distribution of ions along the sensing interface. Therefore, dynamic ionic fluctuation offers the intrinsic fingerprint of the molecules. A biological nanopore-based single molecule sensor features high spatial and temporal resolution, meeting the requirements of label-free single molecule sensing in the biological environment.[7–14] A biological nanopore sensor has been commonly adapted from easy nature sourcing such as MspA,[15] CsgG,[16] Aerolysin,[17,18] *etc.*, whose diameter of the constriction region reaches as narrow as *ca.* 1 nm. The temporal resolution was promoted towards sub-milliseconds thanks to a fine-tuned electrochemical acquisition tool.[19] A high throughput manner has also been achieved by assembling hundreds of nanopores in the chip array, providing high throughput sensing to produce megabyte data.[20]

2.2 Design a Single Biomolecule Sensing Interface

The biological nanopore sensing interface is derived from various pore-forming biomolecules, which determine the two important sensing characterizations of the nanopore technique: sensitivity and selectivity.[4] Taking the example of DNA sequencing, first, the energy barrier exists at the pore opening to capture or sieve the bypass molecule.[22] After entering the nanopore lumen, the interaction between the nanopore lumen wall and the single DNA will induce either translocation or binding and bounce back to where it entered. Therefore, the capture rate of the analyte towards the nanopore should be increased by an enhanced electrical driven force, optimized biochemical environment at the entrance of nanopore, *etc.* Rational design of the molecule sensing interface of nanopore electrochemistry offers exploration such as a powerfull label-free sensing tool to meet various demands. Such a single molecule sensing tool should not only tell us the single molecule features, but also own the capacity to differentiate each species from a heterogeneous system.

Regarding the biomolecule nanopore, which is considered as a single biomolecule interface, the known genomic information allows us to handle each amino acid by site-direct mutagenesis so as to boost the performance of such nanopore sensors, *i.e.* sensitivity and selectivity. Here, we use the aerolysin nanopore to illustrate the design principles of the single-biomolecule interface. WT aerolysin gives a high frequency of over 22 nt s^{-1} for capturing $(dA)_4$.[23] Previous experimental results together with molecular dynamics simulation demonstrate the existence of two sensing regions inside the aerolysin nanopore.[24] The first one is around R220 while the second one is located around K238. The substitution of arginine at 220 with glutamine (R220E) results in a higher entropic energy barrier that prevents the capture of negatively charged adenine segments (Figure 2.2a).[25] Specifically, at position 220 located at the nanopore mouth, the mutation from arginine to glutamine alters such a pocket into a more negatively charged region that repels the approaching DNA strand, in terms of electrostatic force. Taking another example such as α-hemolysin, the introduction of cysteine at the nanopore mouth allows the thiol-tailed thrombin aptamer oligonucleotide to be non-covalently bounded to the nanopore opening (Figure 2.2b).[26] This hybridized nanopore sensor enables the selective analysis of thrombin.

After the modification of composition at the nanopore opening, one can manipulate the capture rate of the molecule of interest; the same strategy is also applied to the sensing region inside the nanopore lumen. From the all-atomic simulation of the aerolysin nanopore, the sensing region inside the nanopore has been found at around position 238. Taking the $(dA)_4$ again as the model segment, recent attempts at using this nanopore enlarged the

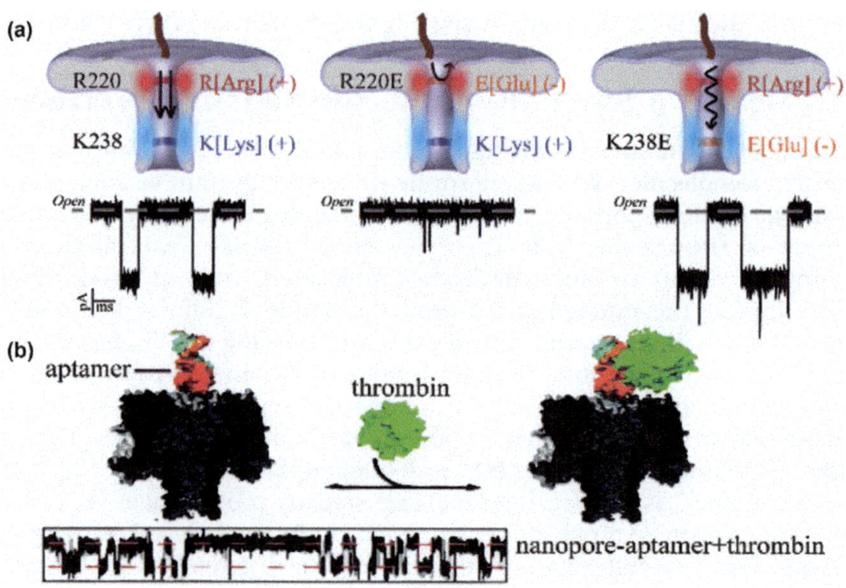

Figure 2.2 (a) Example of a single aerolysin interface for single oligonucleotide analysis. Left: a WT aerolysin with two positively charged amino acids (R220 and K238) produces a distinguishable blockage current. Middle: the R220E mutant generates a high entropic energy barrier for the sensing of negatively charged oligonucleotide molecules. Right: the K238E mutant negatively charged glutamic acid at the lumen leads to a prolonged duration. (b) α-Hemolysin nanopore with a thrombin aptamer attached to the *cis* opening for selective detection of thrombin. (a) Reproduced from ref. 22 with permission from the American Chemical Society, Copyright 2018. (b) Reproduced from ref. 23 with permission from the American Chemical Society, Copyright 2012.

electrostatic repulsion between the charged wall surface and the substrate by replacing K238 in glutamic acid (K238E).[25] The substitution of a negatively charged amino acid breaks the balance created by the positively charged pocket composed of lysine. The alternation prolonged the duration *ca.* seven-fold compared to the WT aerolysin (Figure 2.2a). As a consequence, the single-site mutant of the nanopore sensing interface allows the manipulation of the non-covalent interactions to achieve resolution beyond the single molecule level, *i.e.* the identification of single nucleotide and the monitoring of the translocation process of small molecules, *etc.*[24,27,28]

In order to push the aerolysin nanopore sensor beyond the single base detection limitation, further manipulation of the aerolysin-based single molecule interface at the K238 site was carried out to enhance the interaction between the sensing region and the oligonucleotide. For example, the K238G aerolysin nanopore successfully achieved the discrimination of methylated cytosine and cytosine of a hetero-nucleotide chain from a mixed system.[24] Besides nucleotide sensing for the purpose of DNA lesion or epigenetic modification, peptidomic sensing is more important due to the fact that it correlates to clinically related chronic diseases. The first peptide sensed by nanopore electrochemistry was achieved in 2006.[18] Recently, WT aerolysin has been used to identify single amino acid differences with the homo-polymeric peptides carriers.[29] Further improved sensing ability towards protein sequencing should be carried out by careful design of such a single molecule sensing interface. For example, a T232K/K238Q mutant aerolysin nanopore is designed to achieve direct detection of tau fragment peptide, pS262-, pT263-tau fragment peptide and pS262/T263-tau fragment peptide in the mixture with a nearly 100% identification accuracy for their characteristic peaks.[30] More cases of single nucleotide segment and peptides by using a designed single biomolecule interface are listed in Chapter 3. Both the peptidomic and oligonucleotide sensing examples illustrate the ability of a designed single biomolecule interface to determine and identify target molecules from the heterogenetic impurity presented, which cannot be handled by conventional techniques such as HPLC and mass spectrometry. Therefore, the manipulation of each amino acid at the interface allows one to fine tune the small sensing probe, providing nanopore electrochemistry with superior selectivity and sensitivity. Also, the reader could refer to the reviews in ref. 1, 14 and 31–35 for further perspective and applications.

Commonly used biomolecules, such as α-hemolysin (α-HL),[11,36] mycobacterium smegmatis porin A (MspA),[27,37,38] outer membrane protein G (OmpG),[31,39] phi29 motor,[32,33] cytolysin A (ClyA),[34,35] aerolysin,[40,41] CsgG,[16] lysenin,[42] ferric hydroxamate protein uptake component A (FhuA)[43] and YaxAB,[44] have been applied to the construction of a single biomolecule interface (Figure 2.3). The top view and the side view of each biological nanopore are listed to have a direct view of the protein structure, the smallest diameter of each pore lumen is also indicated as it is critical to the steric effect of the bypass analytes. The critical dimension of the biological

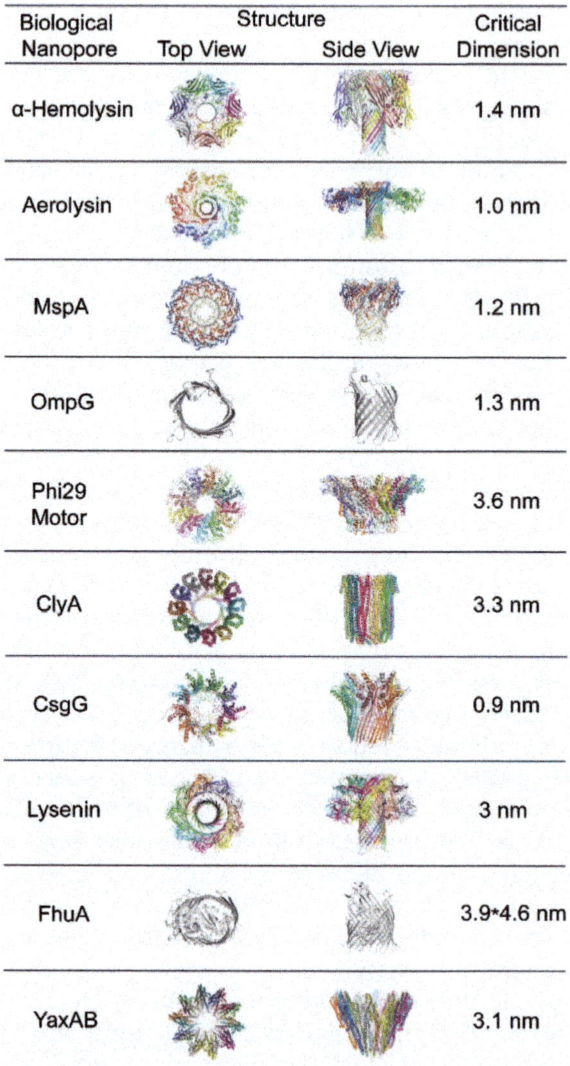

Figure 2.3 Commonly used biological nanopores. The structure of α-hemolysin, aerolysin, MspA, OmpG, phi29 motor, ClyA, CsgG, lysenin, FhuA and YaxAB (PDB 7A-HEMOLYSIN, 5JZT, 1UUN, 2F1C, 1FOU, 2WCD, 4Q79, 5EC5, 1BY3 and 6EL1, respectively) are from the PDB Protein Data Bank (the nanopores are coloured by chains).

nanopores ranges from sub-one-nanometre to nearly four nanometres, which greatly widens the application of nanoscale single molecule analysis technology to different kinds of analytes.

Most of these single-biomolecule interfaces originate from pore-forming toxins. In nature, pore-forming toxins (PFTs) are secreted by a large class of organisms and are evolved in the metabolism mechanisms; PFTs represent the bacterial virulence factors. PFTs disrupt epithelial barriers and interact with the immune system during bacterial infection, which can promote pathogen growth and dissemination. One notable feature of PFTs (*e.g.* α-hemolysin, aerolysin) is that they are in the form of water soluble proteins, which can assemble themselves in the host membrane. There are α and β PFTs that differ in the self-assembly process. The insertion of a pore will lead to cellular ion imbalance and eventually cause cell death (**Figure 2.4**). Taking α-hemolysin or aerolysin as examples, the activated monomer can self-assemble into a heptamer pore and embed the phospholipid bilayer, and many scientific researchers use this characteristic to detect and analyse various molecules, such as nucleic acid, oligosaccharides, peptides and proteins. Here, we mainly introduce α-hemolysin and aerolysin.[18,23,45–48]

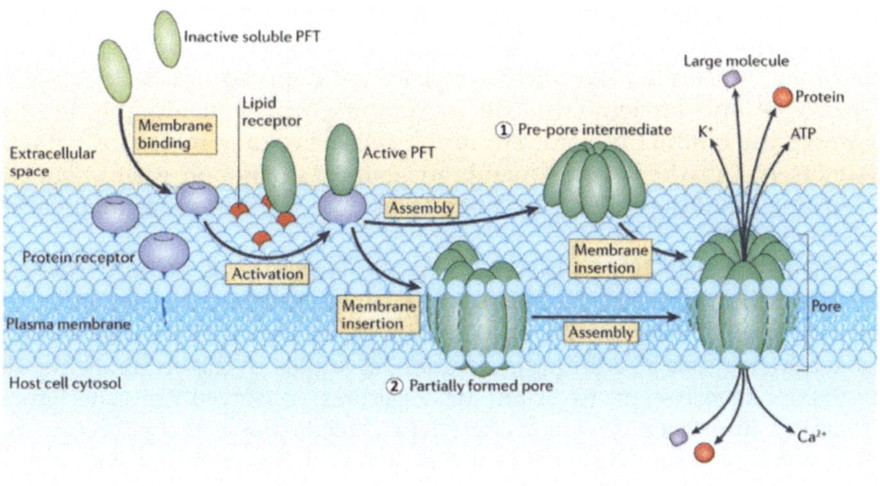

Figure 2.4 Mechanisms of the pore formation of pore-forming toxins (PFTs). Soluble PFTs will bind to the plasma membrane by interacting with protein or lipid receptors when they approach the host membrane. Then, PFTs start to concentrate and oligomerize following one of two pathways. For most β-PFTs, oligomerization usually occurs at the membrane surface, producing an intermediate structure called a pre-pore, which eventually inserts the host membrane *via* conformational rearrangements. For most α-PFTs, the membrane insertion of PFTs is concurrent with sequential oligomerization, which probably results in the formation of partially or completely formed pores. For both mechanisms, PFTs eventually form a transmembrane pore with different size, architecture, stoichiometry and conduction features. Small molecules, ions and proteins flow in or out of the host membrane through the transmembrane pore, which is also involved in various secondary responses and lysis. Adapted from ref. 46 with permission from Springer Nature, Copyright 2015.

2.3 Aerolysin-based Single Biomolecule Interface

2.3.1 The Structure of an Aerolysin Nanopore

According to the secondary structures that assemble in the bilayer, PFTs are classified as α or β PFTs. Aerolysin is a key member of the β-pore-forming toxins produced by *Aeromonas hydrophila* as an inactive precursor, proaerolysin, which can be activated by cleaving the C-terminal peptide (CTP). Aerolysin is composed of a 40% β-sheet structure and a 17% α-helix. The β-structure consists of 20 β-strands that take part in eight antiparallel β-sheets.[38] It contains four domains. Domains 1 and 2 are involved in the binding to *N*-glycosyl phosphatidylinositol (GPI)-anchored aerolysin receptors; domain 3 contains a five-stranded β-sheet and a pre-stem loop; domain 4 contains the CTP.[38] Aerolysin was first determined and named in 1974,[49] which could spontaneously oligomerize and insert into the cellular membrane to form a pore.[46] The sequence of aerolysin indicates its hydrophilicity in terms of monomer water solubility.[50] The water-soluble dimer can oligomerize to a heptamer and subsequently convert to a transmembrane mushroom-shaped pore.[50,51] To activate the aerolysin nanopore in the cellular membrane, enzymes such as trypsin are required to cut the loop between CTP and proaerolysin. Moreover, the glycan core-binding site can promote the binding of aerolysin to the cell membrane receptor.[52,53] Aerolysin inserts into the host cell membrane to form a nanopore with a diameter of ~1 nm, resulting in an imbalance in osmotic pressure that then causes cell rupture.[54,55]

Considering the huge differences in the structure of proaerolysin and aerolysin, the astonishing complexity of their architecture and dynamics has been resolved to reveal the pore-forming mechanism. The water-soluble proaerolysin exists as a dimer in solution as well as in the crystal, while the interactions between hydrophobic residues observed in monomer–monomer interfaces were considered to provide the driving force in the formation of the dimer.[52] As shown in Figure 2.5, the crystal structure of proaerolysin with 2.8 Å resolution exhibits its four domains. The CTP (black) in inactive conformation is water soluble. Domain 1 (red) undergoes a rotated process after CTP removal, which along with the second domain is now responsible for binding the receptor in the target membrane. Domain 4 reorganizes into a β-sandwich, which oligomerizes and forms a β-barrel prepore. The reorganization of domain 4 will lead to the extraction of the prepore and the fold of residues in the stem forming the post-pore structure. The stem β-barrel elongates from the top in a zipper-like pathway.

While the pro-aerolysin structure is available, the crystal structure with high resolution of the aerolysin pore has been unclear up to now, possibly owing to the high protein flexibility of the aerolysin heptamer. As shown in Figure 2.6, combining the quasipore conformation and wild-type pore structure with 7.9 Å resolution, the basic aerolysin pore structure was constructed. It should be noted that the 28 amino acids of the inner interface of

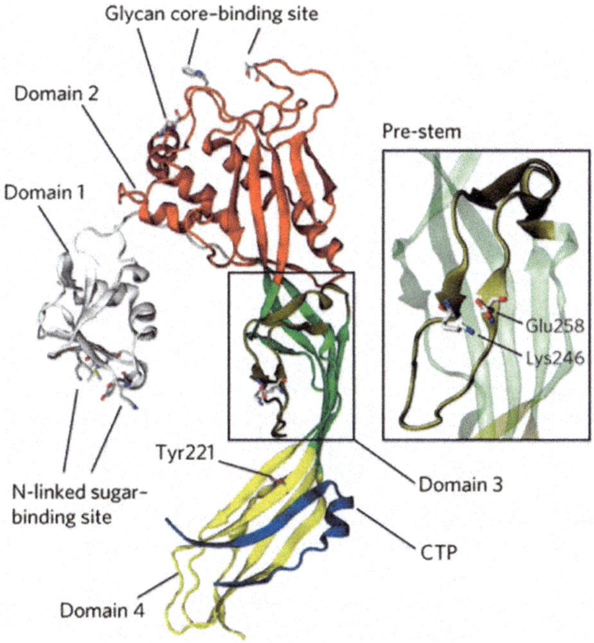

Figure 2.5 Crystal structure of proaerolysin. Reproduced from ref. 45 with permission from Springer Nature, Copyright 2013.

the aerolysin β-barrel were not resolved in the quasipore conformation, which was modelled based on the structural template of a solved anthrax protective antigen transmembrane barrel[17,18] The aerolysin heptamer contains a large cap with a diameter of about 140 Å, while the pore is located at the centre of the cap, owing a diameter of 10–14 Å. The β-barrel of aerolysin is 70–100 Å tall and its constriction region is almost lower than 10 Å.

Based on the analysis of aerolysin formation at different stages with near-atomic resolution, the detailed structure of the proaerolysin monomer was revealed.[17,45] The soluble proaerolysin undergoes pre-pore, post-pre-pore then extending its inner barrel (quasipore) and finally assembled as a nanopore. The aerolysin is folded as a loaded spring, blocked by two steps, the CTP and the pre-stem loop. As shown in Figure 2.7, after removing the CTP domain of proaerolysin, the protein straightens its fourth domain and oligomerizes, which generates a novel structure formed by two concentric β-barrels mainly contributed by the hydrophobic interactions. Then, the pre-stem loop gradually folds in a zipper-like manner using hydrogen bonds and extends the inner β-barrels to a length of 87 Å in the final pore conformation, which is similar to the length of the barrel formed by an anthrax protective antigen and lysenin.

In terms of single molecule electrochemical sensing, aerolysin protein was first used as a biological nanopore to study the structure of α-helix collagens (Figure 2.2).[56] Subsequently, it was used to study the oligonucleotide,[1,57]

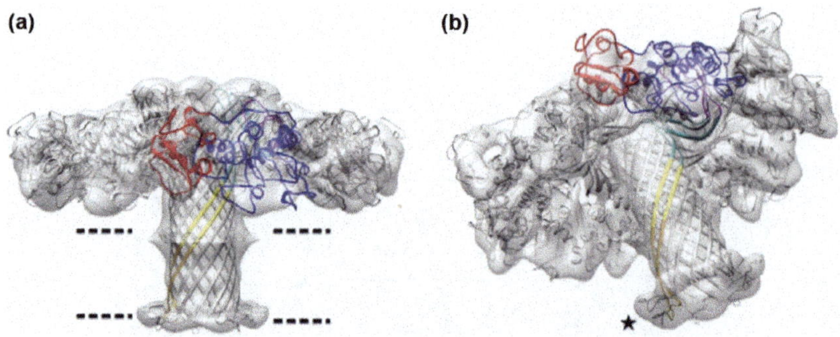

Figure 2.6 (a) The pore structure obtained from the quasipore map was rigid-body docked into the wild-type aerolysin map. The fit confirms that the quasipore structure indeed represents the structure adopted by the wild-type aerolysin in the pore conformation. For clarity, one monomer of the pore structure has been colour coded as in Figure 2.5 with domain 1 in red, domain 2 in blue, domain 3 in purple and domain 4 in turquoise. The part of the pre-stem loop that was resolved in the quasipore map is shown in yellow, while the residues missing from the quasipore map are shown in gold (see text and Figure 2.5). The position of aromatic residues in the β-barrel (shown in a ball and stick representation) hint at the position of the two leaflets of the membrane (dashed lines). (b) Tilted view of (a) showing the fit of the loops at the end of the transmembrane β-barrel (asterisk). As previously reported, upon crossing the bilayer the tips of the β-barrel loops fold back forming a rivet.
Reproduced from ref. 17 with permission from Springer Nature, Copyright 2016.

nucleotide lesion,[58] folding and unfolding kinetics of proteins, polypeptides, protein structures,[59] oligosaccharide aggregation[41] and the kinetics of enzyme degradation.[47] In addition, the aerolysin nanopore can also be used as a single molecule mass spectrometer to achieve real-time resolution of PEG molecular weight. Recently, it has been reported that the application of WT aerolysin can achieve a single molecule level mass resolution based on the blocking current for PEG with repetitive units less than 25 Da.[60]

2.3.2 Expression and Purification of Aerolysin

To obtain aerolysin for single molecule analysis, an *Escherichia coli* expression and purification system of proaerolysin was constructed.[61] The PCR amplification was performed using primers EAPT-F: 5′-TATC GGAA TTAA TTCG GATC CGGA TCCG GCA GAAC C-3′ with a BamH I site and EAER-R: 5′-TCGA GTGC GGCC GCAA GCTT AAGC TTTT GATT TGC-3′ with a Hind III site. In brief, samples were adjusted to a final volume of 50 μL by adding 0.25 μL beyotime Pfu DNA polymerase (5 U μL^{-1}), 5 μL 10×pfu buffer, 4 μL dNTP mixture (2.5 mM respectively), 2 μL of each primer (10 μM),0.5 μL genomic DNA and 38.25 μL ddH$_2$O. Reaction mixtures were incubated at 94 °C for 30 s, 60 °C for 30 s and 70 °C for 1.5 min for 35 cycles; 72 °C for

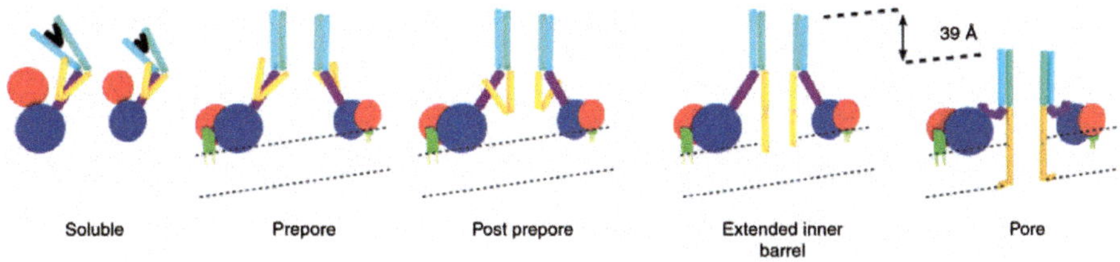

Figure 2.7 Schematics of the structural changes observed during aerolysin mode of action from the soluble inactive monomer to membrane-inserted oligomer sub-unit. In soluble aerolysin, the CTP (black) acts as a peg blocking the protein in its inactive conformation. Removal of the CTP and oligomerization lead initially to the rotation of domain 1 (red), which together with domain 2 is now able to bind the receptor (green), as well as to a reorganization of the domain 4 into a β-sandwich, which through oligomerization becomes a concentric β-barrel (turquoise; pre-pore). The reorganization of domain 4 triggers the gradual extraction of the pre-stem loop (yellow), as more residues are folded in the stem domain (post-pre-pore). Subsequently, torsions at a first hinge located at the interface of domains 2 (blue) and 3 (purple), and at a second hinge at the interface of domains 3 and 4 (turquoise) lead to the collapse of the structure by 39 Å and the insertion of the inner β-barrel through the membrane, resulting in the formation of a pore, while the tips of the β-barrel loops fold back to anchor the pore as a rivet (pore).
Reproduced from ref. 17 with permission from Springer Nature, Copyright 2016.

10 min. The pET-22b plasmids were digested with BamH I and Hind III. Samples were adjusted to a final volume of 20 μL by adding 10 x buffer 2 μL, pET-22b plasmid 2 μL, Bam I 1 μL, Hind III 1 μL and ddH_2O 15 μL. Reaction mixtures were stored at 37 °C for 3 h and purified using the DNA Gel Extraction Kit of Sangon Biotech.

Subsequently, the One Step Cloning Kit of Beyotime was used to link the target fragments obtained by PCR and the purified pET-22b plasmids (Figure 2.8). Samples were adjusted to a final volume of 20 μL, the purified pET-22b plasmid to 1 μL, target fragment 3 μL, 2X Seamless Cloning Mix 10 μL, nuclease free water 6 μL. Reaction mixtures were stored at 37 °C for 1 h and the reaction stopped by incubation for 5 min on ice. Then, reaction products were transformed into DH5α cells; the sequence of clones with correct sizes were further identified by GENEWIZ. Finally, the plasmids pET-22b-proaerolysin were transformed into BL21 [DE3] pLysS *E. coli* cells.

The target gene was introduced to the pET-22b vector and BL21 [DE3] pLysS *E. coli* was transformed with the pET-22b-proaerolysin plasmid, allowing periplasmic expression of the toxin with a His6 tag on the C-terminus. BL21 [DE3] pLysS *E. coli* harbouring the WT aerolysin expression plasmid was grown at 37 °C in 1 L cultures to an OD600 of 0.6. A 0.25 mM final concentration of IPTG was added in the bacterial culture to induce proteins at 16 °C for 4 h. Cells were harvested after about 2 h and reached an OD600 = 1.2. *E. coil* cells were collected from the LB medium by

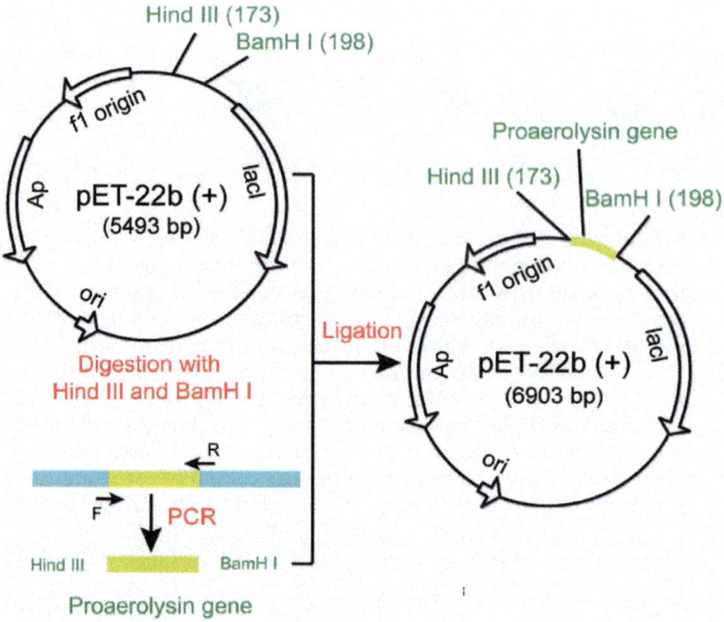

Figure 2.8 Schematic representation of the expression of plasmid pET-22b-proaerolysin.
Reproduced from ref. 62 with permission from Chemical Science China, Copyright 2019.

centrifugation (3000 rpm for 30 min at 4 °C). 50 mL buffer containing 0.5 M NaCl, 20 mM Na_3PO_4, pH 7.4 was added into the sediment and made it resuspended. After sonication on ice for 10 min, samples were centrifuged for 30 min at 4 °C and 10 000 rpm. Then the supernatant was loaded onto a Ni-NTA sorbent using a gravity flow column. The column was washed with buffer (20 mM Na_3PO_4, 0.5 M NaCl, pH 7.4) with a linear gradient of imidazole (0–0.5 M) to remove the unspecific bound proteins. The protein was further purified by size exclusion chromatography using a Superdex 200 16/600 PG column (GE Healthcare). The purity of the protein was confirmed by SDS-PAGE (Figure 2.9) and the precipitate, supernatant, the heterozygous protein and finally the target protein bars are indicated. Equivoluminal 30% (v/v) glycerinum was added into the protein. Finally, the proaerlysin was stored at −80 °C.

It is known that the incubation of proaerolysin and trypsin leads to the release of the CTP from aerolysin (dual chaperone role of the C-terminal propeptide in folding and oligomerization of the pore-forming toxin aerolysin). Trypsin can cut the loop between CTP and the protein body so that aerolysin can be activated. Then CTP can break away from the protein to facilitate oligomerization. We choose trypsin agarose to cut proaerolysin, so as to facilitate the removal of trypsin after restriction enzyme digestion and avoid influencing the subsequent experiments. After obtaining proaerolysin

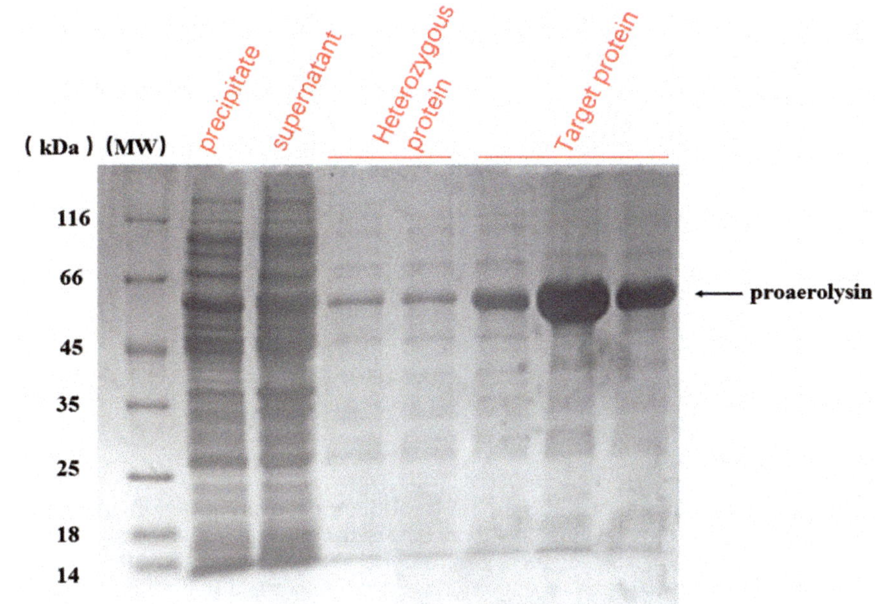

Figure 2.9 SDS-PAGE profile of purified proaerolysin. Lane 1: precipitate; lane 2: supernatant; lanes 3–4: heterozygous protein; lanes 5–7: target protein.

protein with high purity, proaerolysin was activated by trypsin agarose and the reaction was about 4 h at 4 °C. Then, for the enzyme digestion, trypsin agarose was removed by centrifugation. Although the CTP was still attached to the protein body because of the non-covalent effect between CTP and aerolysin, the molecular weight from gel-running decreased after the enzyme digestion. The SDS in buffer destroys this non-covalent effect, so it shows a decrease in molecular weight (Figure 2.10a). Activated aerolysin protein is prone to oligomerization at pH <7.5 at room temperature. Therefore, dialysis buffer (50 mM NaCl, 20 mM HEPES, pH 7) was used to dialyse at room temperature to obtain the polymerized protein (Figure 2.10b).

2.3.3 Manipulation of Sensing Interface of Aerolysin

The biological nanopore was used for single molecule identification and analysis by recording the characteristic blocked current signals by target analytes,[7] which has the advantages of being label-free, high resolution, low cost and high throughput.[63–65] However, the selectivity and sensitivity of WT biological nanopores are of limited value in specific applications in biosensing. Fortunately, protein bioengineering can be used for the precise mutation of biological nanopores. It is possible to try to optimize aerolysin nanopores for high sensing ability. As shown in Figure 2.11, the region above R282 controls the pore formation, the region around R220 determines the pore selectivity, and the region around K238 modulates the sensitivity of the

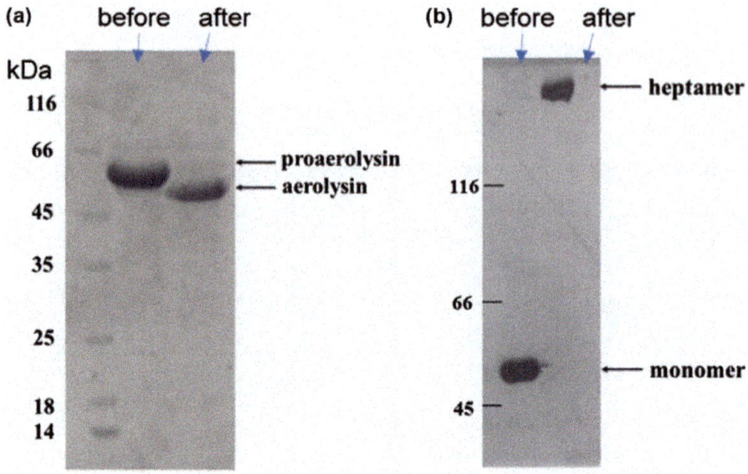

Figure 2.10 Comparison of aerolysin before and after activation/oligomerization. (a) Using trypsin cuts the loop between CTP and proaerolysin to active aerolysin. (b) Purified monomeric α-hemolysin. Activated aerolysin protein is prone to oligomerization to become a heptamer.

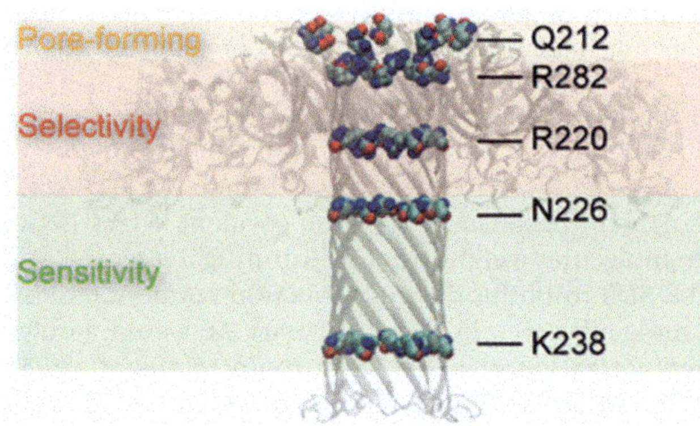

Figure 2.11 The reported position of mutant residues in the aerolysin. Modified from ref. 6, Copyright 2020 Science China Press Co., Ltd.

aerolysin. Generally, the mutant sites of aerolysin are R220 and K238. The single-site mutagenesis of aerolysin could be carried out according to the following steps.

A series of mutations was designed at the K238 site. As shown in Figure 2.12, lysine (K) was substituted by glycine (G), glutamine (Q), cysteine (C), glutamic acid (E), tyrosine (Y) and phenylalanine (F), respectively. Based on the constructed pET-22b-proaerolysin plasmid, we obtained a series of mutant plasmids by the reverse PCR method. These mutant plasmids and primers are used in PCR, as shown in Table 2.1.

In brief, samples were adjusted to a final volume of 50 μL by adding 1 μL KOD-Plus-Neo (1 U μL^{-1}), 5 μL 10× PCR buffer, 5 μL dNTPs (2.5 mM

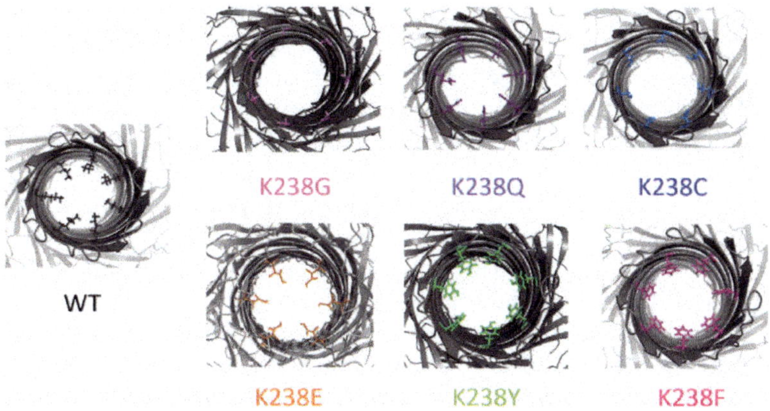

Figure 2.12 The illustration of different residues in the lumen of an aerolysin nanopore.

Table 2.1 Plasmid proaerolysin mutant and primers used.

Mutant plasmid	Forward	Reverse
K238G	GGAG TCAC CACC AAAA ACAA ATTC	CTCG CTCA GACC GTAG GTAT TG
K238C	TGTG TCAC CACC AAAA ACAA ATTC	CTCG CTCA GACC GTAG GTAT TG
K238F	TTCG TCAC CACC AAAA ACAA ATTC	CTCG CTCA GACC GTAG GTAT TG
K238E	GAAG TCAC CACC AAAA ACAA ATTC	CTCG CTCA GACC GTAG GTAT TG
K238Y	TACG TCAC CACC AAAA ACAA ATTC	CTCG CTCA GACC GTAG GTAT TG
K238Q	CAAG TCAC CACC AAAA ACAA ATTC	CTCG CTCA GACC GTAG GTAT TG

respectively), 1.5 μL of each primer (10 μM), 0.5 μL genomic DNA and 34 μL ddH_2O. Reaction mixtures were incubated for 2 min at 94 °C, followed by 15 cycles at 94 °C for 15 s, 68 °C for 2 min, and a final extension at 72 °C for 10 min. Then, we used Dpn I to digest PCR products in order to eliminate template plasmid. The total volume was 20 μL with PCR products 17 μL, buffer tango 2 μL, Dpn I 1 μL at 37 °C for 2h. Finally, we used T4 polynucleotide kinase to connect the linear products that we obtained in the previous step, and this step lasted for 3 h at 16 °C. Then, reaction products were transformed into BL21 [DE3] pLysS *E. coli* cells, and a sequence of clones with correct sizes were further identified by GENEWIZ. Finally, we verified the sequences for each mutant plasmid. Take K238G as an example, the original lysine codon at the 238 site is AAA, while the mutant is glycine and the base is GGA. Figure 2.11 shows the spectrogram of the sequencing results around the mutation point, the sequence of the mutant vector and the original vector with NCBI BLAST. The sequencing results showed that only the codon of amino acid at the 238 site was changed, while other sequences remained unchanged, indicating that our mutant vector was successfully constructed. Other mutations were also compared with their sequences in the same way to confirm that the mutation was completed correctly.

Each mutant plasmid is verified. Take K238G as an example, the original 238 lysine codon is AAA, while the mutant is glycine and the base is GGA. Figure 2.13 shows the spectrogram of the sequencing results around the mutation point, and the sequence of the mutant vector and the original vector with NCBI BLAST. The sequencing results show that only the 238 amino acid codon was changed and the other sequences remained unchanged, indicating that our mutant vector was successfully constructed. Other mutations were also compared with their sequences in the same way to confirm that the mutation was completed correctly.

The performance of mutant aerolysin for nanopore electrochemistry is evaluated by single channel recording. As shown in Figures 2.14 and 2.15,

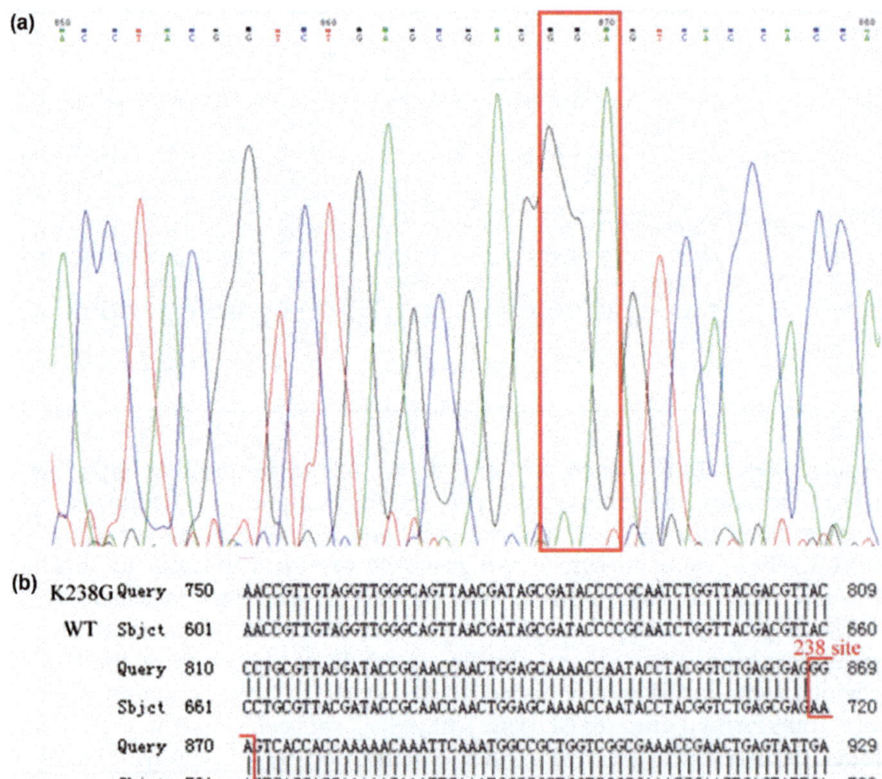

Figure 2.13 The result of K238G sequencing (a) and the nucleotide BLAST (b). Reproduced from ref. 25 with permission from American Chemical Society, Copyright 2018.

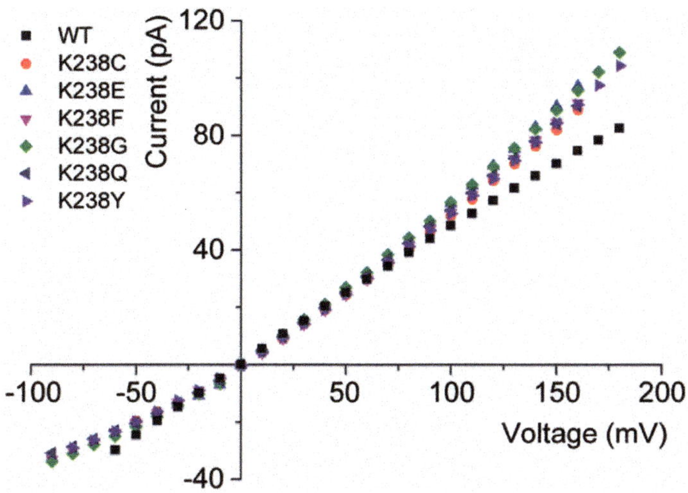

Figure 2.14 The current–voltage curve of wild-type (WT), K238C, K238E, K238F, K238G, K238Q and K238Y mutant nanopores.

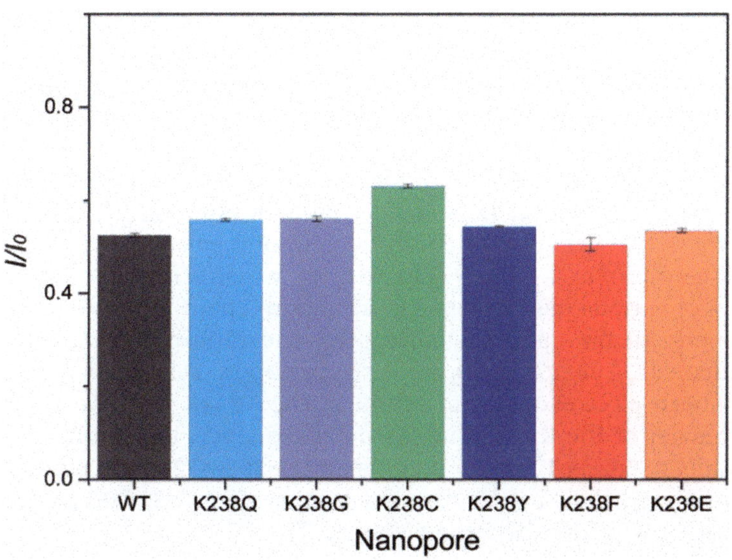

Figure 2.15 I/I_0 of dA_4 by wild-type (WT), K238C, K238E, K238F, K238G, K238Q and K238Y mutant nanopores at the applied voltage of +140 mV. I represents the residual current, whereas I_0 stands for the open pore current.

the mutation on the K238 site would induce a slight change in the I–V curve in single channel recording.[24,25,66] While the positively charged lysine (K) in the 238 site was replaced by neutral glycine (G), glutamine (Q), cysteine (C), tyrosine (Y), phenylalanine (F) and negatively charged glutamic acid (E),

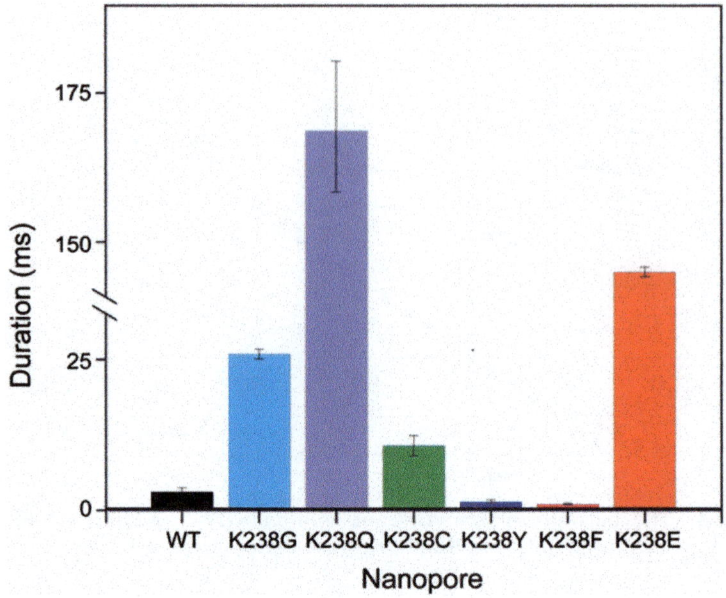

Figure 2.16 Duration of dA_4 by wild-type (WT), K238C, K238E, K238F, K238G, K238Q and K238Y mutant nanopores at +140 mV.

larger I_+/I_- values were observed in higher applied voltage. These observations might be due to ion selectivity changes induced by the charge distribution.

More interestingly, these mutations at the K238 site could significantly modulate the duration of DNA inside the aerolysin. Taking poly$(dA)_4$ as a model, the K238Q exhibits the longest duration, even about 60-fold more than WT aerolysin at +140 mV (Figure 2.16), which allows the simultaneous detection of various base damage in a random chain.[58]

To synergistically optimize nanopores, the multiple sites are mutated simultaneously. T232K/K238Q aerolysin nanopore synergistic incorporates the enhanced electrostatic interaction at the T232K site and the high repelling barrier at the K238Q site. As a consequence, the unphosphorylated tau peptide, pS262-, pT263-tau peptide and pS262/pT263-tau peptide can be directly distinguished by the residual current of a T232K/K238Q aerolysin sensor with 100% identification accuracy.[30]

2.4 α-Hemolysin-based Single Biomolecule Interface

2.4.1 The Structure of an α-Hemolysin Nanopore

α-Hemolysin is secreted by the human pathogen *Staphylococcus aureus*, which can attack human cells (for example, erythrocytes, platelets, lymphocytes, monocytes and endothelial cells). Membrane bound monomers oligomerize at the membrane surface, then insert into target membrane *via* conformational rearrangements and eventually form heptameric transmembrane pores,[67] which cause membrane leakage then destroy the cellular osmotic balance and finally promote cell lysis. In 1996, the nucleotide sequence of α-hemolysin was reported.[36] This protein is composed of 58% β-sheet structure and 4% α-helix. In function, α-hemolysin is similar to bacterial toxins including α-toxin from clostridium sebticum and aerolysin from aeromonas hydrophila. Eukaryotic proteins including perforin and complement component C9 also assemble to oligomeric transmembrane pores.

The crystal 3D structure of the α-hemolysin nanopore was reported in 1996.[36] Each protomer assembles along a crystallographic seven-fold axis. The α-hemolysin nanopore consists of three domains, including the cap, rim and stem region. β-sandwiches from seven protomers compose the cap region. Under the cap region, four β-strands constitute the rim domain. The stem domain forms the transmembrane portion of the channel using a 14-strand antiparallel β-barrel. The heptameric nanopore is approximately 100 Å tall and the stem domain is around 52 Å in height and 26 Å in diameter. The diameter of the cap domain is about 100 Å. The lumen of the α-hemolysin is hydrophilic, and the exterior is about 28 Å wide. In 2015, the effect of residues on the pore-forming process was studied.[67] In this study, protein engineering enabled the formation of an α-hemolysin H35A mutant, and the substitution of His35 caused a marked decrease in oligomerization and nanopore activity was presented (as shown in Figure 2.17).

α-Hemolysin has been the most widely used biological nanopore since it was first used to explore single oligonucleotides in 1996.[7] Then, hairpin structures are constructed to reduce the translocation speed of DNA molecules and achieved ultrasensitive discrimination of single base pair mismatches.[68] The hairpin-type end was found to immobilize DNA in the

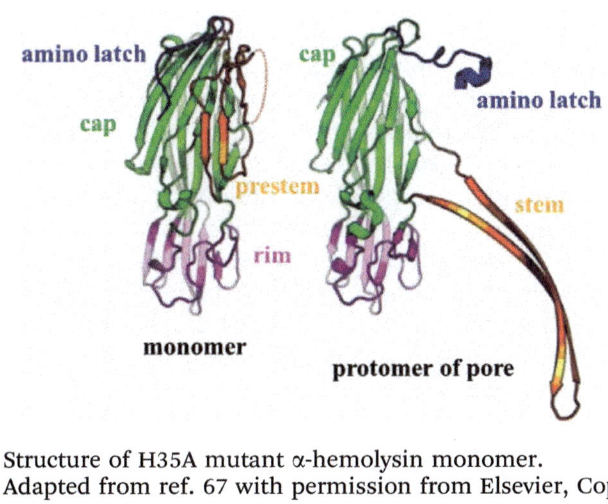

Figure 2.17 Structure of H35A mutant α-hemolysin monomer.
Adapted from ref. 67 with permission from Elsevier, Copyright 2015.

nanopore,[69] and when a cytosine (C) in the cytosine oligodeoxynucleotide was substituted with an adenine (A), the blocking current corresponding to the adenine could be distinguished from cytosine. Later, the same group changed the structure of a α-hemolysin nanopore by site-directed mutagenesis, and embedded the β-cyclodextrin molecule inside the α-hemolysin nanopore with a covalent bond to narrow the lumen.[69] Experiments proved that this improved nanopore can distinguish different kinds of nucleotides and methylated cytosine. Currently, the α-hemolysin nanopore has been used for DNA sequencing, peptide sensing, single molecule reaction, *etc.*, which will be further discussed in Chapter 3.

2.4.2 Expression and Purification of α-Hemolysin

Monomeric α-hemolysin was gathered from cell culture supernatants of *staphylococcus aureus*. Assembly was catalysed by deoxycholate, and the heptamer was purified to homogeneity by size exclusion chromatography.[36] However, this method cannot obtain large purified protein. In 2011, the expression and purification of α-hemolysin was reported.[70] The PCR amplification was performed using primers EAPT-F: 5′-CATG CCAT GGCA GATT CTGA TATT AATA TTAA AACC GG-3′ and EAER-R: 5′-CCGC TCGA GATT TGTC ATTT CTTC TTTT TCCC AATC G-3′. In brief, samples were adjusted to a final volume of 50 μL by adding 0.25 μL DNA polymerase (5 U μL^{-1}), 5 μL buffer, 4 μL dNTP mixture (2.5 mM respectively), 2 μL of each primer (10 μM), 0.5 μL genomic DNA and 38.25 μL ddH_2O. Reaction mixtures were incubated at 94 °C for 3 min; 94 °C for 30 s, 55 °C for 30 s and 72 °C for 2 min for 35 cycles; 72 °C for 10 min. The pET-28b vector was double digested by NcoI and XhoI restriction endonucleases and ligated to the PCR products using the One Step Cloning Kit (Figure 2.18). Then, reaction products were transformed into DH5α cells; the sequence of clones with correct sizes were further identified by GENEWIZ.

The plasmid pET28b-hemolysin was transformed into *E. coli* strain B834 (DE3). Cells grown at 37 °C in 1 L cultures with 50 μg mL^{-1} kanamycin and 34 μg mL^{-1} chloramphenicol to an OD600 of 0.6. A 0.5 mM final concentration of IPTG was added in bacterial culture to induce proteins at 25 °C for 18 h. *E. coil* cells were collected from the LB medium by centrifugation (3000 rpm for 30 min at 4 °C). 50 mL buffer containing 0.3 M NaCl, 50 mM Tris–HCl, pH 8.0 was added into the sediment and made it resuspended. After sonication on ice for 10 min, samples were centrifuged for 30 min at 4 °C and 10 000 rpm. Then the supernatant was loaded on a HisTrap column equilibrated with 20 mM Tris–HCl, 300 mM NaCl, pH 8.0. The column was washed with buffer (20 mM Tris–HCl, 300 mM NaCl, pH 8.0) with a linear gradient of imidazole (0–0.5 M) to remove the unspecifically bound proteins. The protein was further purified using a HiLoad 26/60 Superdex 200-pg column (GE Healthcare). Equivoluminal 30% (v/v) glycerinum was added into the protein. Finally, the proaerlysin was stored at −80 °C.

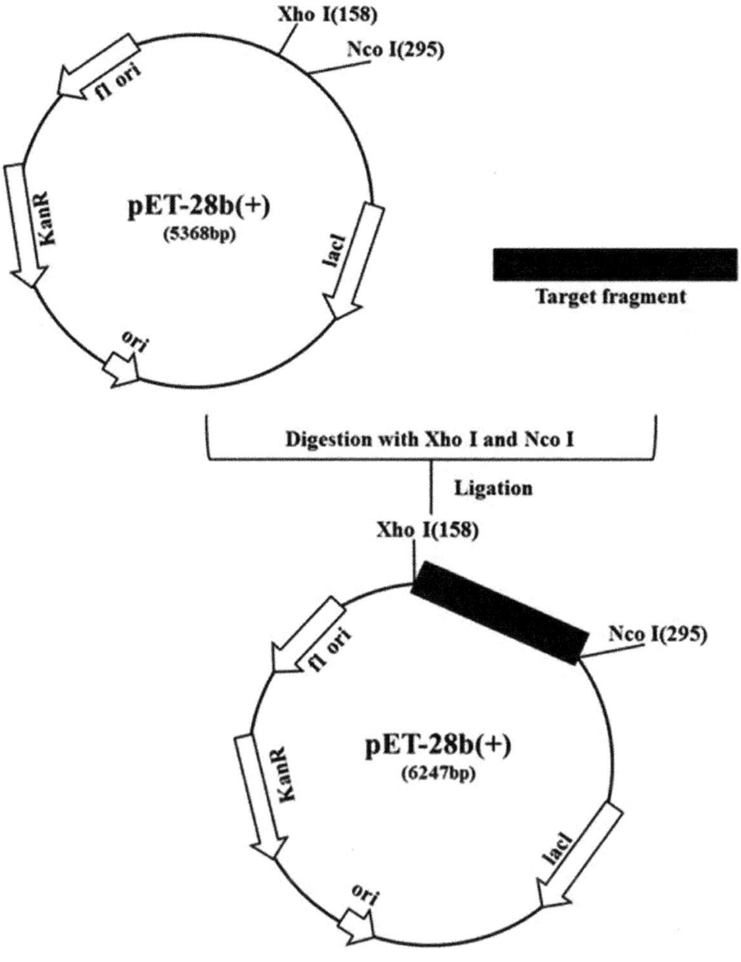

Figure 2.18 Schematic representation of the expression of plasmid pET28b-hemolysin.

2.5 Membrane Formation to Support Pore-forming Molecules

The cell membrane is a thin film structure located at the periphery of the cell (Figure 2.19), and its basic structure consists of a lipid bilayer, a membrane protein and a small number of polysaccharides. The cell membrane acts as a peripheral protective membrane of cells, which is a barrier that hinders the free entry and exit of chemicals to cells. It also maintains the relative stability of the cell environment and ensures that various biochemical reactions can be carried out in an orderly manner. In order to accomplish a specific physiological function, cells must exchange information, energy and substances with the external environment. For harmful substrates, the cell

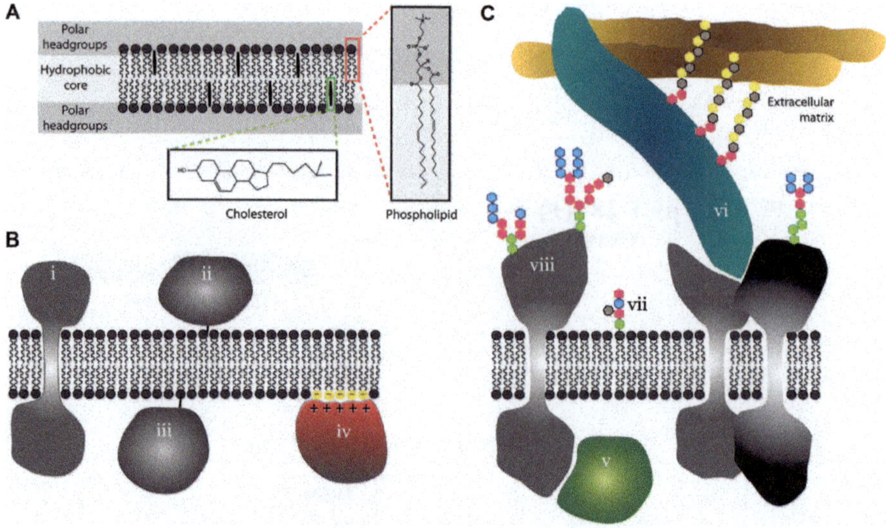

Figure 2.19 Schematic illustration of the basic structure of the lipid bilayer and proteo-lipidic membranes. (a) Cell membranes are lamellar structures with a hydrophobic core and a polar head group space. As examples, phospholipids and cholesterol are shown with almost atomistic detail (red and green boxes). (b) Membrane proteins can integrate into membranes (i), but can use lipid anchors [(ii) and (iii)] or be peripherally associated with membranes *via* electrostatic interactions (iv). (c) Proteins can further associate with membranes *via* protein–protein interactions on the cytosolic side (v) or at the interface between the plasma membrane and the extracellular matrix (vi). Outer leaflet lipid (vii) and extracellular domains of proteins (viii) are often glycosylated. Reproduced from ref. 80 under the terms of a CC BY 4.0 license [https://creativecommons.org/licenses/by/4.0/].

membrane can act as a barrier and avenue to ensure that cells are not invaded. For essential nutrients, the cell membrane can selectively regulate the absorption. Due to the complexity of the cellular membrane, the study of analytes through a nanopore is very difficult. Therefore, an artificial cell membrane system is required to study the structure and function of the nanopore for single molecule sensing. The main component of the artificial cell membrane *in vitro* is a phospholipid forming bilayer, which maintains the fluidity as well as elasticity, and can realize some functions of the cell membrane based on the fluid mosaic model. At present, the artificial lipid bilayer system is used to study nanopore for single molecule sensing as well as various biological processes at the cellular level, including receptor–ligand interaction virus infection, cell signal transduction, *etc.*[71–79]

There are many types of BLM (black lipid membrane), and different types of artificial membrane platforms should be selected for different applications. For example, BLMs supported by a ruthenium base can be used to inlay small fractions to receptors for studying their interaction with extracellular proteins. If it is necessary to study transmembrane proteins or cell membrane pore-forming toxic molecules, a black lipid membrane system (also known as a suspended BLM) or a support membrane system with a polymer mat layer is more desirable to prevent structural deformation of the protein on the solid supporting surface.[81–84] BLMs are named after their real image under an optical microscope. The basic structure is a single-layer BLM structure. Suspended BLMs are the most commonly used artificial BLM models for studying cell membrane electrophysiology. They were first used to study ion channels formed by peptides, proteins, antibiotics or other pore-forming biomolecules. In the 1860s, a BLM system was launched and its electrical properties was determined.[79] In this system, two solution chambers are connected by a 1 mm diameter aperture, surrounded by rouge molecules (Annulus) (Figure 2.20a). Two decades later, Tamm and McConnell built BLMs directly on the surface of solid materials.[85] In 1997, Boxer's group further used lithographic ring technology to prepare complex supporting BLM. This technique has also been developed for the preparation of a sensor array of solid-supported BLMs to study living cells. Artificial cell membranes mainly include suspended BLMs (*i.e.* black lipid membranes) and supporting BLMs. Depending on the type, the preparation methods of BLMs mainly include: the smear method, the Langmuir–Blodgett (LB) membrane method (Figure 2.20b) and the liposome vesicle rupture method (Figure 2.20c).[63,78,86–89]

In a non-solvent system, BLMs can be formed using LB technology: the solution containing pore forming materials is placed in aqueous solvent. First, the solvent interface is gradually increased from the position below the aperture oriented at the water–air interface. With the movement of the liquid surface, a black greasy film gradually forms and extends to the surface of the entire small hole to form a process called Montel–Mueller technology. As with the coating method, the black lipid film prepared by the LB method has poor reproducibility, and its stability is poor due to the lack of a solvent-phospholipid ring at the edge of the pore.[90]

Therefore, the cyclic structure is proposed and often referred to as the Plateau–Gibbs edge whose function is to maintain the stability of the black lipid film. Through the volatilization of the organic solvent on the surface of the hydrophobic material, the intermediate layer will continue to be thinner when the lipid–solvent mixture is thinning out. Then the randomly occurring thermodynamic and mechanical vibration processes will form a phospholipid at the edge of the aqueous layer–non-polar layer. The molecular layer is gradually pulled in and the two layers of monolayers are brought together by van der Waals forces to form a bilayer. The organic solvent used to dissolve the fat affects the fluidity of the artificial membrane and may affect the membrane protein embedded therein. This painting method is a reproducible method for preparing BLMs.

Since the BLM is formed and suspended in a solvent, it is not interfered with by the support, and the embedded transmembrane protein has intact fluidity and protein activity. However, the stability of the BLM is not compatible with the cellular membrane, its lifespan is limited and its use range

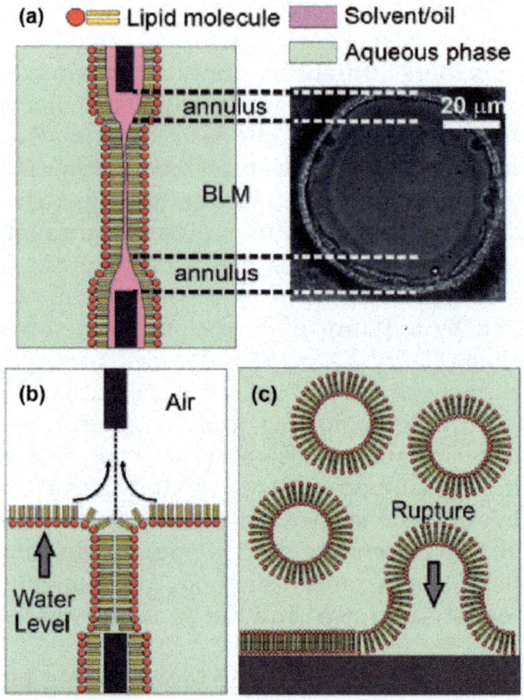

Figure 2.20 Conventional laboratory techniques for lipid bilayer formation. (a) A hydrophobic sheet with a single aperture is clamped between two aqueous reservoirs. A rod or a brush is used to deposit a lipid–solvent mixture across the aperture. (Left) Schematic of a lipid bilayer membrane and (right) a transmission, bright field image of a black lipid molecule (BLM) formed across an aperture using the painting technique. The annulus region, close to the edge of the aperture, appears darker than the membrane in the centre. (b) Schematic of a solvent-free Langmuir–Blodgett film being formed across an aperture. Surfactants self-orient at the water–air interface in each reservoir. Starting from below the height of the aperture, the level of the aqueous solutions is slowly raised above the aperture. (c) A schematic interpretation of a lipid bilayer membrane formed using the vesicle rupture technique. In all these cases, the self-assembly properties of surfactants are exploited, the lipid molecules spontaneously orient themselves at the interface between different phases.
Reproduced from ref. 91 with permission from the Royal Society of Chemistry.

is limited. Therefore, traditional BLM research is often only used for electrical conductivity detection and simple optical microscopy systems. However, in recent years, many researchers have combined microfluidic technology to study sophisticated BLM systems and apply them to the more potential techniques (with prominent examples listed in Figure 2.21). For example, microfluidic film can be used to accurately control the amount of organic solvent, the aqueous solution of the lipid and the passage time. By

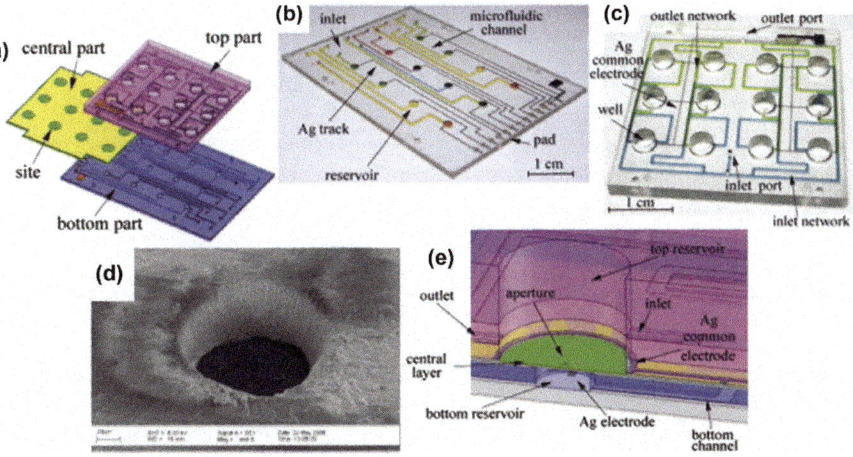

Figure 2.21 (a) Drawing of the three components of the chip system, with the 12 BLM sites placed in the central polymeric membrane. (b) Photograph of the bottom microfluidic part. Food dye is used to highlight the 12 reservoirs and microfluidic channels. Each site is accessed electrically *via* a silver track that is connected to the pads. (c) Photograph of the top microfluidic part. Food dye (light blue and green) highlights the two microfluidic networks, which are connected simultaneously to all 12 wells. A common silver electrode electrically connects to each well. (d) SEM photograph of one of the 12 apertures (100 μm diameter) laser-drilled into the PTFE central layer. (e) Diagram of a cross-section through the structure of each BLM site.
Reproduced from ref. 92 with permission from Elsevier, Copyright 2009.

using this method, the monolayer of the single layer can be assembled into the rigid lipid bilayer by controlling the evaporation time of the solvent. A 12-well independent microfluidic channel array was established (Figure 2.21),[92] which formed a black lipid film by time-controlled solvent evaporation, and integrated Ag/AgCl electrodes in the microfluidic channel used to realize electrical detection. The electrodes are indicated in the bottom part as illustrated in Figure 2.21b while the microfluidic part of the array BLM is illustrated in Figure 2.21c. Each well will support a single layer of BLM whose SEM structure and schematic structure is given in Figures 2.21d and e. This device precisely controls the area of the membrane and the size of the Plateau–Gibbs ring, leading to an enhanced stability of the BLM. An electrophysiological study of potassium ion transmembrane channel protein KcsA was also further performed to show the membrane activity.

Taking the fashion of lab-on-a-chip due to the miniaturization, the lipid membrane could also work in a chip. A gas permeable PDMS material was utilized to prepare a microfluidic chip, which is beneficial to the volatilization of organic solvents to form a suspended BLM (Figure 2.22a). Since PDMS is highly apt to adsorb organic solvents, a water–organic–water structure of a sandwich structure can be formed in an aqueous phase

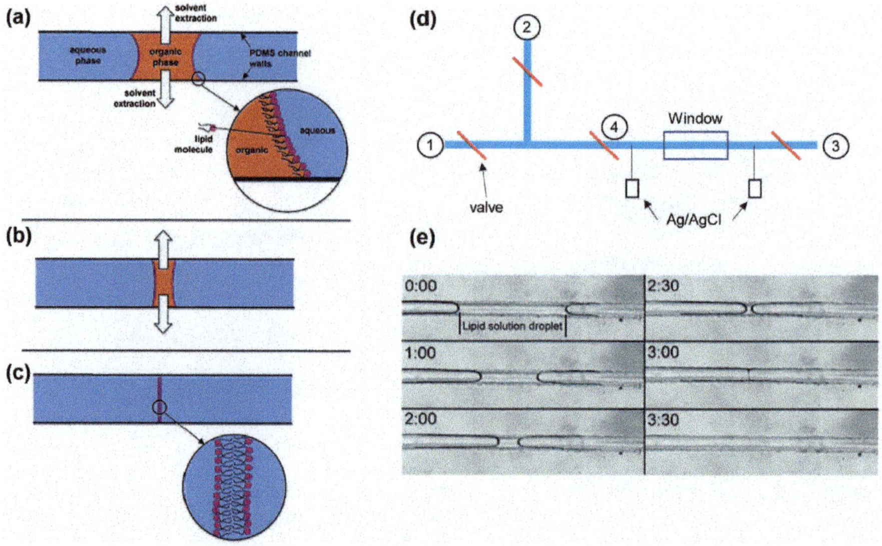

Figure 2.22 Mechanism for lipid-bilayer formation by microfluidic solvent extraction. (a) A droplet of organic solvent with dissolved lipid is formed in an aqueous stream of fluid. Lipids are organized on the hydrophobic–hydrophilic interface (inset). (b) As solvent enters the PDMS, the two interfaces approach one another. (c) Finally, only the lipid layers are left behind, forming a bilayer membrane. Inset: the membrane formation in a microfluidic channel. (d) A photograph of a microfluidic device used for lipid-bilayer membrane formation by solvent extraction. (e) The real time frame of the formation of BLM in the PDMS microfluidic channel.
Reproduced from ref. 93 with permission from American Chemical Society, Copyright 2006.

solution, and an organic solvent can be rapidly discharged by volatilization (Figure 2.22b), and the thinning of the solvent can be easily achieved by using the material (Figure 2.22c). BLM can be formed after complete volatilization due to the PDMS wall can act as the edge of the solvent-free BLM. Figure 2.22d indicates that the channels have been first filled with buffer solution containing food dye. As indicated from Figure 2.22a, the whole structure of the device is constructed by two layers, *i.e.* a fluidic layer and a valve layer, which is pneumatically controlled under microfluidic devices. Inlets 1 and 2 are responsible for aqueous solution as well as organic solvent and towards a final common outlet 3. The measuring electrodes are placed at inlet 4 with extra caution of leakage. Overall, the microchannels are 100 μm wide with an observation window for the formation of BLM. Figure 2.22e gives a real time frame of the formation of BLM in the PDMS microfluidic channel, the process is quite feasible and the BLM could be formed within minute seconds.

The micromachining techniques were adapted to prepare complex black lipid membrane systems on an array microchambers (>100 wells).[94–96] As is indicated in Figure 2.23, the formed BLM is automated formed with the control of aqueous solution from both sides of the membranes. The array system is composed of a syringe and a PDMS chip (Figure 2.23a), the PDMS chip contains arrayed microchannel with 17 μm diameter and 19 μm in depth, as illustrated in Figure 2.23b. The micropattern is connected by an inlet and an outlet. By pumping the solvent, oil and buffer sequentially (Figures 2.23d–f) the lipid membrane is obtained in parallel formation manner. They have used such array membrane system for quantitative detection of non-charged fluorescent molecules across the membrane. For the sake of single molecule detection, such an array BLM enables the detection in a high throughput manner. To form an array membrane in such array system, first, a microwell array was prepared in a hydrophobic polymer flow control chip. After adding a phospholipid solution, a BLM array was prepared using the Montal–Mueller method around the hydrophobic pores. The microwell array was integrated into the microfluidic channel and each well contains a BLM microwell array, and simultaneously records the signals of multiple self-assembled bilayers.

A novel array BLM setup composed of nine Si chips was constructed as given in Figure 2.24a.[97] The assembled setup is pictured in Figure 2.24b. Making use of the conventional Montal–Mueller method, such an approach is the most feasible and easy to prepare among all the BLM forming methods; however, to reach both sides of the BLM, the BLM septum is hardly horizontally placed. In this design, there are nine Si chips with a small hole fabricated and clamped between two Teflon chambers. When forming the BLM array, the setup is first vertically placed and forms the BLM using the conventional Montal–Mueller method. Then the whole setup is turned 90° to be mounted. Due to the robustness of the BLM formed according to the Montal–Mueller Method, 60% of the BLM array could last until the end of the parallel recordings. Such instrumentation is established to achieve the screening of the ion channel proteins.

The array chip could not only be formed on the fabricated cavity arrays, but also by the microcontact printing approach. A feasible method to first array the BLM on a glass support was designed (Figure 2.25).[98] Firstly, the glass support was diced into several regions, making use of the micropatterning, an array BLM could be formed. Regarding the micropatterning, a series of processes were used, including scratching, blotting then stamping the lipid with PDMS onto the desired sites for the formation of the phospholipid bilayer.

Following the previously described array BLM formation, the pattering of the proteoliposome could also be integrated into such a device for parallel nanopore sensing. As indicated in Figure 2.26, the patterning of liposome for the purpose of formation of the array BLM was designed.[99] Similar to the aforementioned approach, a silicon substrate was used. Firstly, a PDMS was taken whose surface is plasma-oxidized. By applying the PDMS layer into the

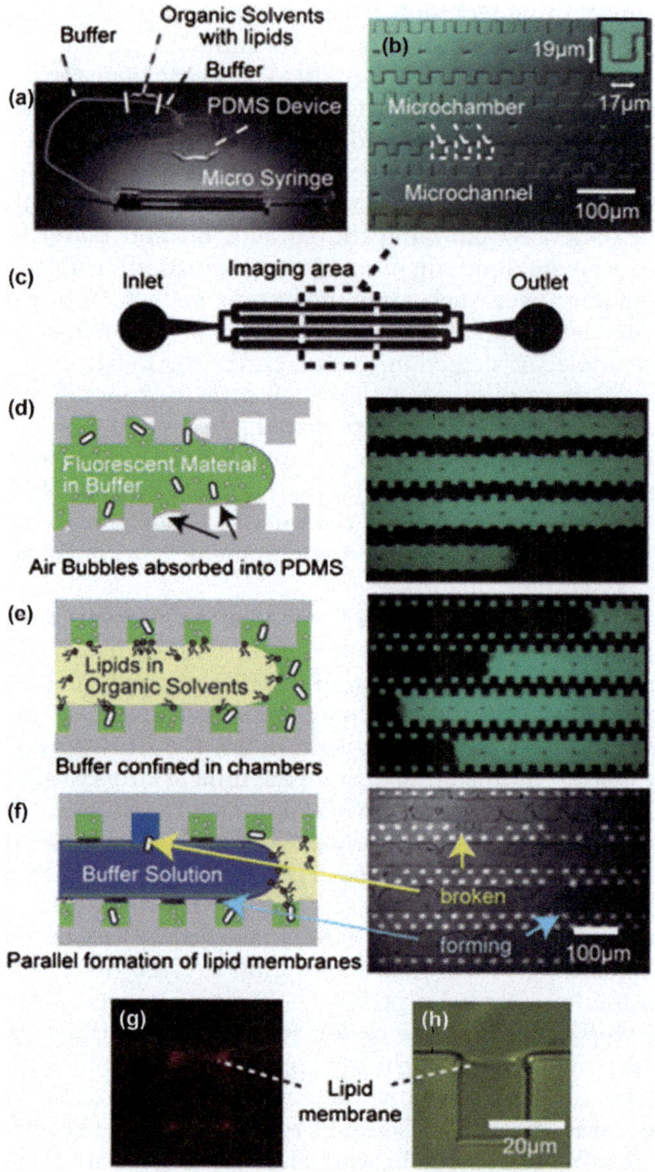

Figure 2.23 (a) An experimental setup composed of a micro-syringe and a PDMS microfluidic device, connected by a tube wherein immiscible solutions are sequentially loaded. (b) Magnified view of the device. (c) Schematic of the microfluidic channels. A procedure to form lipid membranes on microchambers: (d) filling the device with an aqueous solution containing membrane proteins and dyes while pushing air out through PDMS, (e) infusing oil (hexadecane) containing dissolved lipids while confining the first solution in the chamber array, and finally (f) flushing the channels with buffer solution. This process forms oil films with lipids, which then thins down to form membranes. (g) Fluorescent image of the formed lipid membrane and (h) its bright field image. Reproduced from ref. 96 with permission from the Royal Society of Chemistry.

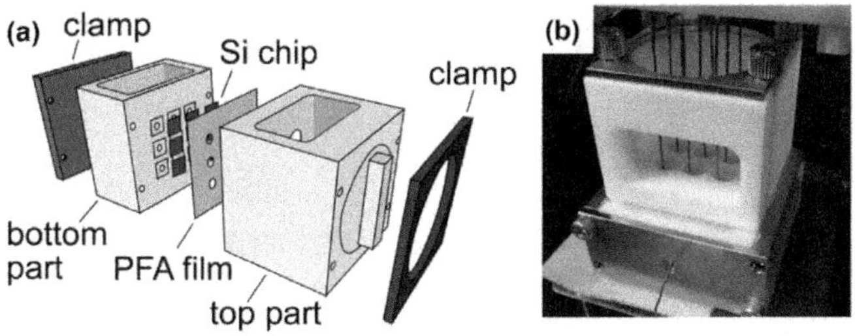

Figure 2.24 (a) Drawing of the components of the present system, with the nine BLM sites placed in the Si chips. (b) Photograph of the experimental chamber in the final arrangement.
Reproduce from ref. 97, Copyright AIP Publishing.

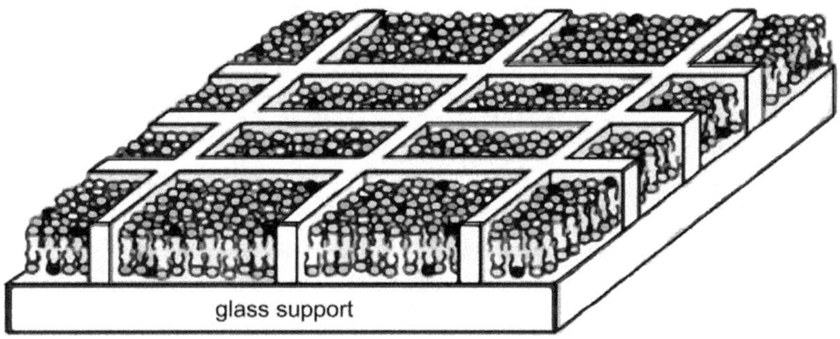

Figure 2.25 Schematic illustration of a micropartitioned fluid lipid bilayer membrane. Patterns of differing surface properties on the solid support restrict the supported membrane, creating barriers to lateral diffusion. The solid support is typically glass and many different materials can be used to fabricate the barriers. The lipid bilayer is separated from the glass substrate by a thin (~10 Å) layer of water that serves to lubricate the interaction and allow lateral mobility. The bilayer itself is approximately ~50 Å thick, while the barrier material is typically 10 ± 1000 nm thick and 1 ± 10 mm wide; the drawing is not to scale. This system serves to partition the bilayer so that its components are free to diffuse laterally within each corralled region, but not between them.
Reproduced from ref. 98 Copyright © 2000 John Wiley & Sons, Inc.

aqueous solution abundant with liposomes, it is possible to cover a large piece of bilayer, thanks to the microstructure over the PDMS. Once the patterning is transferred to the object surface layer, the arrayed BLM could finally be transferred, typically useful for probing and controlling of the interaction happening at the surface of the membrane, which could also be useful for parallel nanopore sensing purposes.

To conclude, nanopore electrochemistry is a unique and powerful technique for single molecule sensing. To achieve high temporal resolution for each analyte, the compatible sensing interface is a must; gifted from nature, the nanopore can already discriminate varieties of individuals and give precise single molecule information. Based on the fundamental nanopore

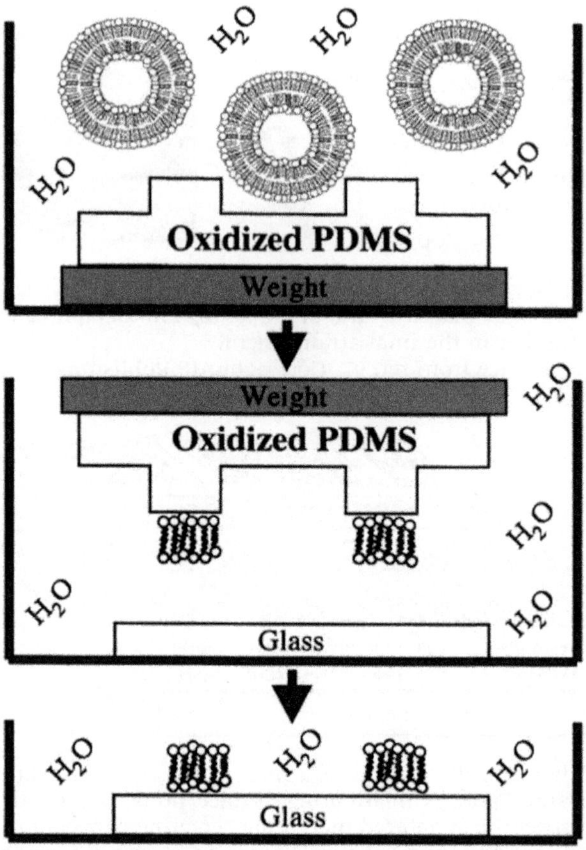

Figure 2.26 Schematic illustration showing the creation of concentration and composition arrays of lipids. Different sized areas of lipid-bilayer-coated PDMS are brought into contact with a pre-patterned surface (the dark vertical bars are fixed barriers; these can be made from a variety of materials). The printed substrate is incubated with vesicles containing lipids of a different composition. The vesicles form a bilayer on the empty areas of the substrate and connect with the printed bilayer, and the two components (dark and light) mix together. This can be used to create concentration arrays if either the printed or vesicle bilayer contains the component that is being diluted or binary composition/concentration arrays where different components are present in the printed part *versus* the added vesicles.
Reproduced from ref. 99 with permission from American Chemical Society, Copyright 2001.

structural information and sensing mechanisms, the sensing interface between the nanopores is explored and designed for the typical purpose of specific analytes. For the sake of much similar chemical structure detection, intermolecular interaction and molecular interaction are considered. Taking examples of aerolysin and α-hemolysin, the biomolecule interface provides compatible sensing domain to the bypass analyte. To functionalize the nanopore sensing, a more robust and parallelized BLM array is desired for real time, real sample, long term and high throughput analysis. The BLM, after several decades' development, still uses the earliest Montel–Mueller methods. Although later the painting method as well as the vesicle rupture approach could increase the BLM robustness, these methods cost a lower effective BLM area and less feasibility respectively. Alternatively, the main hindrance of the conventional BLM methods is their low throughput. Increasing the analysis yield would enable the nanopore single molecule analysis to be used for drug screening and other healthcare or fundamental chemistry research. The development of the array BLM is still young; most approaches are still inspired by the mostly used Montel–Mueller method with an increase of the membrane numbers. The feasibility, accessibility and automation are not gifted from such approach. Recent attempts through the support of the microfluidic approach should be more suitable for the parallelization idea with lower sample consumption and it is more user friendly; most importantly, different materials could be used to decrease the white noise contributors. To sum up, single molecule analysis through the nanopore confining region has great demands of the sensitive and tuneable sensing interface. The BLM as the support of the nanopore should develop in the direction of robustness and low noise. With the development of both systems, then the resolution towards the sub-molecular level is foreseeable.

References

1. C. Cao and Y. T. Long, *Acc. Chem. Res.*, 2018, **51**, 331–341.
2. Y. Lin, Y.-L. Ying and Y.-T. Long, *Curr. Opin. Electrochem.*, 2018, **7**, 172–178.
3. Y. Ying, X. Zhang, Y. Liu, M. Xue, H. Li and Y. Long, *Acta Chim. Sin.*, 2013, **71**, 44.
4. Y.-L. Ying, C. Cao, Y.-X. Hu and Y.-T. Long, *Natl. Sci. Rev.*, 2018, **5**, 450–452.
5. Y. L. Ying, J. Zhang, R. Gao and Y. T. Long, *Angew. Chem., Int. Ed. Engl.*, 2013, **52**, 13154–13161.
6. Y.-L. Ying, J. Wang, A. R. Leach, Y. Jiang, R. Gao, C. Xu, M. A. Edwards, A. D. Pendergast, H. Ren, C. K. T. Weatherly, W. Wang, P. Actis, L. Mao, H. S. White and Y.-T. Long, *Sci. China: Chem.*, 2020, **63**, 589–618.
7. J. J. Kasianowicz, E. Brandin, D. Branton and D. W. Deamer, *Proc. Natl. Acad. Sci. U. S. A.*, 1996, **93**, 13770–13773.
8. D. Deamer, M. Akeson and D. Branton, *Nat. Biotechnol.*, 2016, **34**, 518.
9. J. J. Kasianowicz and S. M. Bezrukov, *Nat. Biotechnol.*, 2016, **34**, 481.
10. C. Dekker, *Nat. Nanotechnol.*, 2007, **2**, 209.
11. B. M. Venkatesan and R. Bashir, *Nat. Nanotechnol.*, 2011, **6**, 615.
12. Y. L. Ying, J. Zhang, R. Gao and Y. T. Long, *Angew. Chem., Int. Ed.*, 2013, **52**, 13154–13161.
13. J. Wang, J. Yang, Y. L. Ying and Y. T. Long, *Chem. – Asian J.*, 2019, **14**, 389.
14. W. Shi, A. K. Friedman and L. A. Baker, *Anal. Chem.*, 2016, **89**, 157–188.

15. M. Faller, M. Niederweis and G. E. Schulz, *Science*, 2004, **303**, 1189–1192.
16. P. Goyal, P. V. Krasteva, N. Van Gerven, F. Gubellini, I. Van den Broeck, A. Troupiotis-Tsaïlaki, W. Jonckheere, G. Péhau-Arnaudet, J. S. Pinkner and M. R. Chapman, *Nature*, 2014, **516**, 250.
17. I. Iacovache, S. De Carlo, N. Cirauqui, M. Dal Peraro, F. G. van der Goot and B. Zuber, *Nat. Commun.*, 2016, **7**, 12062.
18. R. Stefureac, Y.-t. Long, H.-B. Kraatz, P. Howard and J. S. Lee, *Biochemistry*, 2006, **45**, 9172–9179.
19. J. K. Rosenstein, M. Wanunu, C. A. Merchant, M. Drndic and K. L. Shepard, *Nat. Methods*, 2012, **9**, 487.
20. H. Bayley, *Clin. Chem.*, 2015, **61**, 25–31.
21. Y.-L. Ying and Y.-T. Long, *J. Am. Chem. Soc.*, 2019, **141**, 15720.
22. M. Wanunu, W. Morrison, Y. Rabin, A. Y. Grosberg and A. Meller, *Nat. Nanotechnol.*, 2010, **5**, 160.
23. C. Cao, Y. L. Ying, Z. L. Hu, D. F. Liao, H. Tian and Y. T. Long, *Nat. Nanotechnol.*, 2016, **11**, 713–718.
24. Y.-Q. Wang, M.-Y. Li, H. Qiu, C. Cao, M.-B. Wang, X.-Y. Wu, J. Huang, Y.-L. Ying and Y.-T. Long, *Anal. Chem.*, 2018, **90**, 7790–7794.
25. Y.-Q. Wang, C. Cao, Y.-L. Ying, S. Li, M.-B. Wang, J. Huang and Y.-T. Long, *ACS Sens.*, 2018, **3**, 779.
26. D. Rotem, L. Jayasinghe, M. Salichou and H. Bayley, *J. Am. Chem. Soc.*, 2012, **134**, 2781–2787.
27. A. H. Laszlo, I. M. Derrington and J. H. Gundlach, *Methods*, 2016, **105**, 75–89.
28. A. H. Laszlo, I. M. Derrington, B. C. Ross, H. Brinkerhoff, A. Adey, I. C. Nova, J. M. Craig, K. W. Langford, J. M. Samson and R. Daza, *Nat. Biotechnol.*, 2014, **32**, 829.
29. F. Piguet, H. Ouldali, M. Pastoriza-Gallego, P. Manivet, J. Pelta and A. Oukhaled, *Nat. Commun.*, 2018, **9**, 1–13.
30. S. Li, X.-Y. Wu, M.-Y. Li, S.-C. Liu, Y.-L. Ying and Y.-T. Long, *Small Methods*, 2020, DOI: 10.1002/smtd.202000014.
31. M. A. Fahie, B. Yang, B. Pham and M. Chen, *ACS Sens.*, 2016, **1**, 614–622.
32. D. Wendell, P. Jing, J. Geng, V. Subramaniam, T. J. Lee, C. Montemagno and P. Guo, *Nat. Nanotechnol.*, 2009, **4**, 765.
33. J. Geng, S. Wang, H. Fang and P. Guo, *ACS Nano*, 2013, **7**, 3315–3323.
34. V. Van Meervelt, M. Soskine and G. Maglia, *ACS Nano*, 2014, **8**, 12826–12835.
35. L. Franceschini, M. Soskine, A. Biesemans and G. Maglia, *Nat. Commun.*, 2013, **4**, 2415.
36. L. Song, M. R. Hobaugh, C. Shustak, S. Cheley, H. Bayley and J. E. Gouaux, *Science*, 1996, **274**, 1859–1865.
37. T. Z. Butler, M. Pavlenok, I. M. Derrington, M. Niederweis and J. H. Gundlach, *Proc. Natl. Acad. Sci. U. S. A.*, 2008, **105**, 20647–20652.
38. J. M. Craig, A. H. Laszlo, I. M. Derrington, B. C. Ross, H. Brinkerhoff, I. C. Nova, K. Doering, B. I. Tickman, M. T. Svet and J. H. Gundlach, *PLoS One*, 2015, **10**, e0143253.

39. M. A. Fahie and M. Chen, *J. Phys. Chem. B*, 2015, **119**, 10198–10206.
40. M. Pastoriza-Gallego, L. Rabah, G. Gibrat, B. Thiebot, F. O. G. van der Goot, L. Auvray, J.-M. Betton and J. Pelta, *J. Am. Chem. Soc.*, 2011, **133**, 2923–2931.
41. B. Cressiot, E. Braselmann, A. Oukhaled, A. H. Elcock, J. Pelta and P. L. Clark, *ACS Nano*, 2015, **9**, 9050–9061.
42. A. Yamaji-Hasegawa, A. Makino, T. Baba, Y. Senoh, H. Kimura-Suda, S. B. Sato, N. Terada, S. Ohno, E. Kiyokawa and M. Umeda, *J. Biol. Chem.*, 2003, **278**, 22762–22770.
43. M. Krewinkel, T. Dworeck and M. Fioroni, *J. Nanobiotechnol.*, 2011, **9**, 33.
44. B. Bräuning, E. Bertosin, F. Praetorius, C. Ihling, A. Schatt, A. Adler, K. Richter, A. Sinz, H. Dietz and M. Groll, *Nat. Commun.*, 2018, **9**, 1806.
45. M. T. Degiacomi, I. Iacovache, L. Pernot, M. Chami, M. Kudryashev, H. Stahlberg, F. G. Van Der Goot and M. Dal Peraro, *Nat. Chem. Biol.*, 2013, **9**, 623.
46. M. Dal Peraro and F. G. Van Der Goot, *Nat. Rev. Microbiol.*, 2016, **14**, 77.
47. A. Fennouri, R. g. Daniel, M. Pastoriza-Gallego, L. Auvray, J. Pelta and L. Bacri, *Anal. Chem.*, 2013, **85**, 8488–8492.
48. M. Kukwikila and S. Howorka, *Anal. Chem.*, 2015, **87**, 9149–9154.
49. A. W. Bernheimer and L. S. Avigad, *Infect. Immun.*, 1974, **9**, 1016–1021.
50. E. A. Manrao, I. M. Derrington, M. Pavlenok, M. Niederweis and J. H. Gundlach, *PLoS One*, 2011, **6**, e25723.
51. A. H. Laszlo, I. M. Derrington, H. Brinkerhoff, K. W. Langford, I. C. Nova, J. M. Samson, J. J. Bartlett, M. Pavlenok and J. H. Gundlach, *Proc. Natl. Acad. Sci. U. S. A.*, 2013, **110**, 18904–18909.
52. A. D. Tucker, M. W. Parker, D. Tsernoglou and J. T. Buckley, *J. Mol. Biol.*, 1990, **212**, 561–562.
53. M. W. Parker, J. T. Buckley, J. P. M. Postma, A. D. Tucker, K. Leonard, F. Pattus and D. Tsernoglou, *Nature*, 1994, **367**, 292–295.
54. L. Abrami, M. Fivaz and F. G. Van Der Goot, *Trends Microbiol.*, 2000, **8**, 168–172.
55. Y. Tsitrin, C. J. Morton, C. El Bez, P. Paumard, M.-C. Velluz, M. Adrian, J. Dubochet, M. W. Parker, S. Lanzavecchia and F. Van Der Goot, *Nat. Struct. Mol. Biol.*, 2002, **9**, 729.
56. J. He, J. Wang, J. Hu, J. Sun, D. M. Czajkowsky and Z. Shao, *J. Mol. Recognit.*, 2016, **29**, 174–181.
57. C. Cao, J. Yu, M.-Y. Li, Y.-Q. Wang, H. Tian and Y.-T. Long, *Small*, 2017, **13**, 1702011.
58. J. Wang, M.-y. Li, J. Yang, X.-y. Wu, J. Huang, Y.-l. Ying and Y.-t. Long, *ACS Cent. Sci.*, 2019, **6**, 76–82.
59. C. l. Merstorf, B. Cressiot, M. Pastoriza-Gallego, A. Oukhaled, J.-M. Betton, L. Auvray and J. Pelta, *ACS Chem. Biol.*, 2012, **7**, 652–658.
60. G. Baaken, N. Ankri, A. K. Schuler, J. Ruhe and J. C. Behrends, *ACS Nano*, 2011, **5**, 8080–8088.
61. C. Cao, D.-F. Liao, J. Yu, H. Tian and Y.-T. Long, *Nat. Protoc.*, 2017, **12**, 1901–1911.

62. X.-Y. Wu, M.-B. Wang, Y.-Q. Wang, M.-Y. Li, Y.-L. Ying, J. Huang and Y.-T. Long, *CCS Chemistry*, 2019, **1**, 304–312.
63. H. Bayley and P. S. Cremer, *Nature*, 2001, **413**, 226–230.
64. D. W. Deamer and D. Branton, *Acc. Chem. Res.*, 2002, **35**, 817–825.
65. Y. Wang, F. Yao and X.-f. Kang, *Anal. Chem.*, 2015, **87**, 9991–9997.
66. M.-Y. Li, Y.-Q. Wang, Y. Lu, Y.-L. Ying and Y. Long, *Front. Chem.*, 2019, **7**, 528.
67. T. Sugawara, D. Yamashita, K. Kato, Z. Peng, J. Ueda, J. Kaneko, Y. Kamio, Y. Tanaka and M. Yao, *Toxicon*, 2015, **108**, 226–231.
68. W. Vercoutere, S. Winters-Hilt, H. Olsen, D. Deamer, D. Haussler and M. Akeson, *Nat. Biotechnol.*, 2001, **19**, 248.
69. N. Ashkenasy, J. Sánchez-Quesada, H. Bayley and M. R. Ghadiri, *Angew. Chem., Int. Ed.*, 2005, **44**, 1401–1404.
70. Y. Tanaka, N. Hirano, J. Kaneko, Y. Kamio, M. Yao and I. Tanaka, *Protein Sci.*, 2011, **20**, 448–456.
71. A. L. Plant, M. Brighamburke, E. C. Petrella and D. J. Oshannessy, *Anal. Biochem.*, 1995, **226**, 342–348.
72. D. Pum, G. Stangl, C. Sponer, K. Riedling, P. Hudek, W. Fallmann and U. Sleytr, *Microelectron. Eng.*, 1997, **35**, 297–300.
73. T. Yang, O. K. Baryshnikova, H. Mao, M. A. Holden and P. S. Cremer, *J. Am. Chem. Soc.*, 2003, **125**, 4779–4784.
74. T. Yang, S.-y. Jung, H. Mao and P. S. Cremer, *Anal. Chem.*, 2001, **73**, 165–169.
75. A. Ono and E. O. Freed, *Proc. Natl. Acad. Sci. U. S. A.*, 2001, **98**, 13925–13930.
76. L. Xu, P. Frederik, K. F. Pirollo, W.-H. Tang, A. Rait, L.-M. Xiang, W. Huang, I. Cruz, Y. Yin and E. H. Chang, *Hum. Gene Ther.*, 2002, **13**, 469–481.
77. K. Kasahara and Y. Sanai, *Trends Glycosci. Glycotechnol.*, 2001, **13**, 587–594.
78. S. Qi, J. T. Groves and A. K. Chakraborty, *Proc. Natl. Acad. Sci. U. S. A.*, 2001, **98**, 6548–6553.
79. J. Boyle and R. Buckley, *J. Clin. Immunol.*, 2007, **27**, 497–502.
80. J. Bernardino de la Serna, G. J. Schütz, C. Eggeling and M. Cebecauer, *Front Cell Dev. Biol.*, 2016, **4**, 106.
81. E. Bamberg, H. Alpes, H.-J. Apell, R. Bradley, B. Härter, M.-J. Quelle and D. Urry, *J. Membr. Biol.*, 1979, **50**, 257–270.
82. P. Van Gelder, F. Dumas and M. Winterhalter, *Biophys. Chem.*, 2000, **85**, 153–167.
83. T. Nagao, D. Mishima, N. Javkhlantugs, J. Wang, D. Ishioka, K. Yokota, K. Norisada, I. Kawamura, K. Ueda and A. Naito, *Biochim. Biophys. Acta, Biomembr.*, 2015, **1848**, 2789–2798.
84. Y. Yu, J. A. Vroman, S. C. Bae and S. Granick, *J. Am. Chem. Soc.*, 2009, **132**, 195–201.
85. L. K. Tamm and H. M. McConnell, *Biophys. J.*, 1985, **47**, 105–113.

86. P. Mueller, D. O. Rudin, H. Ti Tien and W. C. Wescott, *Nature*, 1962, **194**, 979–980.
87. D. J. Muller, K. T. Sapra, S. Scheuring, A. Kedrov, P. L. Frederix, D. Fotiadis and A. Engel, *Curr. Opin. Struct. Biol.*, 2006, **16**, 489–495.
88. E. Kalb, S. Frey and L. K. Tamm, *Biochim. Biophys. Acta, Biomembr.*, 1992, **1103**, 307–316.
89. M. E. Sandison, M. Zagnoni and H. Morgan, *Langmuir*, 2007, **23**, 8277–8284.
90. M. Montal and P. Mueller, *Proc. Natl. Acad. Sci. U. S. A.*, 1972, **69**, 3561–3566.
91. M. Zagnoni, *Lab Chip*, 2012, **12**, 1026–1039.
92. M. Zagnoni, M. E. Sandison and H. Morgan, *Biosens. Bioelectron.*, 2009, **24**, 1235–1240.
93. N. Malmstadt, M. A. Nash, R. F. Purnell and J. J. Schmidt, *Nano Lett.*, 2006, **6**, 1961–1965.
94. B. Le Pioufle, H. Suzuki, K. V. Tabata, H. Noji and S. Takeuchi, *Anal. Chem.*, 2008, **80**, 328–332.
95. T. Osaki, H. Suzuki, B. Le Pioufle and S. Takeuchi, *Anal. Chem.*, 2009, **81**, 9866–9870.
96. S. Ota, H. Suzuki and S. Takeuchi, *Lab Chip*, 2011, **11**, 2485–2487.
97. A. Hirano-Iwata, T. Nasu, A. Oshima, Y. Kimura and M. Niwano, *Appl. Phys. Lett.*, 2012, **101**, 023702.
98. L. A. Kung, J. T. Groves, N. Ulman and S. G. Boxer, *Adv. Mater.*, 2000, **12**, 731–734.
99. J. S. Hovis and S. G. Boxer, *Langmuir*, 2001, **17**, 3400–3405.

CHAPTER 3

Single-biomolecule Interface for Advanced Single Molecular Sensors

MENG-YIN LI,*[a] ZHENG-LI HU,[a] YA-QIAN WANG,[b] SHUANG LI,[b] JIE YANG,[b] KAIPEI QIU[b] AND YI-TAO LONG[a]

[a] Nanjing University, Nanjing, China; [b] East China University of Science and Technology, Shanghai, China
*Email: mengyinli@nju.edu.cn

3.1 Nucleotide Detection

3.1.1 Nucleotide Detection at the Single-biomolecule Interface

3.1.1.1 Case Study I: First Nanopore-based Nucleotide Detection Experiment

In 1996, it was first shown that an electric field can drive a single-stranded RNA or DNA molecule through an α-hemolysin (α-HL) nanopore[1] in an artificial lipid bilayer. Such a pore works as a single molecular sieve whose steric limitation allows only a single-strand nucleic acid across; each molecule traverses through the pore as an extended chain that could induce partial blockage, which results in the characteristic ionic current blockage. The translocation of each molecule is observed as a transient decrease in ionic current whose duration corresponds to the length. Such a strategy can therefore be used to measure the length of the polynucleotide. In principle,

Detection Science Series No. 17
Confining Electrochemistry to Nanopores
Edited by Yi-Tao Long

Published by the Royal Society of Chemistry, www.rsc.org

with further improvements, this approach could provide direct, high-throughput detection of the sequence of DNA.

In the proof-of-concept experiment,[1] polyuridylic acid was used due to its minimal second structure, which might interfere with translocation of the polymer. Here, polyuridylic acid is denoted as poly(U). The sequence of the ssDNA being tested contained abundant adenine, cytosine, guanine and tyrosine with careful inspection *in silico* to avoid unnecessary base pairing during the experiment. The 150 nt ssDNA used in this demonstration is sequenced as 5′-CTCA CCTA TCCT TCCA CTCA TTTT CCTT AACC ATTT CATT CACC CATC TCTT CACT CCAT CTAT CACC TCCA TACA TACC CTCC ATAT TACA CTCC CATC ACTA TCAT TATC TACA TCCA TTAC ATCA CTAC TCCT CACA CTAC CATA CC-3′. According to the study, the polymer resides longer in the blockade than the homo-polymers. Based on this model 150 nt ssDNA strain, another 120 nt DNA (5′-CTCA CCTA TCCT TCCA CTCA TTTT CCTT AACC ATTT CATT CACC CATC TCTT CACT CCAT CTAT CACC TCCA TACA TACC CTCC ATAT TACA CTCC CA CTAC TCCT CACA CTAC CATA CC-3′) and 100-nt (5′-CTCA CCTA TCCT TCCA CTCA TTTT CCTT AACC ATTT CATT CACC CATC TC ACTA TCAT TATC TACA TCCA TTAC ATCA CTAC TCCT CACA CTAC CATA CC-3′) were also tested and could have successfully been detected by the nanopore electrochemistry.

When there is no substrate applied to the nanopore system, the application of −120 mV (*cis* side is grounded) induced a typical single pore current without pore gating transient (Figure 3.1). Then the addition of poly(U) to the *cis* side of the system triggers the pore–substrate interaction, which represents a few short-lived current blockages. As the typical characterization, the current size of each event should occupy 80–100% of the full pore open current; moreover, the length of the substrate (*i.e.* the polymer, DNA, RNA, *etc.*) determines the duration of the events lasting from hundreds to thousands of microseconds. Taking polymer as an example, the frequency of the transient blockages should be substrate concentration dependent and should be fitted into a linear relation until the rate-limiting condition is met. As shown in Figure 3.1, no events have been acquired in the absence of any substrate, and are even free of gating, indicating the pore is under good conditions. As long as the polymer is added, blockages with extremely long residence inside the pore is regarded as temporal blockage and could be recovered simply by setting the bias voltage to zero or reversed. These long-lived gating events are supposed to be induced by the secondary structure of the polymers that could not transverse through the pore. ‘Normal’ events that last for several milliseconds can be regarded as normal events that could also been observed with other DNA long polymers such as polyadenine (poly(A)), polythymine (poly(T)) and polycytosine (poly(C)). Suppose that the current intervals are caused by the translocation of the polynucleotides in linear formatting through the nanopore, the duration of the blockages should be in a linear relationship with the length of the substrate as well as be inversely proportional to the bias voltage. As a promising case, the 210-nt poly(U) was used to be probed by the nanopore. The analysis through the

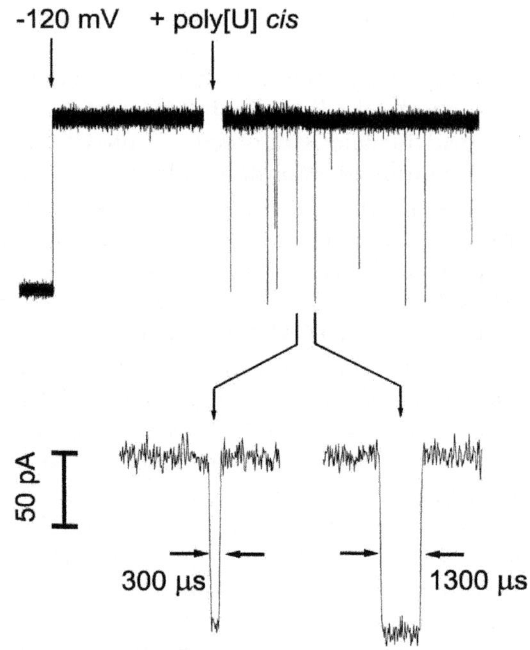

Figure 3.1 Oligomers of poly(U) caused transient blockades in the α-hemolysin single pore current. At the first arrow, a potential of −120 mV was applied across the membrane (*cis* side negative). This voltage caused a continuous current of −120 pA to flow. At the second arrow, poly(U) of mean length 210 bases was stirred into the *cis* compartment to a final concentration of 0.1 mg mL^{-1}. The polynucleotides caused short-lived current blockades. The inset (expanded time scale) shows two typical blockades with lifetimes of 300 and 1300 ms. For the purposes of illustration, the low time resolution current recordings (a total of 4 s is shown here) were digitally filtered using an exponential smoothing moving average algorithm.
Reproduced from ref. 1 with permission from National Academy of Sciences, U.S.A., Copyright 1996.

histogram distribution of the current duration demonstrated three distinct peaks (Figure 3.2). The similar histogram analysis for poly(U) with different size illustrated that the average duration of the second and third peaks should be correlated to the size for the poly(U) length as well as inversely related to the bias voltage. Moreover, the remaining peak, *i.e.* the first peak, refers to the events with swift residence. Such events are attributed to the simple collision of the substrates with the nanopore, inducing unsuccessful translocation. For sequencing, only events that are distributed in the range of the second and third peaks should be regarded as 'real' statistical translocation. It is noteworthy that the unbranched polymers could generate the translocation events with two distinct sojourn distributions, even the area of the event distribution is almost similar. One of the assumptions is that either 3′-end or 5′-end could enter the nanopore under the same

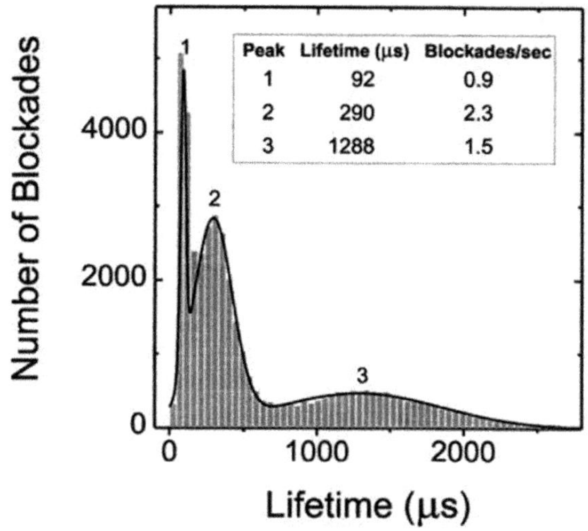

Figure 3.2 The characteristic lifetimes of pore blockades caused by poly(U) (0.1 mg mL^{-1}; mean length, 210 nt) fell within three well-defined peaks. The mean lifetimes corresponding to the three peaks were determined by fitting the sum of three Gaussians to the data.
Reproduced from ref. 1 with permission from National Academy of Sciences, U.S.A., Copyright 1996.

statistics. Since the nanopore lumen is commonly less than 2.6 nm, it is apparently too small to provide the steric effect to double-strand DNA to transverse the nanopore. Consequently, the double-strand DNA could only 'hit' the nanopore mouth, creating a unique peak denoted as peak 1 in Figure 3.2, which represents events with relatively short sojourn and high frequency.

Following the previous outcome, then single-strand DNA with diverse composition but the same molar quantity [*e.g.* poly(A) and poly(U); poly(C) and poly(G)] are mixed. Such a sample should yield many base pairs that hovered by unpaired 'free' single-stranded ends that should interact with the nanopore but hindered in a region of paired regions. These paired regions could produce ultimately long interaction events that represent blocking the nanopore. Such blockages make it impossible for the substrates to move outwards simply by passive diffusion; alternatively, a bias voltage with inversed polarity should be applied, *i.e.* a negative voltage is applied from the nanopore *trans* side to repel the negatively charged DNA strand towards the nanopore in this typical case. Consequently, the ions then moving through the nanopore represent a nanopore open state. In order to correlate the acquired blockages to the DNA or other substrates translocated through the nanopore, work was performed to pair the acquired signals falling into the second and third peaks with the actual single-strand DNA translocated across the bilayer from the *cis* (ground side or sample addition side) to the *trans* side (headstage side or voltage applied side) along through the nanopore path. Based on the previous assumption that the double-strand DNA could not translocate through the nanopore reaching the *trans* side of the nanopore due to the sizing limitation, alternatively only binds at the nanopore *cis* side mouth, the double-strand DNA was set as the internal control for the unintended contamination from the *trans* side, then the single-strand DNA that translocated from the *cis* side to the *trans* side could be accumulated and counted. As illustrated in Table 3.1 and Figure 3.3, the number of the estimated single-strand DNA accumulated at the *trans* side of the nanopore is correlated with the acquired nanopore transient events.

In order to quantitively determine the correlation between DNA concentration and blockage frequency, especially for the DNA strands that vary in size, polymers in different sizes with various concentration through the nanopore sensor were analyzed. The results demonstrated that the relatively

Table 3.1 DNA transport through α-hemolysin pores. Reproduced from ref. 1 with permission from National Academy of Sciences, U.S.A., Copyright 1996.

No. of pores	Blockades per min	Molecules per min
35	1300	1250
8	220	280
6	150	80

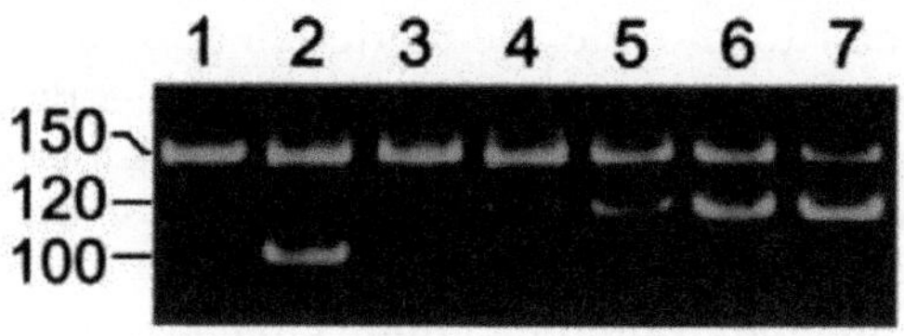

Figure 3.3 Single-stranded DNA traversing the α-hemolysin pore. The PCR-amplified product (lane 1) of a 40 ml sample of the *trans* chamber compared with (lane 2) the similarly amplified product of a 109-fold diluted sample of the *cis* chamber. All samples were taken following an eight-pore experiment in which a mixture of 150 nt single-stranded DNA and 100 nt double-stranded DNA was added to the *cis* chamber. Material in the *trans* chamber sample reflects accumulation during application of 2120 mV (*cis* chamber negative) for 42 min. Lanes 3–7: example of competitive PCR analysis (25) to show accumulation of single-stranded molecules in the *trans* chamber. The results show that the 40 ml sample used here contained 400 of the 150 nt single-stranded molecules. To each of five 40 ml samples of the *trans* chamber were mixed (lanes 3–7) 0, 100, 200, 400, or 800 molecules of a competing 120 nt DNA that was PCR amplified using the same primers as the 150 nt DNA. The position of the amplified 150, 120, and 100 nt polymers is shown on the left. Lanes 1 and 3 show independently amplified 40 mL samples of the same *trans* chamber.
Reproduced from ref. 1 with permission from National Academy of Sciences, U.S.A., Copyright 1996.

short strands (*i.e.* 200 nt) yield much more significant stochastic frequency than the longer strands (*e.g.* 1000 nt) at the same molar concentration. Such an outcome suggested that the nanopore single molecule events could be regarded as an extra sensitive probe, which could allow monitoring of the polynucleotide hydrolysis. This observation was then applied to monitor the enzymatic activity. The hydrolysis of DNA strands with certain size by ribonuclease, catalyzed by the degradation of RNA into smaller components, was monitored by the nanopore sensor. By increasing the substrate concentration, else decreasing the substrate size, could significantly raise the single molecule event occurrence. As a promising example, the addition of ribonuclease A to the poly(U) solution from the *cis* side of the measuring chamber generated an astonishing increase in the single-molecule event frequency, and eventually the enzymatic degradation makes the fragment smaller than the nanopore steric resolution, which is represented as lower single-molecule event frequency on the temporal axis.[1] Alternatively, the replacement of poly(A) showed no such effect on ribonuclease A, which demonstrated the better selectivity of ribonuclease A to poly(U) than poly(A). This nanopore based single molecule monitoring and analyzing tool could be useful to demonstrate many basic biological processes due to its label free single molecule resolution.

3.1.1.2 *Case Study II: Translocation Behaviors of Homopolynucleotide with Different Base Types*

Nucleotides must pass through the pore in sequence when the polynucleotide translocates because the confined space of the nanopore can only one strand of DNA or RNA entering at a time. Deamer *et al.* detected homopolymers of polycytidylic acid (poly(C)), polyadenylic acid (poly(A)) and polyuridylic acid (poly(U)) to distinguish poly(C) and poly(A) segments.[2] The outcome showed that the residual current of poly(U) was 18 ± 3 pA and the residual current of poly(A) was 17 ± 1 pA, which were distinguishable with poly(C) (Figure 3.4a). However, it was easy to distinguish poly(U) and poly(A) by duration time, because the durations of poly(U) were shorter than poly(A) (Figure 3.4a). Also, they studied the signal differences between poly(C) and poly dC single-stranded DNA and found that the blockages of poly(C) were greater than the blockages of the other RNA homopolymers (Figure 3.4b, c), meaning that the poly(C) molecule in the conditions of neutral pH and room temperature exists primarily as a single-stranded helix with a diameter of 1.3 nm.[3–5] Finally, RNA block copolymers ($A_{(30)}C_{(70)}Gp$) were detected by an α-HL nanopore and distinguished between poly(C) and poly(A) segments within a single RNA molecule successfully (Figure 3.4d). The study opened the way to sequence very long individual nucleic acid strands.

Inspired by the above work, the translocation duration and current flow during the traversing of individual polynucleotides was investigated.[6] They found that unlabeled DNA molecules could be discriminated and

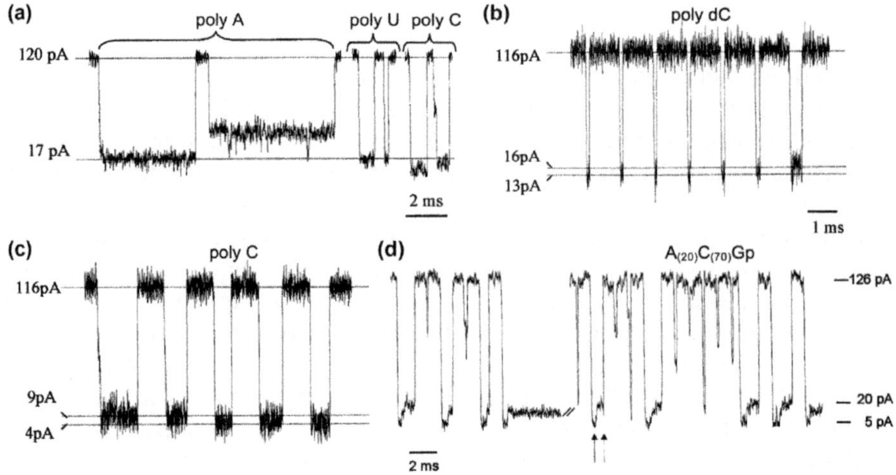

Figure 3.4 (a) Current traces for poly(A), poly(C), and poly(U). (b) Scatter plot in which each point corresponds to the amplitude and duration of a current blockade caused by a single poly(U) or poly(A) homopolymer passing through a single α-hemolysin pore. The data in (a) were digitally filtered at 25 kHz bandwidth. (b) and (c) Typical blockades caused by the poly dC and poly(C) molecules. The data in these experiments were digitally filtered at 50 kHz. (d) Typical blockades of monovalent ion current in the α-hemolysin pore caused by $A_{(30)}$ $C_{(70)}$ Gp RNA.
Reproduced from ref. 2 with permission from Elsevier, Copyright 1999.

characterized by α-HL. Additionally, there is a strong correlated relationship between the temperature dependence of the event properties and the tendency of secondary structure formation of some polymers. The scatter diagram showed that poly(dA)$_{100}$ and poly(dC)$_{100}$ were examined separately (Figure 3.5a). Poly(dA)$_{100}$ had a shorter current blockage and longer translocation duration than poly(dC)$_{100}$ (Figure 3.5b, c). One could also observe this phenomenon through the current trace (Figure 3.5d). Otherwise, they performed a series of experiments with poly(dA)$_{100}$ and poly(dC)$_{100}$ at different temperatures to study the effects of temperature on the translocation process. With the temperature increasing, the translocation duration of polynucleotides decreased (Figure 3.5e) because at lower temperature, the stacked structure of DNA must be broken to translocate through the α-HL; the extra time to disrupt this structure would enhance the translocation durations of polymers. In addition, at low temperatures the differences between polymers are striking. Therefore, it is possible to control the translocation speed and enhance the difference between numerous types of polymers.

Moreover, the nanopore-based artificial lipid bilayer could be ultimately feasible for performing long polymer translocation through a single nanopore or multiple nanopores. Single-nanopore analysis could be applied to probe the length of single molecule such as RNA or DNA strands, which could be

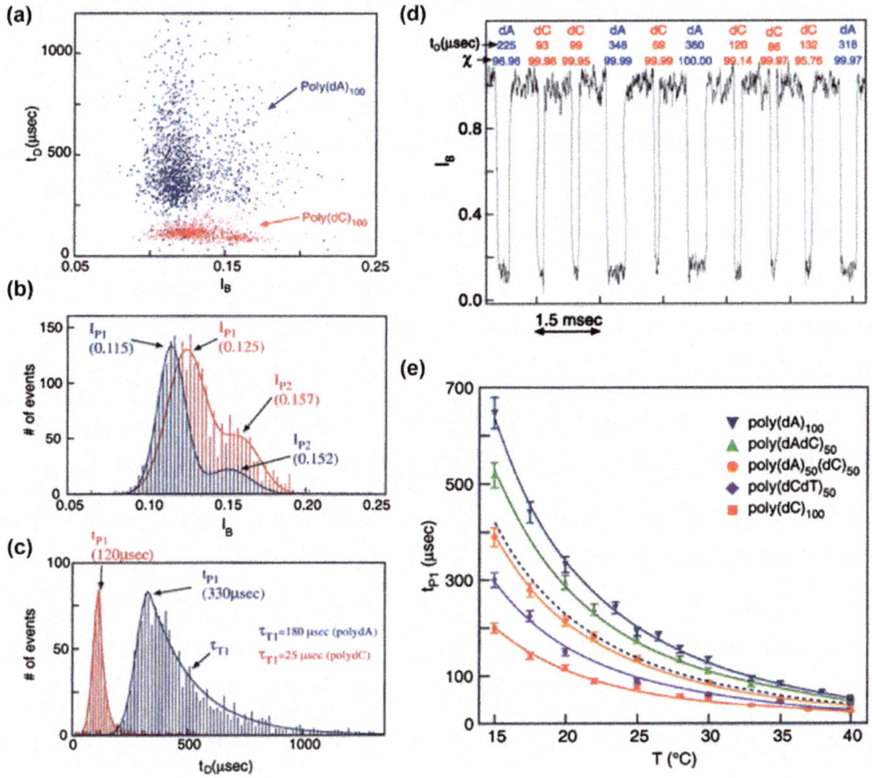

Figure 3.5 (a) Event diagram showing translocation duration *versus* blockade level for poly(dA)$_{100}$ (blue) and poly(dC)$_{100}$ (red) at 20.0 °C. Each point on this diagram represents the translocation of a single molecule that was characterized by its translocation duration, t_D, and blockade current, I_B. (b) Current histogram projected from the above event diagram; the color codes are the same. The two peaks corresponding to the two groups of events are denoted by I_{P1} and I_{P2}. The solid lines are fits of the data to a sum of two Gaussians. (c) Duration histogram projected from (a) for the first group of events. Note the well-defined peak locations, t_P, and the exponential temporal decay constants, t_T. The temporal decay constant associated with the poly(dC) polymers is approximately seven times shorter than the t_T associated with the poly(dA). (d) Representative current trace showing ten events recorded from a mixture of equal molar concentrations of poly(dA)$_{100}$ and poly(dC)$_{100}$. (e) Dependence of t_P for group 1 events for poly(dA)$_{100}$ (blue), poly(dC)$_{100}$ (red), poly(dA$_{50}$dC$_{50}$) (orange), poly(dAdC)$_{50}$ (green), and poly(dCdT)$_{50}$ (purple). Reproduced from ref. 6 with permission from National Academy of Sciences, U.S.A., Copyright 2000.

analyzed to determine the certain biophysical properties of the nucleotides. The molecular descriptors of small molecules could also be sensed through the single-nanopore system with promising examples such as single purine or pyrimidine nucleotides translocated across the artificial lipid bilayer *via* an aperture with confined spatial space. The translocation of analytes through the nanopore generates characteristic current blockages, which contains abundant single-molecule information such as the molecular size, molecular structure, dipole moment and other biophysical and biochemical properties. Moreover, by using such an advanced single-molecule sensor, the polynucleotide sequence could also be ultimately obtained in a label-free manner. In order to meet such a requirement, the following five principles should be adhered to, to meet the sequencing requirements. 1. The unique signature current of the four bases from the nucleotide should be distinct and the probing resolution should be sufficient for harsh conditions. 2. The size of the nanopore should be confined to fit the size of the single-strand DNA so as to ensure that only one strand should enter the nanopore lumen once. 3. The acquisition hardware should also be updated and the nanopore system should also be optimized to boost the current resolution in order to extract nucleotide information from the white noise as well as the substrate fraction inside the nanopore. 4. The movement of the substrate should be optimized to be unidirectional. Movement backward would increase the error rate as well as introduce inevitable trouble during the analysis. 5. The nanopore support should be robust enough to withstand long term measurement as well as sophisticated physiological conditions. While great efforts have been madc to overcome these challenges, the nanopore DNA sequencing have almost been achieved by developing some high spatial resolution biological nanopores combined with DNA polymerase or helicase to slow down and control the translocation speed.

3.1.2 Modulation the Translocation *via* Single-molecule Conformational Changes

3.1.2.1 Case Study III: DNA-binding Antibody Fab Fragment

α-HL was used to examine the conformation induced translocation of DNA polymer *via* the DNA-binding antibody antigen binding fragment (Fab HED10) effect (Figure 3.6).[9] The Fab HED10 exhibits four consecutive thymine residues and it could bind to poly(dT) (Figure 3.6a). As such, the interaction of 45-mer poly(dT)$_{45}$ with α-HL was studied under a potential of 100 mV and it was shown that it traversed the pore from the *cis* to *trans* side. The translocation events of poly(dT)$_{45}$ indicated two states named level I and level II, where level I displayed a partial blockade with a longer duration time, while a deeper blockade was attributed to level II. Poly(dT)$_{45}$ was thus subjected to a random process for exploring the vestibule area, resulting in a more deterministic translocation since the lifetime of level I was found to be much larger than that of level II (Figure 3.6b, top and 3.6c, left). Consequently, the conformational flexibility of DNA was indicated by level I together with the pore and target interactions. Moreover, a mixture poly(dT)$_{45}$:Fab HED10 fabricated *via* equimolar ratios was examined. In addition, it was found that the obtained complex produced similar blockage current with two current levels (level I and level II) comparable with

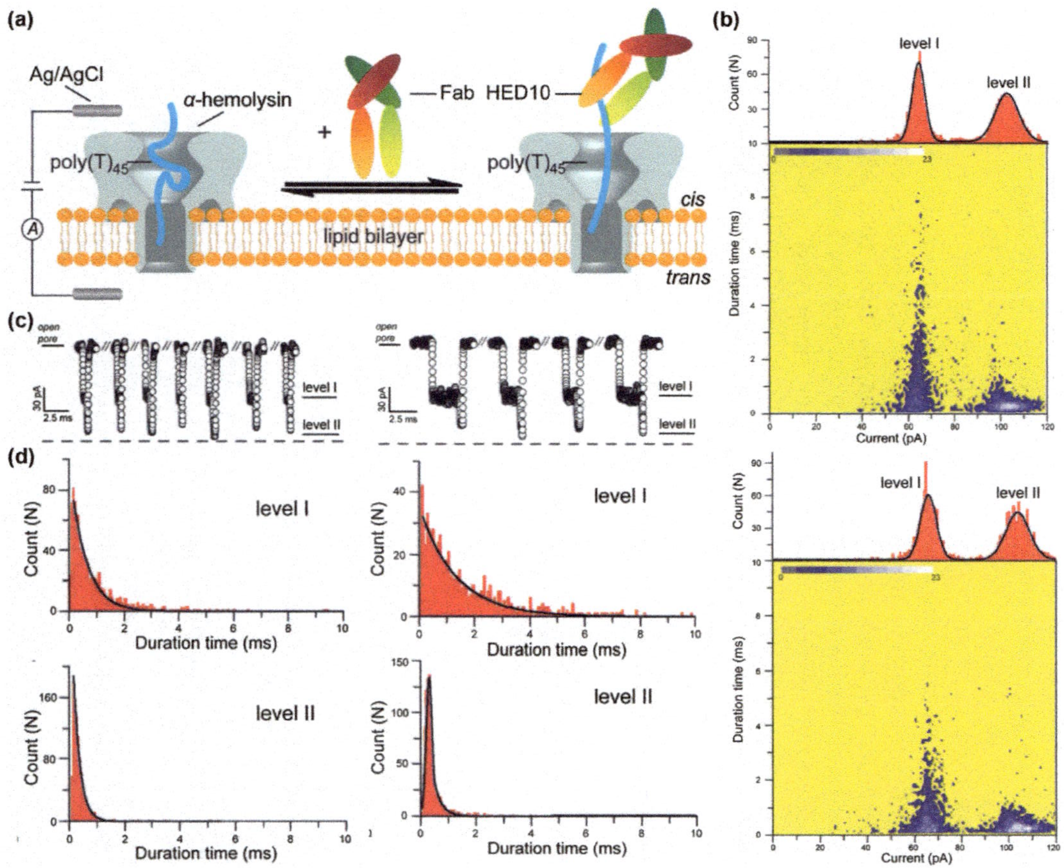

Figure 3.6 (a) Representation of an α-HL pore and the mechanism for the poly(dT)$_{45}$ specific binding with Fab HED10. (b) The contour plots fit Gaussian distributions with two current peaks and contour plots of current transit of the two blockage levels produced by poly(dT)$_{45}$ (top) and poly(dT)$_{45}$:Fab HED10 complex (bottom), respectively. (c) The typical current traces of poly(dT)$_{45}$ translocation events. (d) The duration time histogram for levels I and II.
Reproduced from ref. 9 with permission from the Royal Society of Chemistry.

poly(dT) alone, owing to the fact that Fab HED10 was very large to translocate (Figure 3.6b, bottom and 3.6c, right). Note that the presence of Fab HED10 did not change the current trace; however, it greatly affected the lifetime of level I by reducing it from 1.33 ms to 0.54 ms, but not the second one. According to the Arrhenius relationship related to a target escape over an energy barrier, the translocation analysis for poly(dT)$_{45}$:Fab HED10 revealed two components of the energy barrier, including a free energy cost combined with an entropic recoil force due to the confinement of poly(dT)$_{45}$ and the energy caused by the exploration of the entrance of the β-barrel with the free-jointed poly(dT)$_{45}$ chain. The number of conformational complexes formed by the Fab HED10 and the polymer was reduced because of the existence of only four bases, which offered the binding sites, so Fab occluded as large as 25% of DNA. Also, given that DNA-binding antibodies were positively charged, the translocation process occurred in the opposite sense to DNA. Hence, Fab was considered as a rudder that steered the DNA into the pore. This study allowed understanding the translocation property of polymers for innovation in the fabrication of nanopore biosensors with high specificity.

3.1.2.2 Case Study IV: Peptide–Oligonucleotide Conjugates (POCs)

The fact that the behavior of the peptide–oligonucleotide conjugates (POCs) are generally influenced by structural variations was studied. To do so, the structural properties of two POCs were examined *via* an α-HL nanopore by single-molecule analysis in order to understand their translocation process across α-HL, as shown in Figure 3.7.[7] Both POCs were fabricated and explored with α-HL separately. Thus, by linking a collagen-like peptide (P7) with a 20-mer oligonucleotide *via* flexible disulfide, a compound 1 was obtained, while a noted compound 2 was synthesized by conjugating the same P7 to an oligonucleotide poly(dA)$_{60}$. Under a bias potential of 100 mV, compound 1 was investigated in α-HL through translocation events and it was shown that around 80% of the events exhibited a characteristic blockage current. During that process, the fluctuations of the ion current displayed three segments, namely SI, SII and SIII, with SII the lowest generated by the small and short linker chain. It was found that P7 underwent an unfolding process along with the successive translocation of the additional part when one of the three ssDNA was first driven into α-HL, which was confirmed by the duration time of each segment. As for compound 2, the current blockage traces were the same as compound 1 showed when α-HL was held at 100 mV. Upon comparing the duration times, it was noticed that the translocation process across α-HL increased with increased length of the POCs. The as-investigation allowed a deeper understanding of the translocation properties of biomolecules where α-HL was considered as a powerful tool to distinguish different structures of POCs at the single-molecule level.

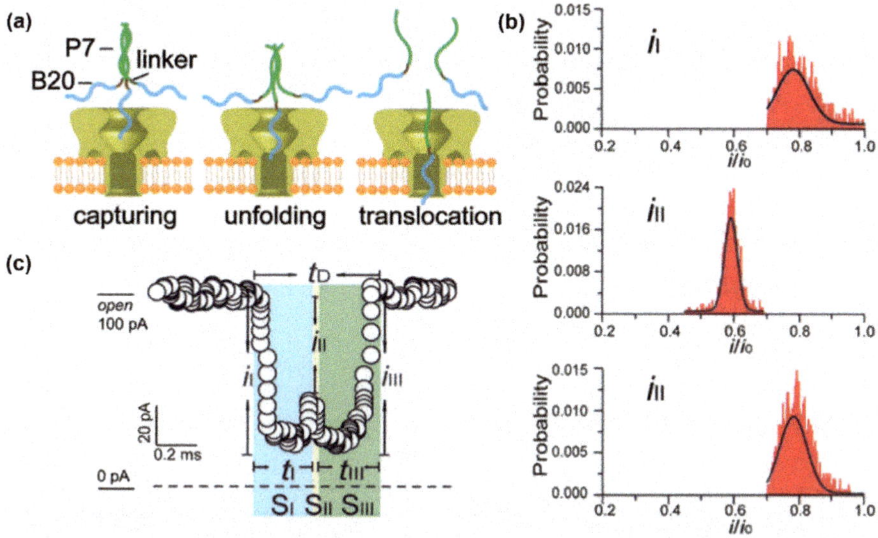

Figure 3.7 (a) A model illustrates the translocation of compound 1 through the α-HL pore in Tris–HCl (pH = 7.80). (b) Histograms of each part of the blockage current were fitted to a Gaussian function. (c) Typical current trace of a translocation event for compound 1.
Reproduced from ref. 7 with permission from the Royal Society of Chemistry.

3.1.2.3 Case Study V: ATP-binding Aptamer (ABA)

The aptamer is a specific oligonucleotide that can form secondary structure. The secondary structure falls into the interest of nanopore conformational detection. ATP-binding aptamer (ABA) is then designed and nanopore sensing is applied to detect its conformational changes. The ABA has a stem–bulge–stem structure, which is a typical secondary structure. The aptamer consists of two G-quartets and two small Watson–Crick helices form a linear chain to adopt conformation changes that can be regulated according to the environmental conditions (Figure 3.8). Four conformations of DNA, folded, linear, duplex strand and complex with ATP were studied *via* the α-hemolysin nanopore. According to the analysis of current behavior, the linear formation of ABA provides a more unique signature largest blockage current than any other conformation. Regarding the residence time, folded ABA, ABA:reporter and ABA:ATP are ranked from long to short. In addition, the outcomes demonstrated the accessibility of distinguishing the current response of three forms of ABA, *i.e.* ABA, ABA:reporter and ABA:ATP. The monitoring of the competition of ABA target molecules in different composition is also possible *via* a contour plot. This demonstration allows one to use the nanopore technique to real-time monitor the four forms of ABA.[8]

Now, the characteristic events induced by the conformational changes is applied into the various applications. Based on the distinct typical events of

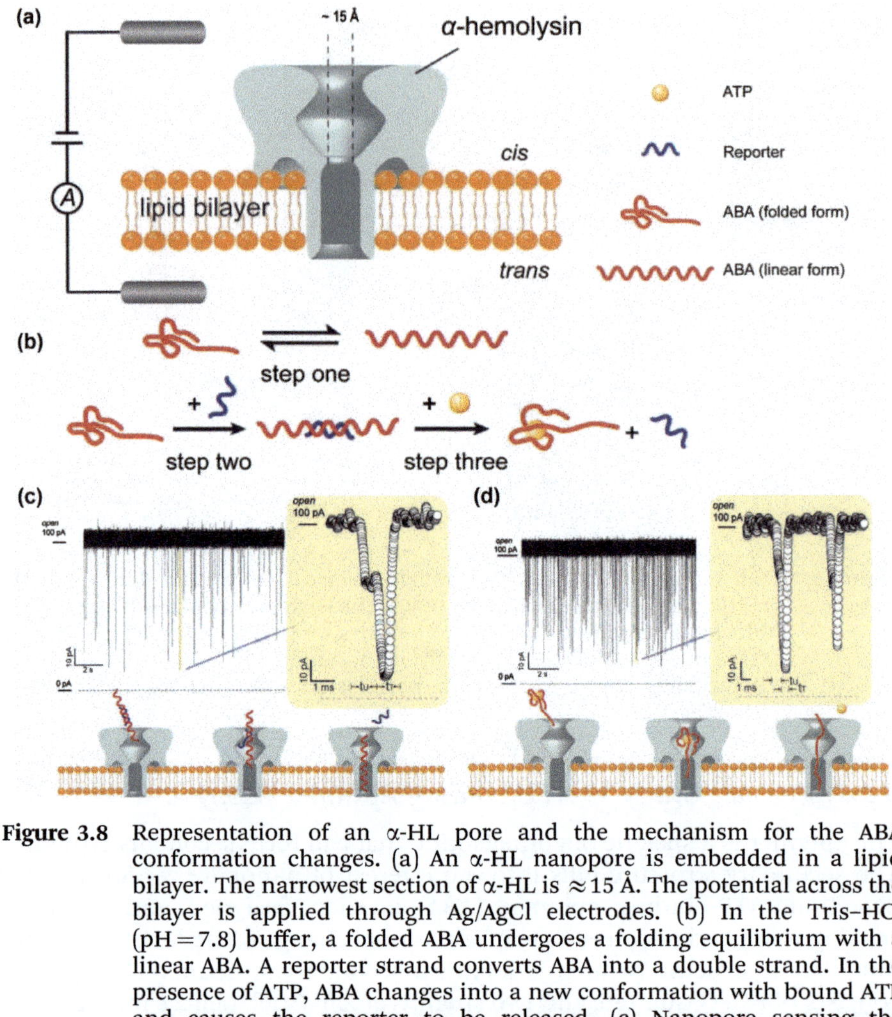

Figure 3.8 Representation of an α-HL pore and the mechanism for the ABA conformation changes. (a) An α-HL nanopore is embedded in a lipid bilayer. The narrowest section of α-HL is ≈15 Å. The potential across the bilayer is applied through Ag/AgCl electrodes. (b) In the Tris–HCl (pH = 7.8) buffer, a folded ABA undergoes a folding equilibrium with a linear ABA. A reporter strand converts ABA into a double strand. In the presence of ATP, ABA changes into a new conformation with bound ATP and causes the reporter to be released. (c) Nanopore sensing the ABA:reporter events. Top: typical translocation signal and raw data; bottom: potential mechanism for the duplex of ABA:reporter unzipping to traverse the pore. (d) ABA:ATP complex detected by the α-HL pore. Top: characteristic events of the ABA:ATP complex and its experimental data; bottom: proposed mechanism of ABA:ATP transiting the α-HL pore. Reproduced from ref. 8 with permission from John Wiley and Sons,

ssDNA and the unfolding process of dsDNA, the detection of MicroRNA with a certain sequence could be detected using a designed complemental DNA chain.[118] Additionally, due to the volume characteristics induced by conformational change, the events with distinct current blockage and duration have been used to detect the protein conformations and ligand binding[119,120] and the DNA G-quadruplex induced by metal ions.[32]

3.1.3 Localize the DNA Targets at the Hot-spot for High Sensitivity

3.1.3.1 Case Study VI: Phi29 DNA Polymerase Decorated Nanopore

Coupling nucleic acid processing enzymes to nanopores allows governed movement of every individual DNA or RNA strands, which is reported as an ionic current *versus* time series. A large number of individual enzyme complexes is examined in single-file order at high temporal and spatial resolution. The phi29 DNA polymerase (phi29 DNAP) that exhibits notable processivity and high affinity for DNA substrate, has been considered as an excellent candidate. Therefore, the phi29 DNAP–DNA complexes are robust when captured under the applied potential across the α-HL nanopore (Figure 3.9). DNA substrates were replicated at the nanopore entrance by exploiting the 3′-5′ exonuclease activity of wild-type phi29 DNAP to excise a 3′-H terminal residue, which yields a primer strand 3′-OH. DNA was synthesized with the presence of deoxynucleoside triphosphates, which allowed measurement of substantial sequential nucleotide additions that was hindered only by the DNA template length in real time. Traversing of phi29 DNAP along DNA substrates was monitored in real time at nanoscale precision as the template was drawn through the nanopore lumen during the replication.[10] Moreover, taking advantage of the high spatial resolution of the MspA nanopore system, this phi29 DNA polymerase method has achieved DNA sequencing and constructed the basics of nanopore DNA sequencing.[121]

3.1.3.2 Case Study VII: Uracil DNA Glycosylase

In an α-HL nanopore biosensor, the DNA hairpin structure is often used to localize the target sequence into the constriction region of the nanopore known as the sensing hot spot. This sensing strategy significantly increases the sensitivity of an α-HL nanopore biosensor for discrimination of the base difference including DNA damage, as shown in cases VIII, IX, X, XII and XIII. Deamination of cytosine to yield uracil is some of the important DNA damage[11] that will lead to the U:G(guanine) mismatch and causes/results in a T:A mutation in replication.[12] Thus, the errors will happen in the messenger RNA and translational process/protein.[13] Then, the DNA repair enzyme and uracil-DNA glycosylase (UDG) will cleave the *N*-glycosidic bond between the uracil base and the sugar of the nucleotide to repair the deamination lesions.[14] The ability to measure UDG activity in label-free and high-throughput is an important goal. Nanopores have been regarded as a simple approach to monitor DNA activity and characterize DNA nucleotides. α-HL, as a kind of protein nanopore, has been widely used at the single-molecule level.[15] Monitoring of DNA repair enzyme activity using a wild-type α-HL nanopore based on base-specific interaction of dsDNA with a sensing

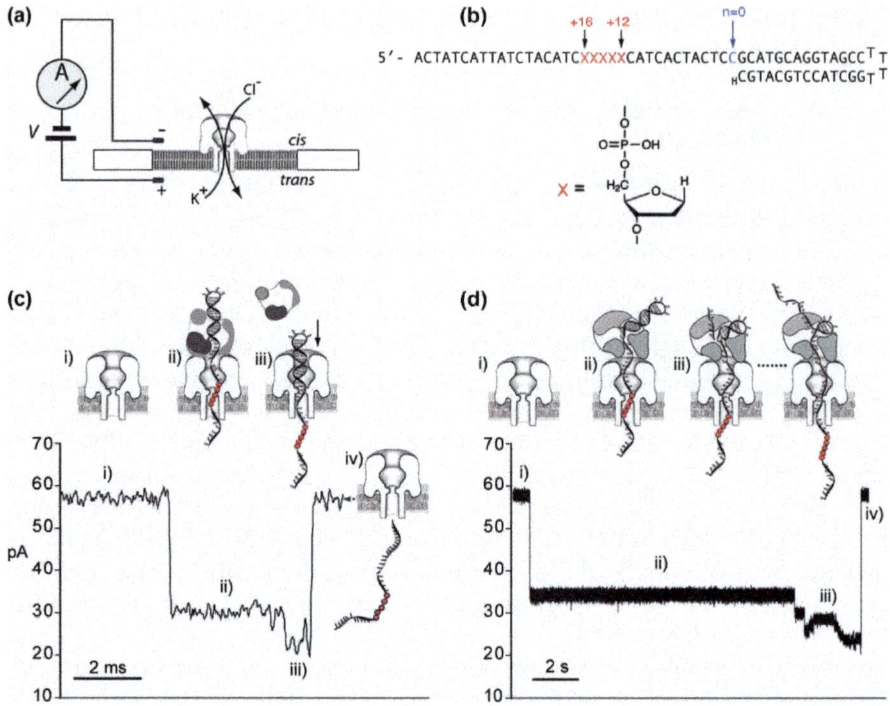

Figure 3.9 Capture of polymerase–DNA binary complexes in the α-HL nanopore. (a) Schematic of the α-HL nanopore setup. An α-HL nanopore inserts into an artificial lipid bilayer that separates two 100 μL chambers containing 10 mM K-HEPES, pH 8.0, 300 mM KCl, 1 mM DTT and 1 mM EDTA at 23 °C. The nanopore buffer contained no added $MgCl_2$. (b) The DNA hairpin substrate used in this experiment. (c) Ionic current signature for capture of a KF (exo-)–DNA complex at 180 mV applied potential. (d) Ionic current signature upon capture of a phi29 DNAP–DNA complex at 180 mV potential. Reproduced from ref. 10 with permission from American Chemical Society, Copyright 2010.

zone 'latch' is specific for a dsDNA structure in α-HL. The accurate quantitation of the kinetics of base repair reactions is achieved by a large increase in the ion pore current resulting in the presence of a single abasic site that is in the proximity of the latch zone. In general, the nanopore is a powerful enzymology tool that realizes the identification of structural change for single base in ds-DNA.[16]

3.1.3.3 *Case Study VIII: Cytosine: Cytosine Mismatch*

Gene regulation in human cells can be controlled by epigenetic modification of the nucleobase cytosine, which have an impact on the development of a lot of diseases.[24] The most universal modification is ref. 25. Recent research has discovered three other epigenetic modifications to cytosine. Hydroxymethyl-(hC),8-formyl-(fC);[26] and carboxy-(caC), which together constitute active reversal of cytosine methylation by sequential oxidation, base excision and subsequent repair.[27] Discriminating between C, mC and the oxidized derivatives of mC is of vital importance in understanding how genes regulate cell function and development. The kinetics of base flipping at the mismatch site is monitored by holding the mismatch at the latch construction of α-hemolysin (Figure 3.10). Thus, the modification can be identified unambiguously without labelling and complicated statistical analysis.

3.1.3.4 Case Study IX: γ-Hemolysin for Isoguanine Detection

The γ-hemolysin (γ-HL) produced by Staphylococcus aureus, has a larger vestibule and a wider β-barrel than the commonly used α-HL. It has been investigated as a potential nanopore for single molecule analysis (Figure 3.11). It has been reported that the γ-HL is a cation-selective pore and voltage-dependent gating. Therefore, the dsDNA is translocating through this nanopore at optimized experimental conditions. The translocation of the dsDNA through the nanopore is at the rate *ca.* 20 bp ms^{-1} and allows guanine:cytosine and a mismatch case such as isoguanine:cytosine to be distinguished. This nanopore opens an area where the biological nanopore could be synthesized into a novel nanopore that allows specific dsDNA monitoring purposes.[36]

3.1.3.5 Case Study X: 8-Oxo-7,8-dihydro-2′-deoxyguanosine

Oxidative stress in the cell is related to multiple diseases, such as cancer, and neurological and heart stress.[17] Reactive oxygen species (ROS) result in

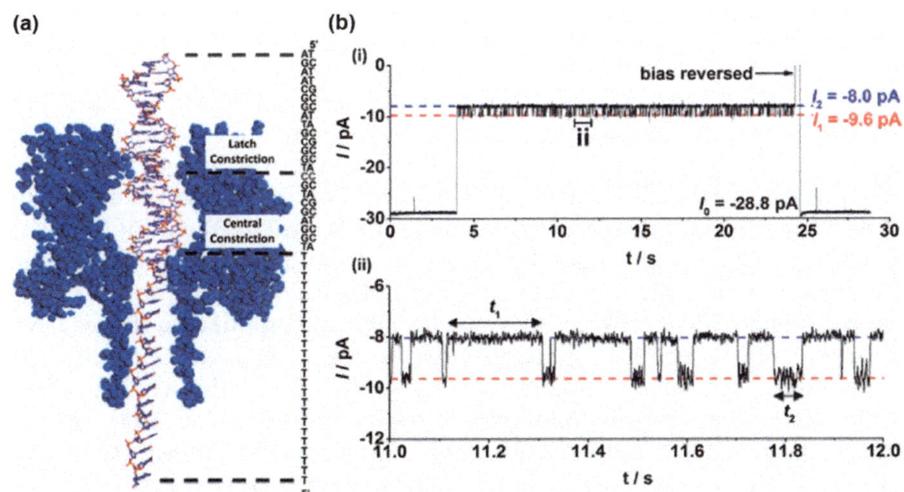

Figure 3.10 Trapping of DNA and analysis of base flipping at a C:C mismatch site within a DNA duplex.
Reproduced from ref. 28 with permission from American Chemical Society, Copyright 2017.

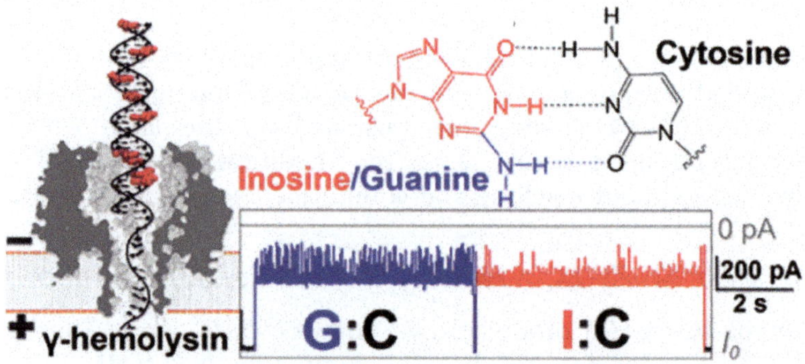

Figure 3.11 Cross section structure of the γ-HL nanopore showing the dimensions of the lumen. Structure of G:C and I:C base pairs. I – t trace of I:C-containing duplexes captured by the protein pore (post filtered to 2 kHz).
Reproduced from ref. 36 with permission from American Chemical Society, Copyright 2018.

oxidation of many components in cells and have potential to cause mutation in DNA bases.[18] The most important of this DNA base damage is 8-oxo-7,8-dihydroguanine (OG), which is an oxidized base that exists in 1 in 10^6 bases at normal cellular conditions, and exists at much higher levels in stress or certain disease conditions. Present methods for detecting OG can neither provide sequence information,[19] nor provide data about multiple lesions per strand, which is highly detrimental to DNA function.[20] In contrast, nanopore ion pore detection and many other single molecule sequencing methods provide the potential to obtain the information of sites on DNA strands based on the passage of DNA through ion pores. However, these sequencing methods have obstacles due to insufficient current resolution.[21] It has previously been shown that this obstacle can be resolved with the immobilization technique, which was introduced by the Bayley and Schmidt laboratories, capturing and immobilizing the ssDNA in an α-HLpore by linking a large molecule to the DNA. This technique results in a longer time of the DNA inside the α-HL and makes it able to discriminate the entering orientation of the DNA and a single change in the base sequence. Furthermore, this mechanism was applied to detect DNA damage, such as the detection of a single oxidative base lesion like ^{O}G using a wild-type α-HL nanopore combined with a membrane suspended across the orifice of a glass nanopore membrane (GNM). The GNM with an orifice diameter of 1–2 μm could provide a strong lipid bilayer that supported low electrical noise. The oligomers were immobilized inside the ion pore in α-HL with a tethered biotin–streptavidin complex[22] because the translocation of ssDNA is too far to identify accurately. Then, the biotin–streptavidin DNA complex was driven by voltage into the ion pore. The complex was held for 1 to 2 s in the pore to collect current blockage event data. The open pore current is restored by reversing the voltage bias, which drives the complex into the bulk solution. Figure 3.12 shows a typical current–time trace when the pore is unblocked and blocked, respectively. The schematic in Figure 3.12 also shows the different conditions in the biotin streptavidin immobilization experiment, including open pore capture and release.

This strategy in α-HL shows the power for the measurement of DNA damage and especially the detection of a single lesion within both hetero- and homo-polymer DNA.

3.1.3.6 Case Study XI: Peptide Labelled MicroRNA Recognized by an Engineered Nanopore

A nanopore sensor has recently been designed to electrically detect microRNAs (miRNAs),[37,38] a class of tiny but extremely important regulatory RNA molecules,[39–42] providing a useful tool in medical diagnostics. However, there are still some problems such as the presence of numerous other nucleic acid species found in biofluid extracts that can non-specifically interact with the nanopore and contaminate the target signals so to reduce the detection accuracy. To utilize a polycationic peptide-PNA probe as the carrier to achieve the selective miRNA detection in the nucleic acid mixture (Figure 3.13).[43] The probe can specifically hybridize with target miRNAs, which can be captured by the α-hemolysin and induce characteristic current events. Therefore, only the target miRNAs can be detected and distinguished from the probe. The study selected the Let-7 tumor-suppressing microRNA family[44–46] as the target and P_{7b} containing a 10-base PNA sequence designed to specifically hybridize with Let-7b as the probe (Figure 3.13). There were no current signals with the addition of Let-7b. In addition, the current trace of P_{7b} showed a large number of blockages. When the Let-7b/P_{7b} mixture was added, a large number of blockages occurred that had a longer duration and shorter blockage than P_{7b} (Figure 3.13). This novel approach offers a useful method for clinical detection.

3.1.4 Direct Detection of Hetero-nucleotide *via* Ionic Current Discrimination

3.1.4.1 Case Study XII: Direct Readout of Single-Nucleotide Base Discrimination

Recent studies have found that aerolysin has an excellent capacity for unmodified oligonucleotide detection without any additional approaches. Driving a oligonucleotide of four nucleotides length through aerolysin occludes nearly 50% amplitude of the open pore current. Further, the total internal reflection fluorescence (TIRF) measurements have provided direct evidence to confirm the translocation of the single-stranded DNA through an aerolysin nanopore.[122] The inherent advantages of aerolysin reveal its

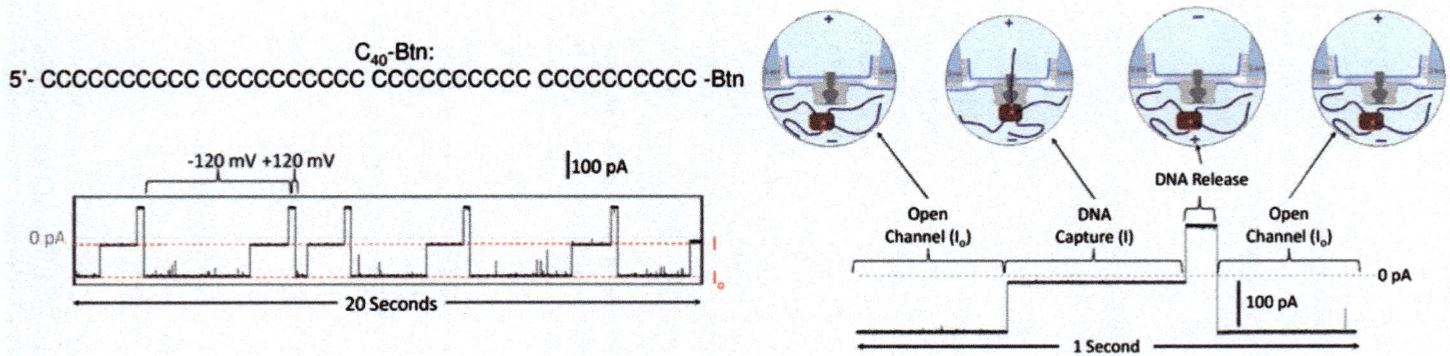

Figure 3.12 Strep-Btn-C40 structure, example current–time trace and molecule capture/release schematic. A C40 ssDNA strand is tethered to a Btn molecule on the 3′ end, and for strand immobilization within the α-HL pore, the Btn-DNA is bound to Strep. Schematic not drawn to scale.
Reproduced from ref. 23 with permission from American Chemical Society, Copyright 2010.

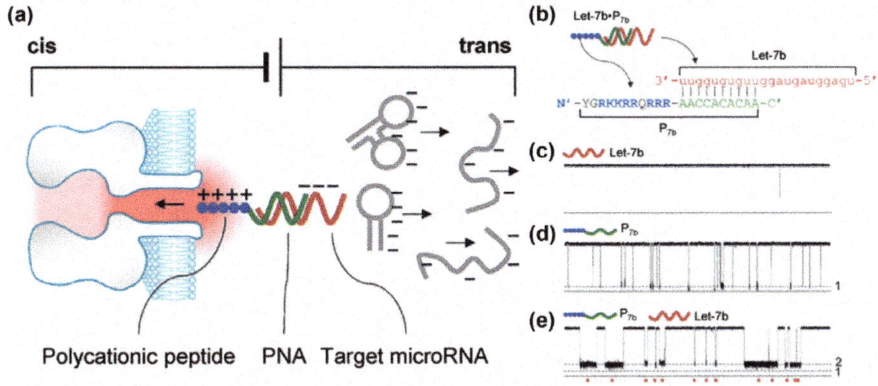

Figure 3.13 (a) Cationic probe-enabled interference-free detection of miRNAs in the nanopore. The probe comprises a capture domain (PNA, green) attached with a polycationic polymer lead (peptide, blue). The capture domain hybridizes with the target miRNA (red). Under a transmembrane voltage, the miRNA• probe complex is drawn into the nanopore by the electric field gradient at the pore opening, while any free nucleic acids without the probe binding carry negative charges (gray) and electrophoretically move away from the pore. (b) Sequences of the target miRNA Let-7b and the polycationic peptide-PNA probe P_{7b}. (c)–(e) Current traces for the K131D pore in the presence of (c) Let-7b alone, (d) P_{7b} alone and (e) Let-$_{7b}$/P_{7b} mixture.
Reproduced from ref. 43 with permission from American Chemical Society, Copyright 2013.

superiority in the discrimination of polydeoxyadenines (dA_n, $n = 2$, 3, 4, 5 and 10) based on high current resolution and sufficient temporal sensitivity (Figure 3.14).[29] For instance, the separation of residual current level between dA_3 and dA_4 can be up to 17%. The observed high spatial and temporal sensitivities result from the special geometry and the strong electrostatic interactions between aerolysin and the oligonucleotides.[29] Moreover, a set of short DNA oligomers with single-base variation in a dA_3 background could also be resolved based on the two-fold information of the residual current levels and the duration simultaneously (Figure 3.14).[29,30]

Moreover, aerolysin nanopore exhibits high sensing capability nucleic acid detection. The first direct characterization of single native RNA of $(A)_4$ and $(U)_4$ shows that $(A)_4$ induces about 10% large blockade current level compared with $(U)_4$ (Figure 3.15). Moreover, the statistical results display that duration time of $(A)_4$ is one hundred times longer than $(U)_4$. According to previous studies, the energy barrier for RNA traversing through aerolysin is lower than that for DNA. Moreover, the strong RNA interaction is due to the hydroxyl in pentose, which increases the translocation time. This study opens up a novel insight into an aerolysin nanopore detecting RNA, such as discrimination of RNA epigenetic modification and selective detection of miRNA.[47]

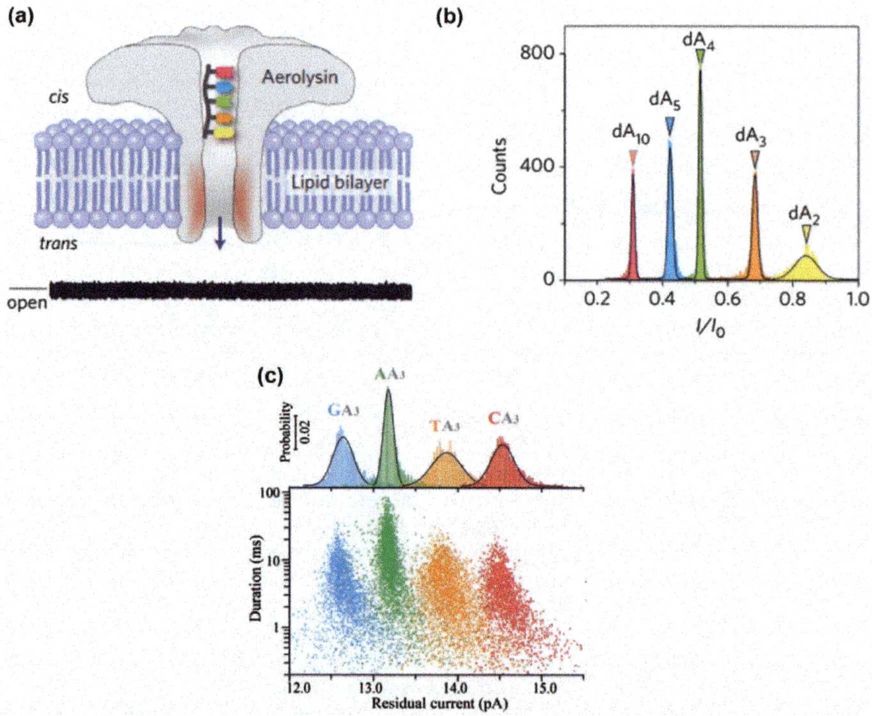

Figure 3.14 (a) Illustration of a short oligonucleotide traversing through an aerolysin nanopore, which is shaped like a rivet and embedded in a lipid bilayer. The region of positively charged residues positioned in the β-barrel is shown by the red shading. The two compartments of the bilayer chamber are termed *cis* and *trans*. (b) Blockade current histograms for dA_n ($n = 2$, 3, 4, 5 and 10), including Gaussian fits. Here, I represents the blockade residual current and I_0 is the open pore current, therefore, I/I_0 reflects the residual current depth. The current separation is the difference between two adjacent Gaussian peak values of dA_n ($n = 2$, 3, 4, 5 and 10). The current separations between dA_2 and dA_3, dA_3 and dA_4, dA_4 and dA_5, and dA_5 and dA_{10} are 15, 17, 10 and 12%, respectively. (c) The residual current histograms (top) and scatter plots (bottom) of duration *versus* residual current of CA_3, TA_3, AA_3, and GA_3, respectively. All residual current histograms were fitted to Gaussian distributions. (a) and (b) Reproduced from ref. 29 with permission from Springer Nature, Copyright 2016. (c) Reproduced from ref. 30 with permission from John Wiley and Sons, © 2017 WILEY-VCH Verlag GmbH & Co. KGaA, Weinheim.

3.1.4.2 *Case Study XIII: Methyl-cytosine Discrimination from Homo-nucleotide*

Single-nucleobase identification allows the possibility for personalized diagnosis, diverse responses to treatments, and better understanding of genomes in the future. In addition, the subtle difference between methylated cytosine (^{m}C) and cytosine (C) can be directly discriminated by wild-type

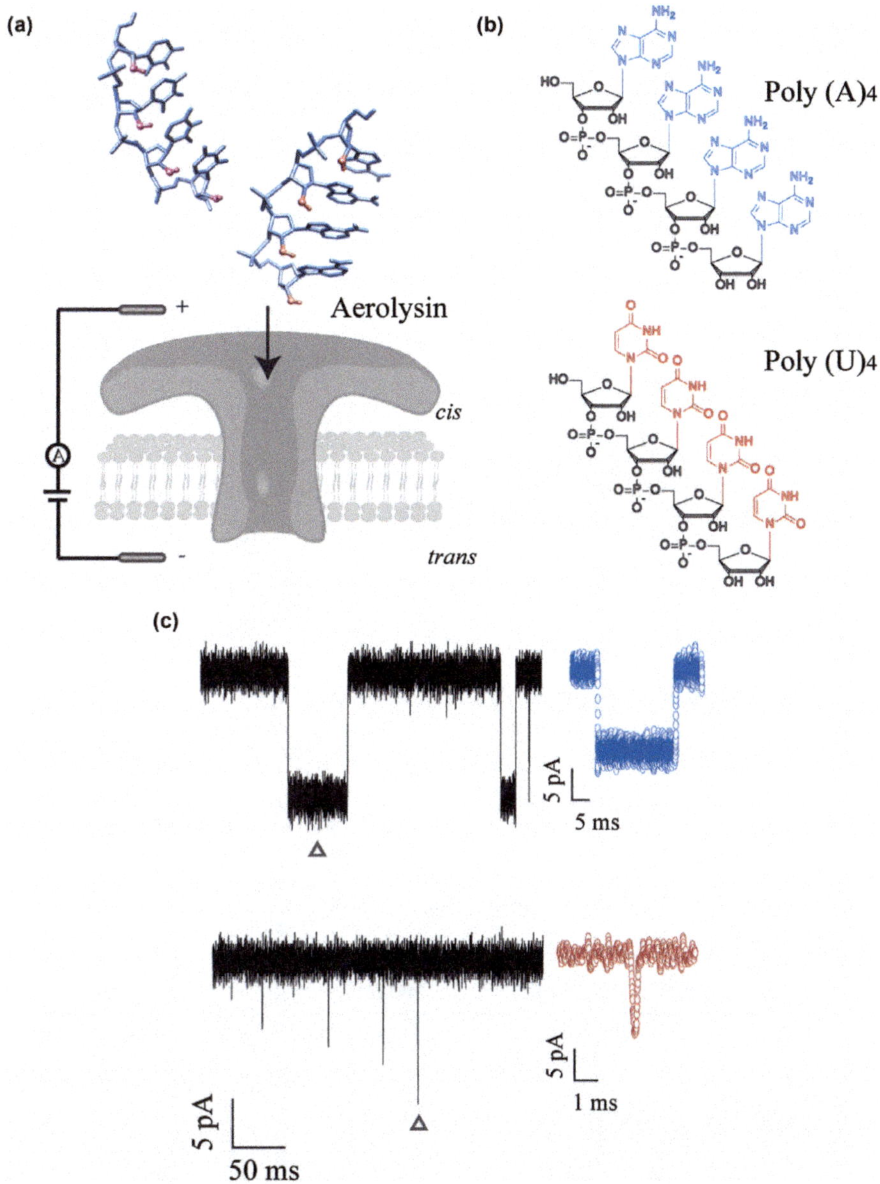

Figure 3.15 (a) Illustration of RNA traversing through an aerolysin nanopore. (b) Structures of Poly(A)$_4$ (upper) and Poly(U)$_4$ (below). The adenine and uracil are shown in blue and red, respectively. (c) Continuous current recording traces of (A)$_4$ and (U)$_4$ and corresponding typical events, respectively. All data were acquired in 1.0 M KCl, 10 mM Tris, 1.0 mM EDTA, pH 8.0 at +60 mV. The ultimate concentration of each RNA in the *cis* chamber is 2 μM.
Reproduced from ref. 47 with permission from American Chemical Society, Copyright 2018.

(WT) aerolysin under human serum conditions.[31] Considering that the DNA methylation is a fundamental epigenetic modification in human genes, detection of DNA methylation in human serum is of vital importance to promote the development of clinical research. Taking advantage of the high stability of aerolysin structure in a harsh environment, methylated cytosine and cytosine by aerolysin nanopore in human serum can be distinguished. As for WT aerolysin showing high sensitivity for detection of DNA, the subtle difference between ^{m}C and C could be observed directly (Figure 3.16). To

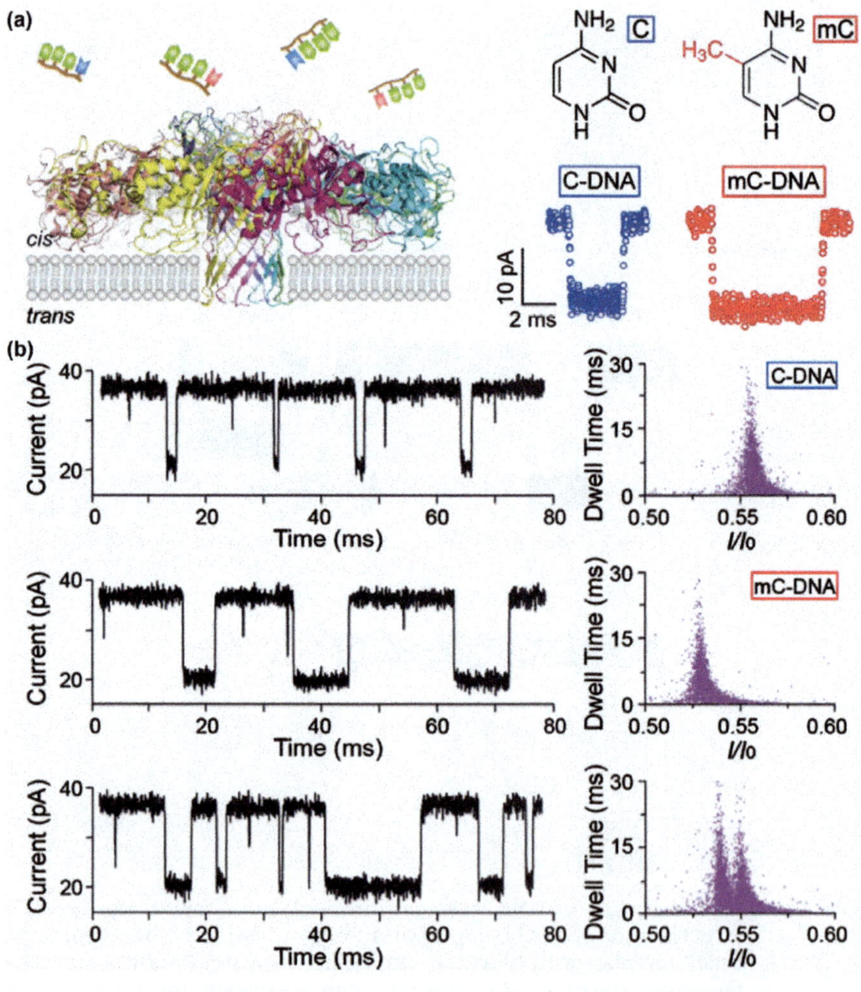

Figure 3.16 (a) Nanopore detection of methylated and unmethylated DNA. (b) Single-pore recordings of methylated and unmethylated DNA. Continuous raw current traces and corresponding scatter plots of C-DNA, ^{m}C-DNA, mixtures of C-DNA and ^{m}C-DNA.
Reproduced from ref. 31 with permission from American Chemical Society, Copyright 2017.

further determine the methylated cytosine in complex samples and to explore the possibility of aerolysin nanopore for clinical diagnosis, methylated cytosine in the presence of human serum was detected.[31]

3.1.4.3 Case Study XIV: Monitoring Exonuclease I Activity

Taking advantage of the excellent ability for analyzing oligonucleotides and detecting oligonucleotides as short as two nucleotides in label free, it is possible to apply an aerolysin nanopore to directly study enzymatic reaction processes. In fact, the real-time monitoring of stepwise cleavage of dA_5 by Exonuclease I (Exo I) has been achieved, and the current levels of every intermediate during the entire digestion process has been ambiguously determined and resolved (Figure 3.17). In the initial state, the majority of the oligonucleotides in the solution are dA_5, which generated a Gaussian peak of $I/I_0 \approx 0.37$ (I represents the residual pore current, while I_0 represents the open pore current). Then the Gaussian peaks attributed to dA_4 and dA_3 successively increased as the dA_5 peak decreased significantly, revealing the cleavage of dA_5 by Exonuclease I. About 8 h later, the majority of dA_5 was cleaved into dA_4 and dA_3. Meanwhile, dA_4 began to decrease and dA_3 continued to increase. Finally, nearly all the species were cleaved into dA_3. Since the operational conditions were not optimal for Exo I activity, after 20 h of recording little dA_3 was cleaved into the terminal dinucleotide. The above sequential changes in the current histograms are consistent with the stepwise cleavage of dA_5 by Exo I. Therefore, the whole digestion process of dA_5 by Exo I can be monitored in real time by identifying every individual cleavage intermediate from the current–time traces.[29]

3.1.4.4 Case Study XV: DNA Secondary Structure

Meanwhile, further studies could be accomplished by the virtue of pH, temperature and buffer concentration. By lowering the pH on one side of the nanopore, aerolysin is immediately 'activated' and able to capture target analyte from the opposite side.[8] Owing to the exceedingly narrow diameter (~1 nm), aerolysin generally squeezes a linear oligonucleotide into its lumen, but cannot accommodate a structured oligonucleotide with large volume. The oligonucleotides with the secondary structure are created by the cation-regulation mechanism without the requirement of high temperature or chemical denaturation. A series of oligonucleotides 5′-(TTAGGG)$_n$-3′ from a human telomere sequence, which tend to fold into G-quadruplexes, were designed as the model molecules. The structured oligonucleotide is effectively unfolded into a linear form by regulating the cation type in the buffer solution, which allows the identification of these oligonucleotides by aerolysin nanopore. The length-dependent duration time demonstrated that the secondary structure was indeed modulated to unfolded state, revealing the achievement of the direct analysis of structured DNA (Figure 3.18).[32]

3.1.5 Uncovering the Sensing Region of an Aerolysin Nanopore

The ability to sense target molecules with high sensitivity and selectivity lies at the heart of a biochemical sensor's performance. With the assistance of an electric field, an aerolysin nanopore can use its confined space and highly charged pore lumen to capture and accommodate individual molecules. The current feedbacks respond to the movements of molecules through the inner channel of aerolysin. However, a central question arises: how to make an aerolysin single-molecule interface feature a higher specificity and sensitivity to the molecules of interest. Our group first demonstrates the sensing regions

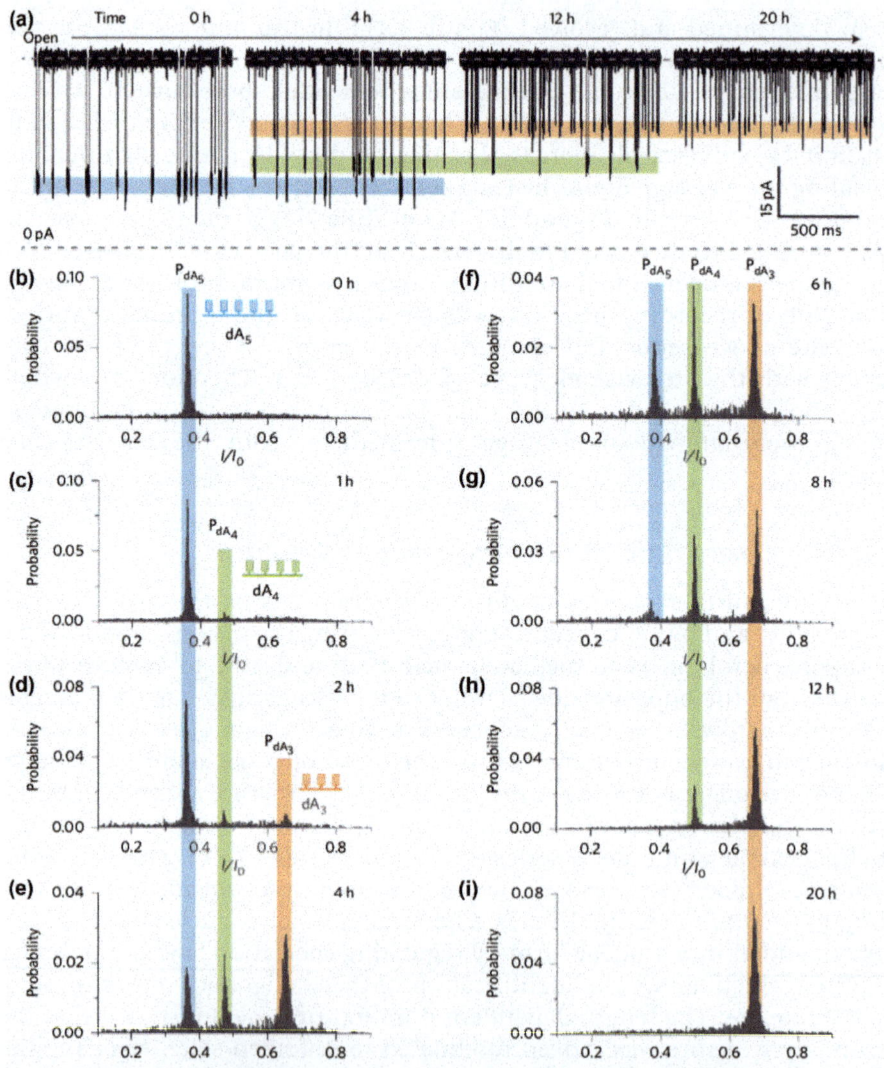

inside aerolysin, which direct the rationally design of mutant aerolysin nanopores with chemical bond precision. The experimental results integrating with the molecular dynamic simulations based on a recent pore structural model was used to precisely map the sensing regions of this pore-forming toxin for ssDNA translocation (Figure 3.19). Rationally probing ssDNA length and composition upon pore translocation provides new insights for molecular determinants of the aerolysin nanopore. The findings revealed two clear sensing regions located at the two main constriction points across the pore lumen, located at amino acid residues R220 and K238, which could help the future design of aerolysin variants with improved capabilities for sequencing.[33]

3.1.5.1 *Case Study XVI: Chemical-biological Engineering of R220 and K238*

After probing the sensing region inside aerolysin, it is necessary to precisely and controllably tune the selectivity and sensitivity of nanopores by modulating the interactions between pores and analytes. Therefore, aerolysin can be modified by a specific group at its internal surface *via* site-directed mutagenesis. A recent study has replaced the positively charged R220 located at the entrance of the pore with negatively charged glutamic acid (E), leading to few current blockages occurring by negatively charged oligonucleotides. Moreover, the substitution of the positively charged lumen-exposed amino acid K238, located at the bottom of the β-barrel to glutamic acid, led to a surprising prolonged duration time, indicating the stronger interactions between pore and analyte. Therefore, it is a feasible and powerful strategy to design a specific biological nanopore for single molecule measurement with high selectivity and sensitivity (Figure 3.20).[34]

Figure 3.17 Real-time monitoring of the stepwise cleavage of dA_5 induced by Exo I with WT aerolysin. (a) The continuous current traces for the stepwise cleavage process of dA_5 at 0, 4, 12 and 20 h. The widths of the bands were calculated from the peak widths of the individual Gaussian fits. (b)–(i) Current histograms for the blockades recorded at 0 h (b), 1 h (c), 2 h (d), 4 h (e), 6 h (f), 8 h (g), 12 h (h) and 20 h (i). The blockades with durations less than 0.13 ms were excluded from the histogram of I/I_0. The colored bands represent the blockade current distribution of each stepwise intermediate during the cleavage process. Exo I degrades oligonucleotides in the 3′ to 5′ directions, liberating 5′-mononucleotides in a stepwise manner. The data were acquired in 1.0 M KCl, 10 mM Tris, 10.0 mM MgCl2 and 1.0 mM DTT at pH 9.5 and +100 mV in the presence of 1.0 μM dA_5 and 2000 U Exo I. The reaction temperature was set to 24 ± 2 °C. The presence of Exo I in a high salt concentration at low temperature provides a relatively low digestion rate, which allows the WT aerolysin to discriminate the stepwise intermediates of the 3′-end oligos.
Reproduced from ref. 29 with permission from Springer Nature, Copyright 2016.

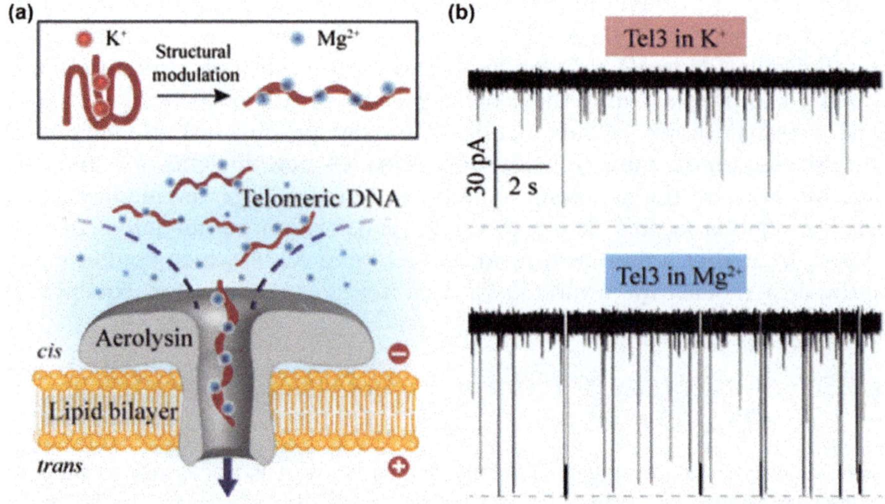

Figure 3.18 Analysis of the oligonucleotides with the secondary structure by the cation-regulation mechanism. (a) Illustration of Tel_n translocating through an aerolysin nanopore in Mg^{2+} buffer. By the strategy of cationic regulation, a Tel_n is unfolded from G-quadruplex to a linear form, which can be driven through an aerolysin nanopore from the *cis* to *trans* side. The solution was in the conditions of 1.0 m $MgCl_2$, 10×10^{-3} m Tris, pH = 7.4. (b) Example current traces for the addition of Tel_3 to the *cis* solution in K^+ (upper) or Mg^{2+} (bottom). Compared to the short-lived current blockades with low amplitude in KCl solution, Tel_3 produced long-lived and deep current blockades in $MgCl_2$ solution. Reproduced from ref. 32 with permission from John Wiley and Sons, © 2018 WILEY-VCH Verlag GmbH & Co. KGaA, Weinheim.

3.1.5.2 *Case XVII: Methyl-cytosine Discrimination from Hetero-nucleotide by Engineering K238G Aerolysin*

A further study investigated the aerolysin sensing regions for oligonucleotide analysis by a series of rationally designed mutant aerolysin nanopore experiments combined with molecular dynamics simulations results. The positively charged lumen-exposed K238, inducing a deep valley in the electrostatic potentials, was identified as one of the key sensing regions. The results for K238 replaced by different charged and sized amino acids exhibit that the translocation time of oligonucleotides through the nanopore can be effectively manipulated by the choice of the target amino acid at the 238 site (Figure 3.21). The results also displayed that both the electrostatic properties and geometrical structure of the aerolysin nanopore are critical to its sensing ability. These insights open an avenue for designing the sensing capability of aerolysin nanopore, further providing a novel paradigm for nanopore technology.[35]

Moreover, by using the K238G mutant with high spatial resolution, the discrimination of the mixtures of methylated cytosine and cytosine in a four-base length hetero-nucleotide was achieved (Figure 3.22). ACGA yields a wide

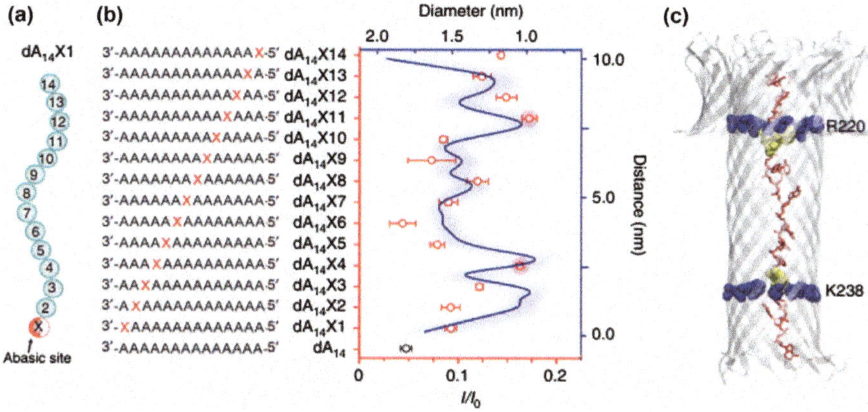

Figure 3.19 Mapping the sensing regions of aerolysin by abasic site scanning of dA_{14}. (a) Schematic representation of $dA_{14}X1$ oligonucleotide; the red sphere of X represented the single-nucleobase deletion. (b) Comparison between the percentage of increase in residual current for each abasic oligonucleotide (red circle with error bar) and the pore diameter (blue line) calculated by the PoreWalker server. Data were obtained in 1.0 M KCl, 10 mM Tris and 1.0 mM EDTA, pH = 8.0, 24 ± 2 °C at the bias potential of +140 mV. The errors of the data were based on at least three separate measurements. (c) Selected MD snapshot of dA_{14} fully inserted in the pore, showing the two main constriction points (R220 and K238, in blue van der Waals representation) and the DNA in red, with bases at positions 11 and 4 in yellow van der Waal representation. Reproduced from ref. 33, https://doi.org/10.1038/s41467-018-05108-5, under the terms of a CC BY 4.0 license, https://creativecommons.org/licenses/by/4.0/.

distribution of signals, by WT aerolysin, while displaying a sharp peak distribution with about five-fold prolonged duration time by K238G. The addition of AmCGA to the solution containing ACGA showed that the populations were almost all overlapped by WT aerolysin. However, these two populations in the K238G mutant aerolysin were well distinguished with peak separation of ~2 pA. This is the first time that aerolysin nanopore achieved the discrimination of single base modification in random sequences (Figure 3.22).

3.1.5.3 *Case Study XVIII: Direct Readout of DNA Damage from Heteronucleotide*

Our studies show that the K238 region is capable of effectively governing the translocation speed of oligonucleotides *via* pore–analyte electrostatic interactions. To further enhance the sensitivity of aerolysin, K238 is latterly replaced by a series of amino acids with different size and charge. Among all the mutants at the K238 site, K238Q shows the strongest ability in providing the ultra-high energy barrier for the DNA translocation. This mutant aerolysin nanopore prolonged the duration time of homonucleotide inside the pore lumen for more than 200 fold. More interestingly, this prolongation of

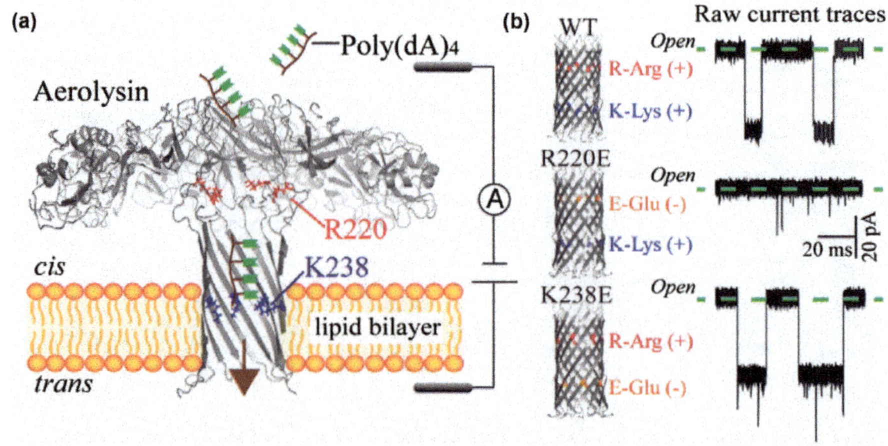

Figure 3.20 Illustration of poly(dA)$_4$ traversing through wild-type and mutant aerolysin pores. (a) Monomeric aerolysin self-assembled and then inserted into the membrane; R220 is shown in red, and K238 is shown in blue. (b) Examples of analyzing poly(dA)$_4$ through a single aerolysin nanopore. The WT aerolysin with positively charged amino acids Arg and Lys produced distinguishable current blockages (top). R220E mutant aerolysin with negatively charged Glu is located at the entrance of the pore, which barely produced the current blockages with the continuous time–current recording (middle). K238E mutant aerolysin with negatively charged Glu at the trans-ward third of the β-barrel stem lead to a prolonged duration. The data were acquired in 1.0 M KCl, 10 mM Tris, and 1.0 mM EDTA at pH 8.0 and +100 mV in the presence of 2.0 μM poly(dA)$_4$.
Reproduced from ref. 34 with permission from American Chemical Society, Copyright 2018.

duration can be extended to hetero-nucleotide, giving a such narrow distribution of current blockages, which further allows the analysis of structural differences on heteronucleotides by using mutant aerolysin nanopore. As expected, the K238Q mutant aerolysin nanopore could detect the single-base variation of heteronucleotide with spatial high resolution. Moreover, K238Q mutant exhibits a high sensing capability to achieve the discrimination of much single-base damage in the mixture simultaneously, including ^{o}G and I, which are the epigenetic modifications in G and methyl cytosine mC in the 5′-ACGA sample. Even the modification quantity in a heteronucleotide chain can be measured by K238Q, with the results indicated in Figure 3.23.[114]

3.2 Peptide Sensing with Biological Nanopores

3.2.1 Case Study I: Peptide/Protein Conformation and Folding/Unfolding Detection

The first attempt for peptide detection was with an α-HL nanopore utilizing the peptides of three different structures.[48] Specifically, peptides containing

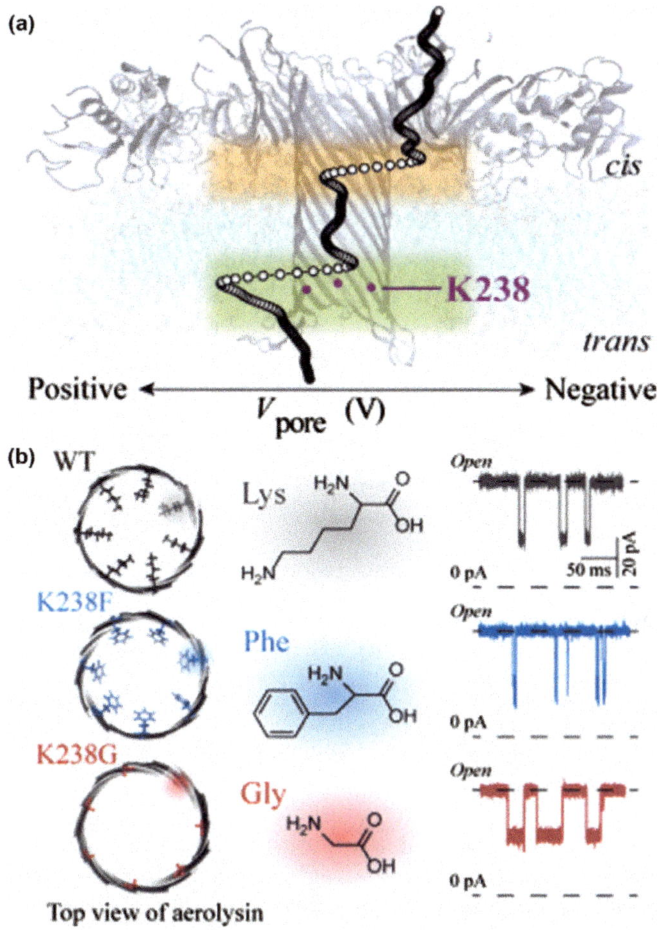

Figure 3.21 Identification of sensitive regions of the aerolysin nanopore for oligonucleotides detection. (a) All-atom model of full-length aerolysin nanopore system. The black line is electrostatic potential distribution along the WT aerolysin at +120 mV. The green and orange bands represent the sensing regions. The 238 sites are shown in purple. (b) Left: top views of WT (black), K238F (blue) and K238G mutant (red) aerolysin. Middle: structure of amino acids, Lys, Phe and Gly. Right: current traces of poly(dA)$_4$ detected by WT (black), K238F (blue), and K238G mutant (red) aerolysin, respectively.
Reproduced from ref. 35 with permission from American Chemical Society, Copyright 2018.

repeat units of the collagen-like sequence (Gly-Pro-Pro)$_n$ ($n=1$ **P1**, $n=2$ **P2** and $n=3$ **P3**) that end in ferrocene (Fc) were synthesized (Figure 3.24). Three repeats of this sequence with a crystal graphically determined diameter of 12 Å can form a constant collagen triple helix. The shorter repeats form dimers and monomers, exhibiting shorter transit times and smaller

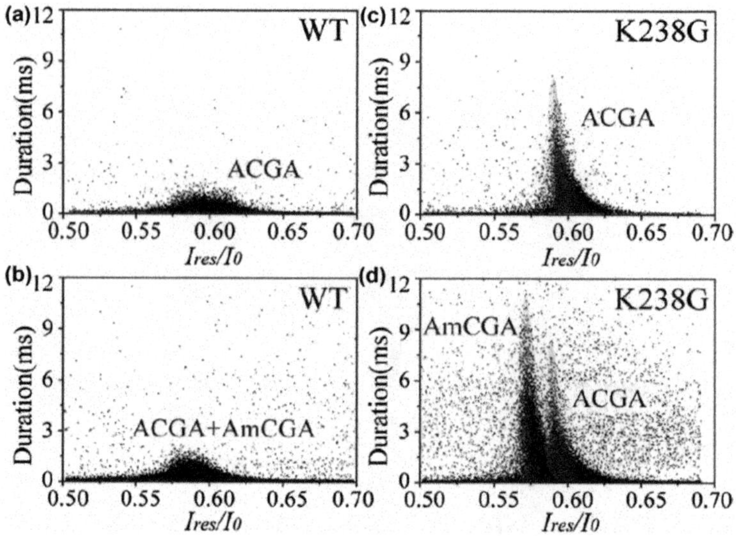

Figure 3.22 Discrimination of methylated cytosine and cytosine in a four-base length heteronucleotide with WT aerolysin and K238G mutant aerolysin. The sequencings are 5′-ACGA-3′ and 5′-AmCGA-3′. Scatter plots of ACGA by WT (a) and K238G mutant (c) aerolysin. Scatter plots of addition of AmCGA in the ACGA solution by WT (b) and K238G mutant (d) aerolysin. The data were acquired at +130 mV in the presence of 4.0 μM ACGA and AmCGA.
Reproduced from ref. 35 with permission from American Chemical Society, Copyright 2018.

blockade currents. As shown at the bottom of Figure 3.24, there are four major structures that the peptides might adopt, which can be recognized based on blockage current and duration time. For **P1**, peak 1 is acknowledged as a linear peptide. The major peak 2 of Figure 3.24d may be caused by a monomeric peptide folded into a U-shape during translocation. For **P2**, there are various minor populations with blockade current ranging from −45 pA to −55 pA, relating to linear and folded monomeric peptides. The major peak for **P2** (population 3) has a smaller increase in blockade current than that for **P1**; however, the duration time significantly increased, which is uniform with a dimeric linear molecule. For **P3**, the major population (population 4) exhibits a larger blockade current of −75.4 pA and a longer duration time of 591 μs, which is consistent with a linear collagen-like triple helix structure. Furthermore, many events with shorter durations can be assigned to dimeric and monomeric linear-extended structures or even folded monomers, as in the instance of **P1** (populations 1, 2, and 3). Moreover, the mixture for single peptide molecules of (Gly-Pro-Pro)$_n$ ($n = 1$, **P1**; $n = 2$, **P2**; $n = 3$, **P3**) with the same concentration can be discriminated by their distinct blockade current and duration. With this attempt of peptide detection, the biological nanopore has been demonstrated that it can be used to analyze the peptide structures and resolve the presence of intermediate

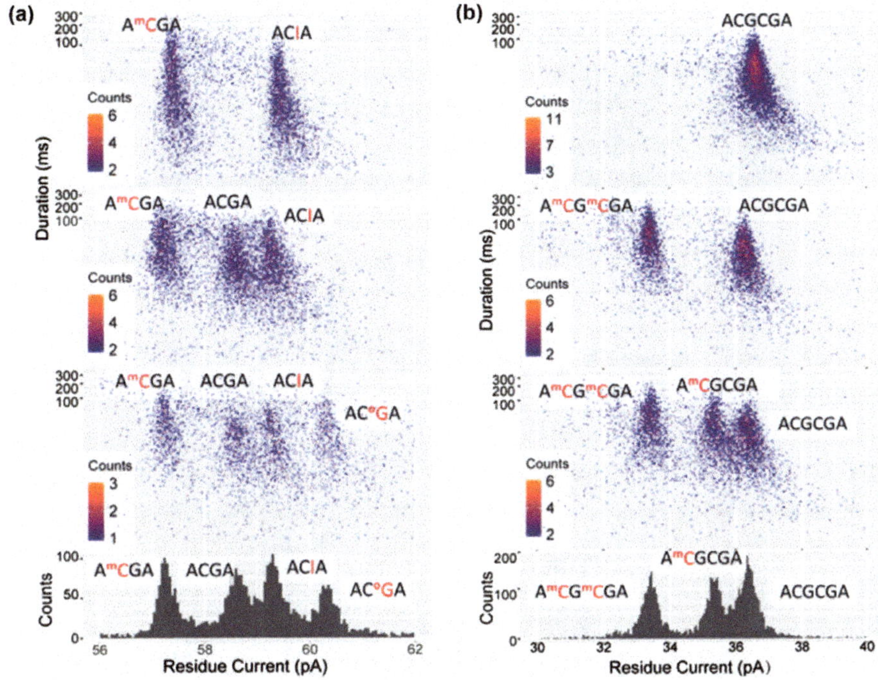

Figure 3.23 (a) Distribution of current and duration from K238Q interaction with ACGA and the modified segments. The color bar illustrates the accumulation of each DNA segment. From the top figure, A^{m}CGA was first introduced, and then ACGA, ACIA and AC^{o}GA were added successively with different peaks that correlate to each segment. A histogram of the current distribution of the mixture was plotted in the panel to demonstrate the current distribution and separation ability of K238Q. (b) Distribution of current and duration from K238Q interaction with ACGCGA and the modified DNA segments. The color bar illustrates the accumulation of each DNA segment. From the top figure, A^{m}CG^{m}CGA was first introduced, and then A^{m}CGCGA and ACGCGA were added successively with different peaks raised correlate to each segment. A histogram of the current distribution of the mixture was plotted in the bottom panel to demonstrate the current distribution and separation ability of K238Q. The data were acquired in 1.0 M KCl, 10 mM Tris, and 1.0 mM EDTA at pH 8.0 and +140 mV in the presence of 2.0 μM ACGA and the modified segments or ACGCGA and the modified segments.
Reproduced from ref. 114, Copyright © 2020 American Chemical Society.

conformations, which were not apparent from the CD spectra. The α-HL was also applied to analyze other peptides in 2005.[73]

Due to the narrow diameter of biological nanopores, it is hard to directly detect protein translocation. Therefore, a remarkable number of studies with protein detection have been stimulated by the virtue of chemical denaturant, ion gradient, pH or temperature change. One of the representative

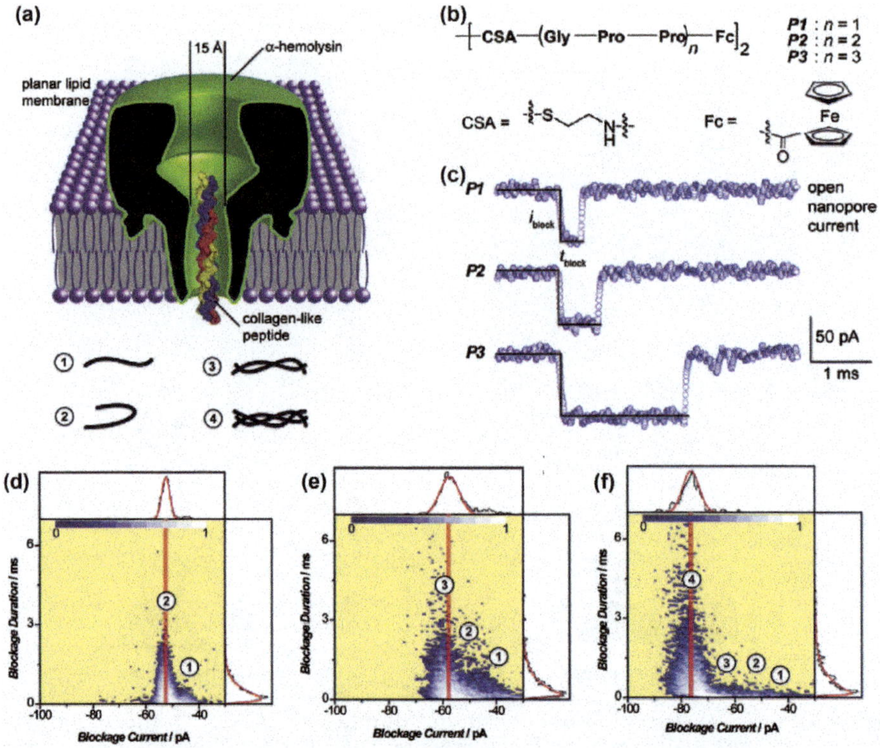

Figure 3.24 Representation of the α-HL pore, the structure of the peptides tested and current transients as the peptides interact with the pore. (a) A schematic of the structure of the α-HL inserted into a planar lipid bilayer with a linear collagen-like peptide traversing the α-HL pore. (b) The structure and labeling scheme of the collagen-like peptides used in this study. Typical current transients of peptides P1, P2, and P3 are shown by the open blue circles. (c) Idealized blockage behavior represented by the black line. The blockade current is equal to the difference between the baseline open current and the average amplitude of the blockage current. Similarly, the duration of the blockage is found from the idealized current trace. Bottom: contour plots of current transients of P1 (d), P2 (e), P3 (f).
Reproduced from ref. 66 with permission from American Chemical Society, Copyright 2004.

works on folded peptide analysis was the detection of the unfolding state of the 370-residue maltose-binding protein (MBP) at different concentrations of Gdm-HCl (Figure 3.25).[49] The experiments with α-hemolysin nanopore were accomplished in a higher concentration of guanidinium hydrochloride (Gdm-HCl, 1.5 M), a commonly used chemical denaturant in protein-folding studies, to unfold the proteins. In the absence of the denaturing agent, no current blockades could be recorded, revealing a large energetic penalty hindering the protein entering into the pore lumen. Remarkably, with the

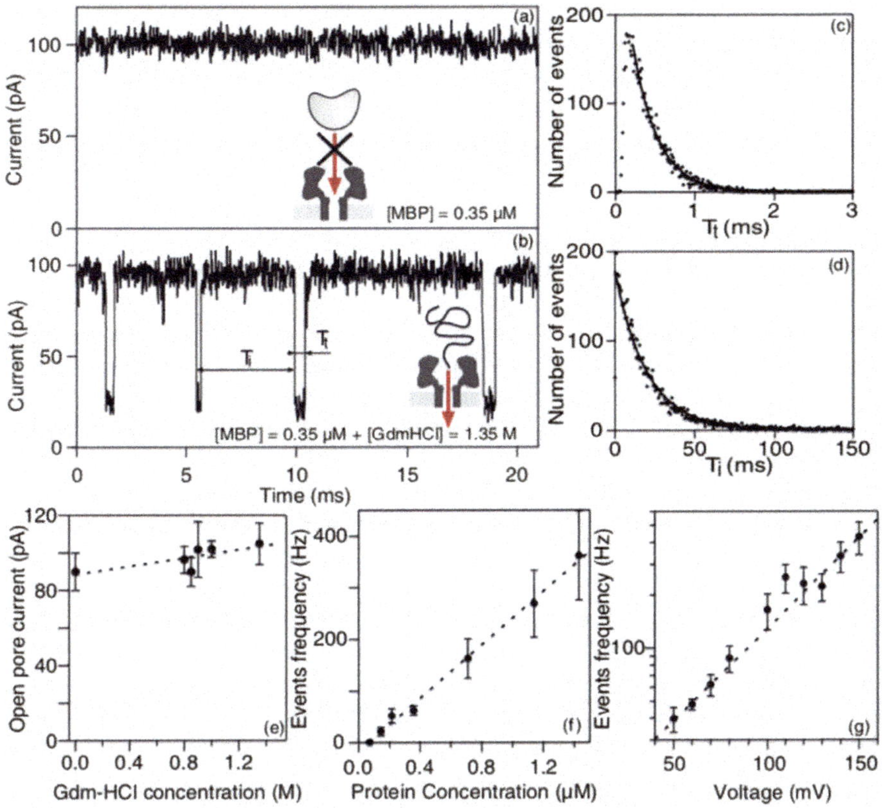

Figure 3.25 Parts of current traces through α-HL (at 100 mV) in the presence of MBP, examples of statistical analysis, open pore conductance *versus* Gdm-HCl and events frequency *versus* MBP and applied voltage. (a) Without Gdm-HCl the current is constant at 100 pA, MBP cannot enter to the pore; (b) with Gdm-HCl 1:35 M well above the unfolding transition, the current trace decreases down to 20 pA when a MBP molecule is in the pore. The RMS noise of the open pore current does not change at this Gdm-HC value. (c) Distribution of blockade time; (d) distribution of time intervals T_i between two events of current. The blockade time and interval both were fitted with exponential function. (e) Open pore current *versus* Gdm-HCl. (f) Frequency of events. The dashed line is a linear fit with a slope of 260 ± 10 Hz μM^{-1}. (g) Frequency of events *versus* applied voltage on a semi-log scale. The dashed line is an exponential fit.
Reproduced from ref. 49 with permission from American Physical Society, Copyright 2007.

increased concentration of Gdm-HCl, protein-induced current blockades were observed. The duration and frequency of the events were correlated with the concentration of Gdm-HCl in the detection system, which indicated that the partitioning of the MBP protein into the nanopore lumen is associated with its folding state in solution.

In addition, there were different stabilities for the entry and translocation of unfolded proteins through an aerolysin pore at the single molecule level, associated with the applied potential and protein concentration.[50] The frequency of ionic current blockades as a function of the applied voltage shows an exponential dependency, and as a function of protein concentration exhibits a linear dependency. The duration time of unfolded proteins decreases exponentially with the increasing applied voltage. The duration time of a double-sized protein is longer than that assessed for a single protein asserting the transport phenomenon (Figure 3.26). These results are consistent with the theory of confined polyelectrolyte and with some experimental results about DNA or synthetic polyelectrolyte translocation through protein nanopores with increasing applied voltage. Since unfolded protein is transported slowly through aerolysin nanopores, it could be used as a sensor for unfolding studies. In the future, aerolysin nanopores may be widely used to study the unfolding of WT and variant proteins and probe protein–protein or protein–DNA interactions in different denaturant environments.

In addition to denaturant, temperature also has been studied as a controllable factor affecting peptides/proteins translocation through biological nanopores (Figure 3.27).[51] The native folded proteins cannot automatically enter into thc porc at room temperature. When the temperature was increased from 25 to 50 °C, the molecules unfolded and the event frequency of current blockade increased. The normalized event frequency for both nanopores can be fitted by a similar sigmoid function, thus the unfolding curve is independent of the net charge and the structure of the nanopore. These transportation events arise with increasing temperature and are therefore

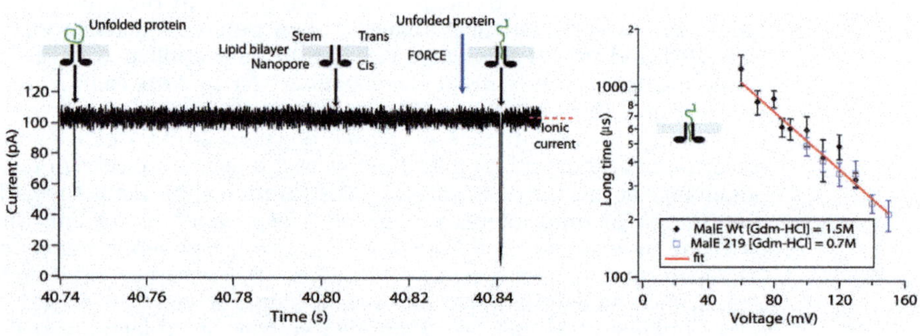

Figure 3.26 Single nanopore current traces of unfolded protein transport through an aerolysin pore as a function of applied voltage for wild-type protein (black color), MalEwt and mutant protein (blue color) and MalE219. The stem domain of the aerolysin pore is placed into a lipid bilayer on the *cis* side, the proteins enter by the stem side. Experiments were made at 1 M KCl, 5 mM HEPES pH 7.4. The final guanidium concentrations are respectively: 1.5 M for MalEwt and 0.7 M for MalE219. The protein concentration is 0.35 μM.
Reproduced from ref. 50 with permission from American Chemical Society, Copyright 2011.

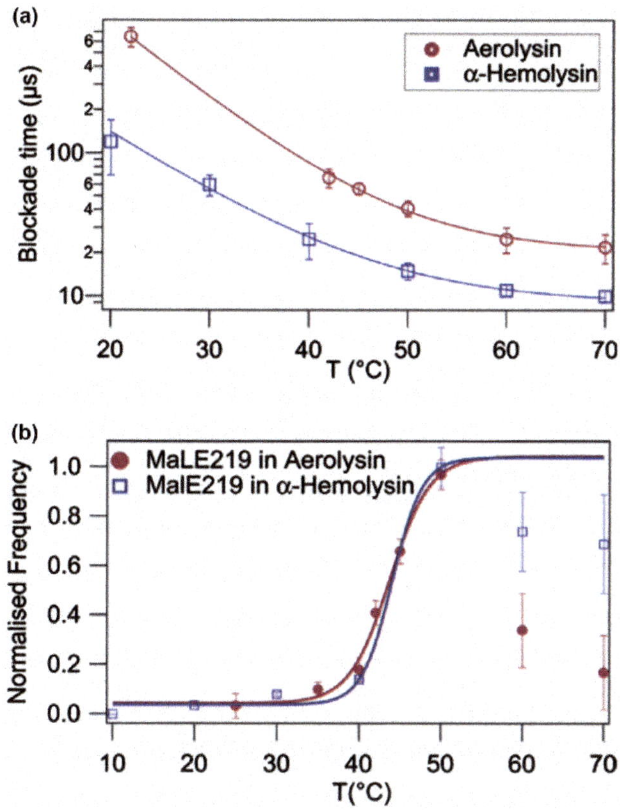

Figure 3.27 (a) Blockade time as a function of temperature for α-hemolysin and the aerolysin pore. The points at 20 and 22 °C for α-HL and aerolysin, respectively, correspond to MalE219 unfolded in the presence of 1 M guanidium chloride. The lines are drawn to guide the eye. (b) Thermal unfolding transition curves of MalE219 detected by, respectively, α-HL and aerolysin pores. The curves represent the normalized event rate (equivalent to the proportion of unfolded conformations) as a function of temperature. Below 50 °C, the data are fitted to a sigmoid. Above 50 °C, the frequency decreases due to the detection limit of short duration time events.
Reproduced from ref. 51 with permission from American Chemical Society, Copyright 2012.

associated with the thermally unfolded proteins translocating the nanopore. Obviously, transport dynamics are accelerated by thermal motion. Likewise, for both nanopores, the event frequency, linearly correlated to the proportion of unfolded proteins, varies as a function of temperature as a growing study thermal unfolding of proteins by electrical detection of a single molecule going through nanopores.

The examples presented above demonstrate that biological nanopore technology is a useful tool in studying peptide/protein conformation and folding, paving the way to experimentally exploring the unfolded and partially folded conformations,[49,51] unfolding intermediates, to obtain unfolding curves or rate of proteins[52] and intrinsically disordered proteins.[51]

3.2.2 Case Study II: Peptide/Protein Translocation Analysis

The behavior of proteins or peptides (*e.g.* membrane protein, functional, peptide enzyme) plays an essential role in biological processes. It is quite necessary for many proteins to be transported across membranes through nanopores to execute their function. The misfolding of protein may lead to the disorder in the physical features. Most studies about protein behavior *in vitro* are performed in bulk and probe the average behavior of an ensemble of molecules. Therefore, the understanding of protein folding and translocation remains a great challenge.[53–56] Recent developments in nanopore analysis allow the real-time observation of individual molecules along their folding pathway and exhibit a potential application on exploring regions of the energy landscape that are inaccessible to bulk measurements.[57,58] Taking advantage of stable biological and physical properties in various environments, high sensitivity and excellent reproducibility of biological nanopores, much research has been devoted to investigate peptides,[59–61] proteins,[62–64] or protein – ligand complexes[65] through nanopore biosensors. As one of the most powerful approaches at single-molecule level, the biological nanopore sensor has been employed for the investigation of peptides/proteins as well as DNA analysis. However, in contrast to DNA, which is considered as strongly and homogeneously charged biopolymers, the charge of a protein is irregularly distributed and can be positive, neutral or negative, depended on the solution pH and temperature. To simplify the peptide/protein detection with biological nanopore, some deliberately designed peptides containing an α-helix, β-sheet structure have been investigated as the models.

A series of negatively charged α-helical peptides of the general formula fluorenylmethoxy-carbonyl (Fmoc)-DxAyKz through two pores either aerolysin or α-HL were investigated in 2006.[94] Compared to α-HL (~100 pA at +100 mV in 1.0 M KCl in 10 mM phosphate buffer, pH 8), aerolysin produced a smaller open pore current (~50 pA) (Figure 3.28). However, both pores have similar percentage block for a translocation event. For instance, the shortest peptide Fmoc-$D_2A_{10}K_2$ has a percentage block of 62% with α-HL and 63% with aerolysin, whereas the values for the longest peptide Fmoc-$D_2A_{22}K_2$ are 78% and 76%, respectively. The results demonstrate that the net charge and dipole moment of each peptide have the main influence on the transport characteristics. The ratio of translocation events/bumping events increases with increasing dipole moment and uncharged peptides mostly gave the bumping events. Notably, the vestibule of α-HL helps to orient the peptides for translocation. Following fundamental studies include characterizing different charged and length peptides with aerolysin nanopores.

Negatively charged peptides with different length composed of repeat unit $(YEQ)_n$, ($n=2$, 3, 4) can be discriminated by aerolysin based on I/I_0 and

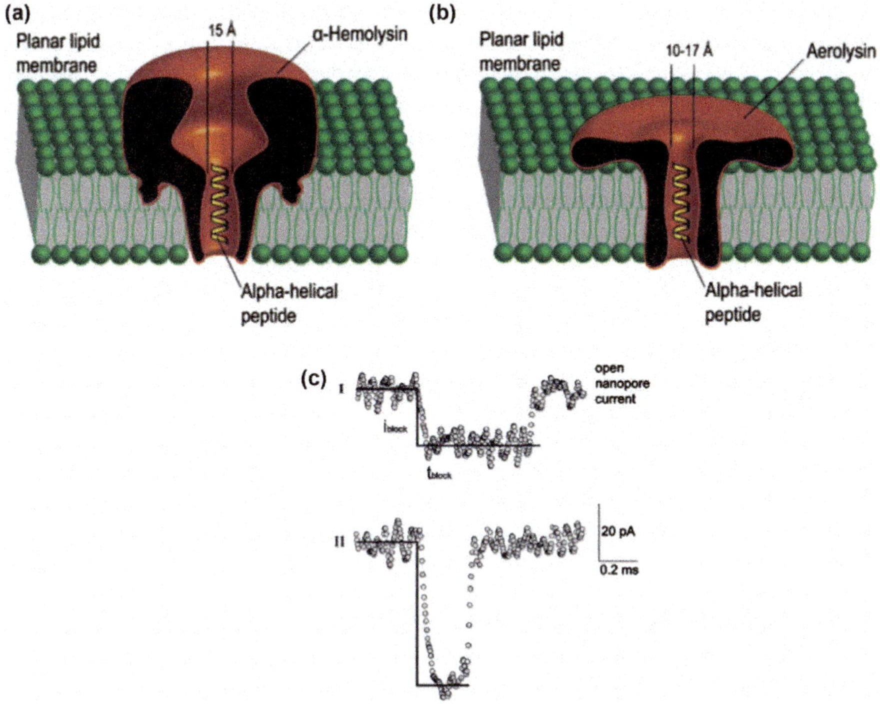

Figure 3.28 (a) Schematic illustration of the α-HL and (b) aerolysin pores inserted into planar lipid bilayers with linear α-helical peptides traversing the angstrom-sized nanopore. (c) Two types of current blockade events on the aerolysin pore are shown by the open circles. Idealized blockade behavior is represented by the black line.
Reproduced from ref. 94 with permission from American Chemical Society, Copyright 2006.

duration time. Under the applied voltage of 100 mV, I/I_0 of $(YEQ)_n$ has been characterized for n from 2 to 4. The histograms of I/I_0 showed a clear single peak distribution that $(YEQ)_2$ is centered at 0.68 ± 0.01; $(YEQ)_3$ and $(YEQ)_4$ displayed a centralized peak at 0.59 ± 0.01 and 0.50 ± 0.01, respectively. Moreover, the location of the peak correlates to the peptide length and is represented by the following linear equation: $I/I_0 = 0.68 - 0.1 \times (n-2)$ ($n = 2$, 3, 4), the increase of one repeat unit will cause the decrease of I/I_0 around 0.1. Referring to the peptide translocation time, $(YEQ)_2$, which is composed of six amino acids, has the fastest translocation time of 0.15 ± 0.02 ms at a transmembrane voltage of +100 mV. The duration time for a repeat unit (YEQ) of 3 and 4 is 0.27 ± 0.01 ms and 0.30 ± 0.01 ms under the same conditions. As expected, a longer peptide needs more time to translocate across the aerolysin nanopore. The results showed that both charge and length of peptides affect the translocation process simultaneously.

Meanwhile, a WT aerolysin nanopore enables the size discrimination of several short uniformly charged homopeptides $(Arg)_n$ ($n = 5$–10), mixed in solution.[62] The addition of the mixture of arginine peptides with six different lengths that differ by a single amino acid in length induced well-resolved current blockades of different amplitudes and durations under a negative applied voltage (Figure 3.29). The high current blockade depth resolution is combined with the high temporal resolution up to 10 ms/amino acid at the single amino acid level, paving the way for the construction of a sequence library of all the different amino acids. Overall, the results demonstrate that the nanopore technology can provide useful structural information, while peptide sequencing still requires further improvements.

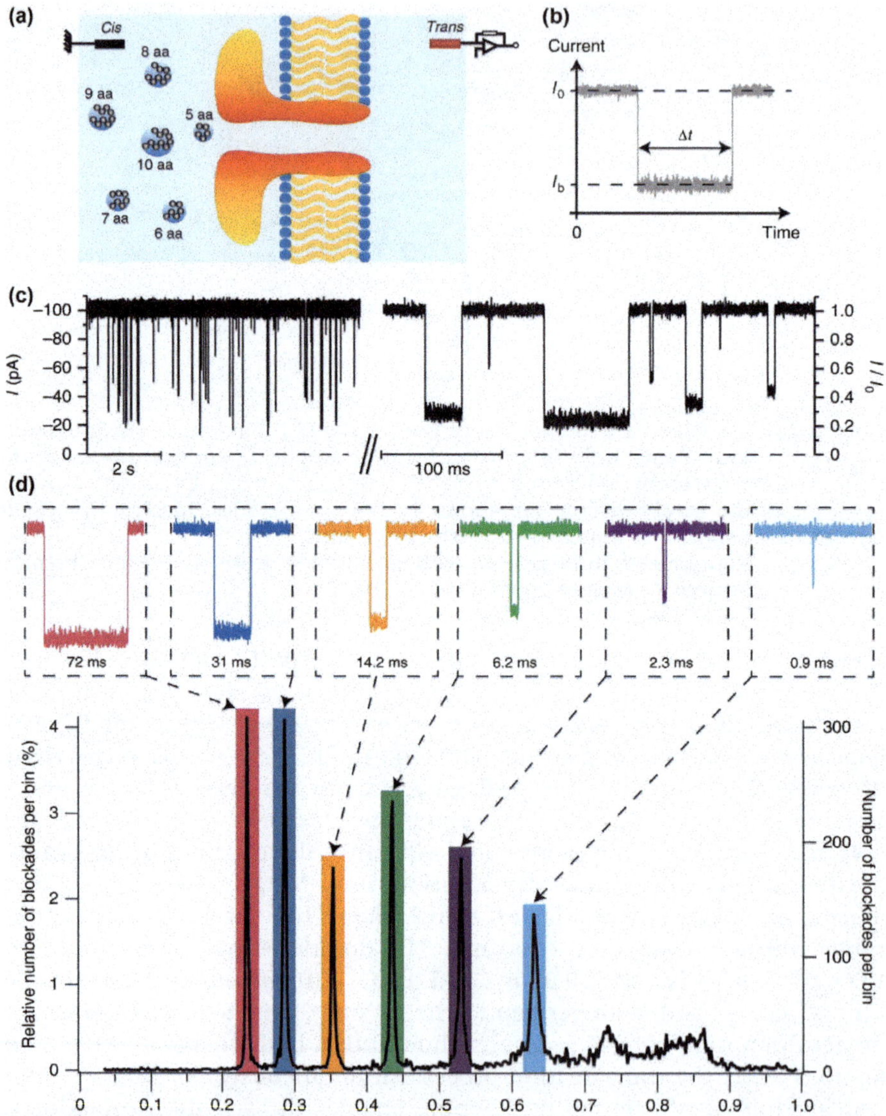

Above-basic understanding of simple model peptides through biological nanopores paves the way for further investigation on peptides/protein with biological nanopore. Much research has utilized α-hemolysin nanopore for monitoring the peptide translocation profile and the interaction of model peptide with a pore wall. One method to decrease the high energy barrier for a peptide to translocate the nanopore is to set a 'traps' in the pore lumen.[64] As shown in Figure 3.30, the entry and exit of the β-barrel part of the α-hemolysin have engineered negatively charged binding sites. Such a negatively charged pocket is achieved by mutagenizing lysin at position 147 into aspartic acid and forming a heptamerous ring. First, how positively charged peptides interact with these negatively charged electrostatic traps has been explored. The engineered electrostatic trap at the *trans* opening had a major impact on the association rate while it had little influence on the dissociation rate. However, nanopore with the *trans* and *cis* traps were concurrently engineered at the lumen, significantly increasing the association and dissociation rate. Similar to ion-conducting protein nanopores, these electrostatic traps worked as binding sites for the peptide. The two traps would catalyze the net flow of polypeptides more efficiently than either the WT α-hemolysin nanopore or a single-trap-containing protein nanopore from the *trans* to the *cis* side of the membrane. The considerable increase in the association rate occurred with the single-trap-containing α-HL nanopore was caused by the over-lap of the transition state of the entry barrier with the energetic minimum formed by the trap engineered on the *trans* side of the pore.

In another case, the CAMA P6 (KWKLFKKIGIGKFLQSAKKF-NH_2) peptide was chosen as the model peptide to analyze the peptide translocation process through α-hemolysin (Figure 3.31).[68] When adding the CAMA P6

Figure 3.29 Detection of a mixture of arginine peptides with different lengths. (a) Schematics of the experimental setup (not to scale). (b) Illustration of a typical current blockade event. The interaction of a peptide with the nanopore is detected as a temporary blockade of the electrical current flowing through the nanopore from its open-pore current value I_0 to the blockade current value I_b. The current blockade duration Δt corresponds to the duration of the peptide/nanopore interaction. Portion of a typical current *versus* time recording through aerolysin nanopore in presence of an equimolar mixture (1 μM) of arginine peptides of different lengths (5, 6, 7, 8, 9, and 10 amino acids) in 4 M KCl, 5 mM HEPES, pH = 7.5 at −50 mV and at 20 °C. (c) Histogram of the relative blockade current I_b/I_0 values corresponding to the current trace. The histogram reveals the existence of at least six distinct populations corresponding to preferred relative blockade current values. Inserts on the top show a typical current blockade event and indicate the mean blockade duration for each population.

(a)

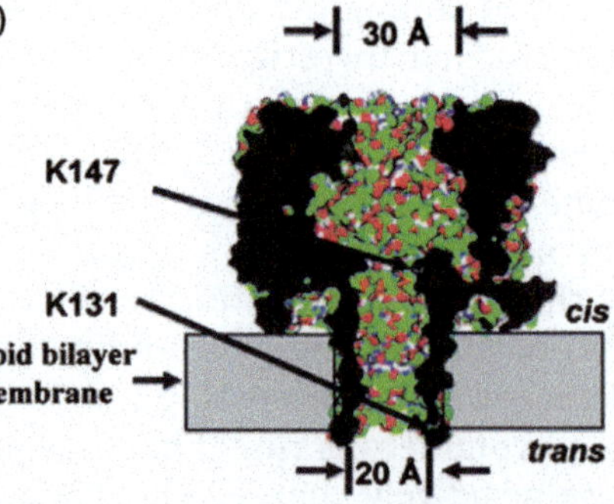

(b)

Analyte constructs	Name	Charge	Charge density
pb$_2$(35)–N barnase C	pb$_2$(35)-Ba	+7	0.20
pb$_2$(65)–N barnase C	pb$_2$(65)-Ba	+13	0.20
pb$_2$(95)–N barnase C	pb$_2$(95)-Ba	+16	0.17

Figure 3.30 (a) Molecular model of the α-hemolysin protein pore highlighting the engineered acidic binding sites (in magenta) made of rings of aspartic acid (K131D$_7$ and K147D$_7$). (b) Design of the pb2-Ba proteins. Different fragments of yeast pre-cytochrome b2 (pb2) were fused to the N-terminus of barnase (Ba).
Reproduced from ref. 64 with permission from American Chemical Society, Copyright 2008.

peptide on the *trans* entrance of the membrane (*i.e.* on the β-barrel side of the pore) with the final concentration of 30 mM, the peptide translocation process is studied at different pH conditions. In relation to the amperometric response at physiological pH, decreasing the pH value induced the occurrence of a second level, thus denoting the larger current drop level as B1 and the latter B2.

To provide a better understanding of this result, Tudor and coworkers performed molecular dynamics simulations in neutral pH to approve this sequential trip and show a larger transversal available area when the peptide traverses through the entrance than when through the β-barrel. When the peptide is added into the *trans* side, level B1 always appeared ahead of level B2, suggesting the impossibility that a peptide threading traverse the constriction region to the entrance returns into the protein β-barrel. Therefore, two consecutive pore translocation steps of passing through the β-barrel region and through its vestibule has been distinguished with tunable pH condition. This provides a facile possibility for further modulation of the biomolecule translocation process, and improves the spatio-temporal resolution of α-HL nanopore for the investigation of folding and translocation of biological polymers.

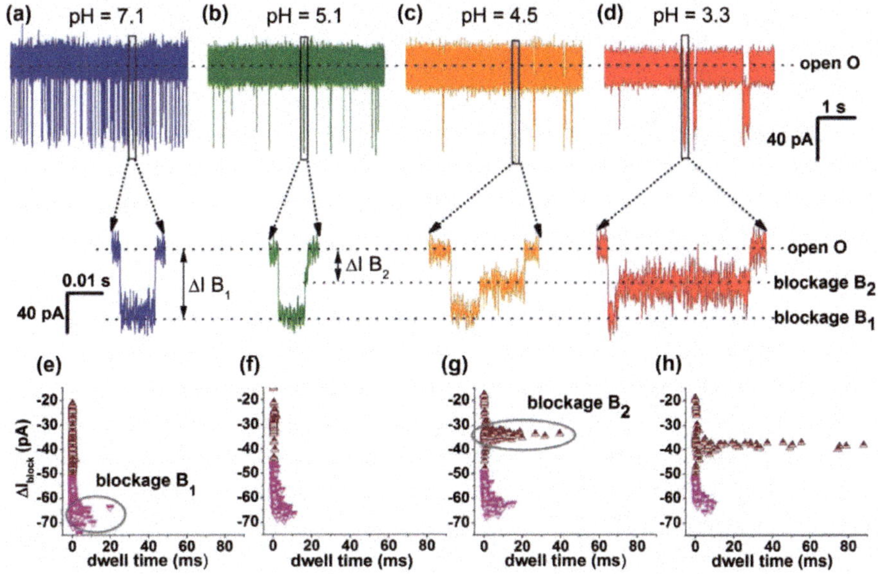

Figure 3.31 Typical single-nanopore current recordings reflecting peptide interaction with the α-HL pore. All traces were recorded at 150 mV, in symmetrical 2 M KCl at pH 7.1 (a), pH 5.1 (b), pH 4.5 (c) and pH 3.3 (d) with 30 mM peptide applied to the *trans* side of the membrane. The dotted line in (a)–(d) shows the level of open-pore current before a peptide partitioned within the β-barrel; downward spikes reflect stochastic reduction of pore current induced by the reversible association of a peptide with an open protein pore. The zoomed-in trace segments under (a)–(d) reveal the distinct blockage sub-states, B1 and B2 (see text), associated with a single peptide in the pore at various pH values. The distinct sub-states are visualized in (e)–(h), scatter plots of dwell time *versus* relative blockage amplitude of blockage events. For exemplification, we highlight by encircling the B1 state at neutral pH (e), and the B2 state at pH 4.5 (g).
Reproduced from ref. 68 with permission from Springer Nature, Copyright 2014.

3.2.3 Case Study III: Pathogenic Related Peptide/Protein Investigation

The above studies on the current blockages analysis involving transit events of peptides have showed us the ability of biological nanopores to identify the secondary structure and even get information about folding events.[61] An increasing number of neurodegenerative diseases involve the misfolding or unstructuring of proteins, such as the amyloid protein in Alzheimer's disease, huntingtin in Huntington's disease and prion proteins in transmissible spongiform encephalopathies (TSEs), and α-hemolysin could provide a sensitive approach for disease-related protein analysis. Recently, many researchers have concentrated on investigating the modulation of the folding state to achieve early diagnosis or therapy of the relative disease.

A series of studies for amyloid peptide analysis has been performed. One of the most common analyzed the amyloid peptide Aβ42, which is crucial in the pathogenesis of Alzheimer's disease. Aβ42 is toxic to neuronal cells in its aggregated form fibrils and plaques. An important study has achieved the monitoring of the aggregation transition of Aβ42 induced by two different drug molecule, β-cyclodextrin (β-CD) and Congo red (CR), which are a promoter and inhibitor of Aβ42 aggregations, respectively.[62] Analysis of the characteristic blockade current and duration time suggests that β-CD and CR have contradictory effects on the aggregation of Aβ42. For the translocation events of the monomeric Aβ42 peptide, its ion current through the α-hemolysin pore decreased from about 100 to 5 pA (Figure 3.32). After adding an equimolar amount of Aβ42 and CR, the main population exhibits a blockage current of 18.06 pA (named long-lived events), and the other population has a blockage current of 25.47 pA (named short-lived events). The short-lived events with τ_2 might be due to monomeric Aβ42 translocated through the pore as the distribution is more focused than for the long-lived events. The long-lived events may be caused by horizontally captured Aβ42 monomer organizing itself in the entrance before transport happens. It demonstrated that the CR binds to Aβ42 and its peptide fibrils by reducing the aggregated fibril formation. The use of a nanopores suggested the

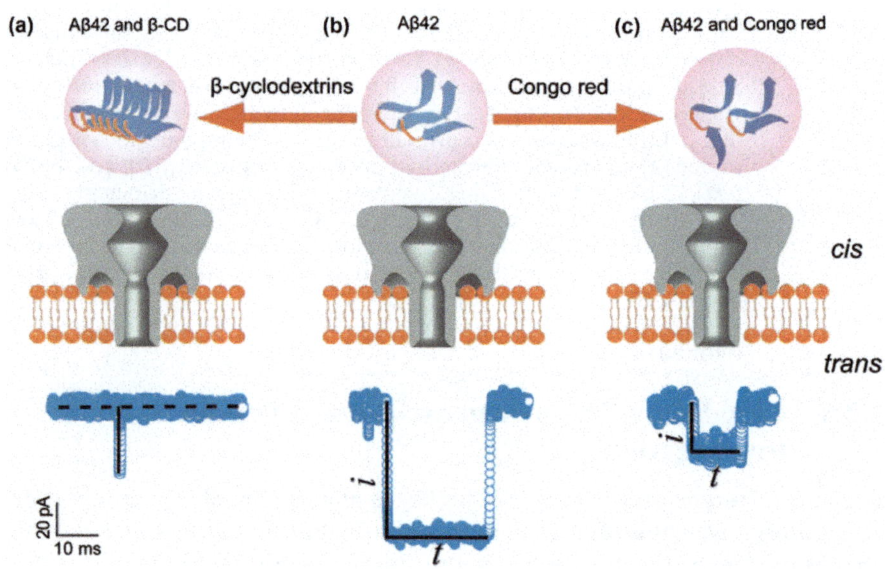

Figure 3.32 The representation of the α-hemolysin protein inserted into the planar lipid bilayer. Aβ42 in the presence and absence of β-cyclodextrin (Aβ42-CD) or Congo red (Aβ42-CR) were added from the *cis* chamber, respectively. (a) Representation of the collision behavior of Aβ42-CD. (b) Representation of the blockage behavior of Aβ42. (c) Representation of the blockage behavior of the translocation event of Aβ42-CR. Reproduced from ref. 62 with permission from American Chemical Society, Copyright 2011.

possibility of using α-HL nanopores to assess the influence of small drug molecules on the peptide aggregation process and structure, which is of great importance for drug screening.

The sensitive distinguishing of different aggregate states for amyloid peptide using an α-HL nanopore promotes further analysis of aggregate dynamics. Previous studies demonstrated that monomeric intermediate of Aβ can influence the aggregation kinetic process. Nevertheless, the correlation between primary structural differences of single monomeric Aβ and the assembly process of monomeric Aβ peptide over time remains to be explored. Therefore, there currently exists an essential necessity to understand the primary structural effects of a monomeric Aβ peptide on the dynamic behaviors of a single Aβ aggregate in solution.

The α-HL nanopore has been used to monitor the primary structure features of two β-amyloid peptide fragments, Aβ25-35 and Aβ35-25, in the time-dependent aggregation process (Figure 3.33).[53] With different charged states of terminal residue, Aβ25-35 and Aβ35-25 show opposite aggregate properties. The freshly prepared Aβ25-35 and Aβ35-25 were driven toward the *trans* side of the lumen, respectively, as shown in Figure 3.33. Most of the events for monomeric Aβ35-25 are centered on the population I (PI) with a low current blockade and a short residue time (Figure 3.33). By contrast, Aβ25-35 causes characteristic distributions named as PII with a high current blockade at a center of $\Delta I/I_0 = 0.63 \pm 0.02$. By analysis of the voltage dependence of duration time corresponding to both these two peptide fragments, the PI distribution has been attributed to the translocation of Aβ35-25 in the conformation of a random coil through the α-HL from its *trans* side in aqueous solution. The PII events could be attributed to the strong interactions of a β-sheet structure with the pore lumen before leaving from either the *trans* or *cis* entrance, bringing a volume exclusion of the ionic current. With the different initial structures, the Aβ25-35 and Aβ35-25 present different aggregate dynamics, revealed by the event frequency obtained with incubation time period of 84 h. With this attempt, these works may demonstrate the potential of nanopore for monitoring the conformational transition of peptide from monomeric amyloid β-peptide into large aggregates, which would benefit the understanding of the amyloidosis in AD.

Besides the conformational transition from the monomer to aggregations, the cleaving of monomer amyloid peptide to a small fragment can also be real-time monitored by an α-HL nanopore.[70] Trypsin is a serine protease that could cut peptide bonds after Arg (R) or Lys (K) amino acid residues. Trypsin (38 Å) is too large to translocate hence it cannot cause current blockage events that might identify the target peptide(s). Therefore, this system demonstrates the ability to monitor the peptide cleavage process and quantify enzymatic chemical kinetics information.

To further analyze the pathogenesis of misfold-related diseases, many approaches have been applied to analyze or modulate the misfolded protein state. Luchian's group investigated the interaction between different metal ions and Aβ-peptide fragment Aβ1-16.[71] The α-hemolysin nanopore makes it

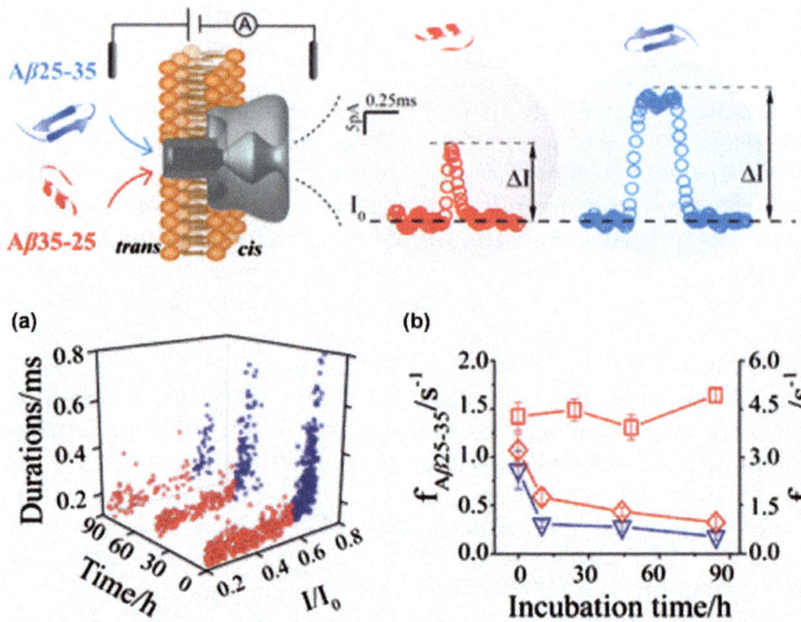

Figure 3.33 Illustration of Aβ35-25 and Aβ25-35 traversing through the α-HL pore. Aβ35-25 and Aβ25-35 adopt random coil (red) and β-sheet (blue) as their initial structures, respectively. The potential was applied using a pair of Ag/AgCl electrodes. The current traces were recorded in solutions containing 1.0 M KCl, 10 mM Tris, and 1 mM EDTA buffer (pH = 8.0) at −60mV. (a) Scatter plots of Aβ25-35 blockades within 5 mins recording for the incubation times of 0 h, 44 h and 84 h, respectively. (b) Frequency of blockades for Aβ25-35 (PI in red diamond and PII in blue triangle) and Aβ35-25 (red square) *versus* incubation time. Data of the errors were based on three separate experiments. The peptides were added from the *trans* chamber containing 1.0 M Tris-KCl at an applied potential of −60 mV.
Reproduced from ref. 53 with permission from the Royal Society of Chemistry.

possible to recognize the bound and unbound states at the single-molecule level. Further, the quantification of dissociation constants of metal ions binding to peptide fragments provide a way to understand the metal ion modulated amyloid aggregation kinetics. Besides the metal ions, the folding behavior of the β-hairpin kind, histidine containing peptide, could be altered by modulating the pH and ionic strength through the inter-amino acid residue electrostatic interactions. The reversible changes of the folded peptide structure in space could be monitored by peptide and α-HL nanopore interaction events, which brings support for revealing the distinct folding sub-populations of individual peptides that are usually invisible, induced by chemical and physical factors. Apart from Aβ peptide related to Alzheimer's disease, α-synuclein is the most important clinical assessment of neurodegenerative problems. The formation of an α-synuclein fibril is critical in the pathogenesis of Parkinson's disease. Therefore, some researchers strive to investigate the intermediate and aggregate process of α-synuclein fibril based on biological nanopores.

Under experimental conditions, it was confirmed that NaCl can induce partial folding of native unfolded α-syn protein, an intramolecular process without involving self-association.[63] After about 10 min, we can frequently observe a blockade for several minutes. In the probe step, there exist three current levels from the open current to block current as a ternary complex. As shown in Figure 3.34, the open pore current level is named as L_o, the block current level is named as L_b and is about 5 ± 0.5 pA, and the capture blockade current level is named as L_c is about 20 ± 1.0 pA. In the primary

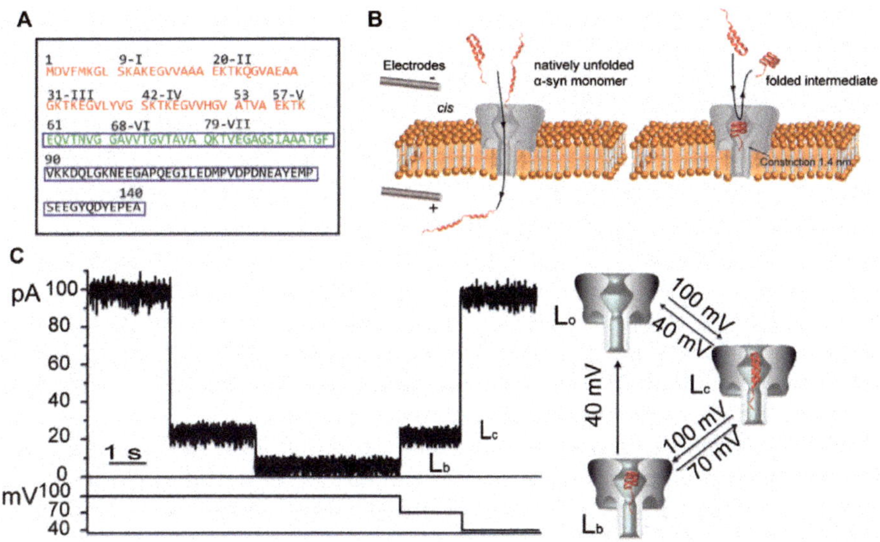

Figure 3.34 (A) The amino acid sequence of WT α-syn. The difference between A53T α-syn and WT α-syn is the residue 53 mutation from A to T. Residues 1 – 64 labeled in orange color are positive in neutral solution and along with residues 65 – 90 (labeled in green color) form the N-terminal part of the polypeptide. Residues 65 – 140 enclosed with a blue frame are negative in neutral solution (PBS, pH = 7.4). The C-terminal region of α-syn (residues 90 – 140) is very rich in acidic residues and thus highly negatively charged at neutral pH. (B) Schematic of the α-syn monomer and intermediate interaction with the α-HL nanopore. (C) Characteristic current blockade event structure control by voltage change of a single α-syn (negative charge) molecule captured inside the vestibule. At an applied voltage of 100 mV, α-syn monomer translocates through the pore as an unfolded chain. When an electronic field is applied to the system, unfolded monomer is transformed to a partially folded intermediate and is captured inside the vestibule. α-syn monomer explores the exit nanopore and translocates through the pore or it curls up inside the vestibule until the voltage is lower than 40 mV.
Reproduced from ref. 63 with permission from American Chemical Society, Copyright 2013.

stage of the formation of the ordered fibril structure, an unfolded chain is folded into a monomeric native structure through one or more partly folded intermediates. The capture of a partially folded WT α-syn may decrease L_c. The ordering of intermediate in destabilized states inside the vestibule may cause transient blockade signals from L_c to L_b, Once the voltage decreases down to 70 mV, the L_b returned to L_c. The L_c current returned to L_0 upon exit of intermediate until the voltage lower of 40 mV. The technique showed here can be used to real-time detect the α-syn monomer elongation with the partially folded intermediate under physiological conditions *in vitro*, paving a new way for PD pathogenesis and intervention.

3.2.4 Case Study IV: Enzymatic Kinetics Monitoring

Enzymatic catalysis processes are involved in many fundamental biological human diseases. Aerolysin nanopore can be used to monitor the enzyme depolymerization of glycosaminoglycan polysaccharides without label.[72] The polysaccharide degradation is catalyzed by hyaluronidase kinetics with increasing substrate concentration by the aerolysin. The enzyme depolymerization dynamics through detecting reaction products were determined on the foundation of current blockade frequency measurements (Figure 3.35). And the recording of the enzymatic cleave of high molecular weight polysaccharides with a nanopore can be qualified by a Michaelis – Menten model. Based on a model relating the degradation process and considering the theoretical chain size distribution and comparing them the experimental translocation time distribution can be evaluated. Modeled and experimental distributions are well connected. This suggests that the nanopore analysis could be extended to any other enzymatic cleavage and not only for small size substrate molecules. Considering the size distributions of various degraded polysaccharides, the nanopore is worked as a 'nano' mass spectrometer.

In the present work, the current recording of peptide translocation through a nanopore in the presence and the absence of an enzyme is a great way to observe peptide cleavage. The method allows real-time investigation of protease kinetics through monitoring the ionic current fluctuations caused by the interactions between enzyme and peptide.[70] Without enzyme, the current blockades only originate from the substrate. However, after adding the enzyme into the solution, there exist two fragments derived from enzyme digestion. Thus, we can see new blockage events with distinct duration times (τ_{off}) and/or amplitudes. To confirm this concept, $(M113F)_7$ α-HL nanopore was used to investigate the trypsin cleavage process of peptide Aβ (10-20). The results suggested that without trypsin ($t = 0$ min), the solution including peptide Aβ (10-20) exhibited only a kind of blockade with a small mean residual current of -1.8 ± 0.3 pA, and residence time of 0.63 ± 0.06 s (Figure 3.36). In sharp contrast, after adding trypsin there were two new kinds of blockades, which exhibited obviously shorter residence times (0.85 ± 0.08 and 0.75 ± 0.05 ms) and larger residual blockage currents

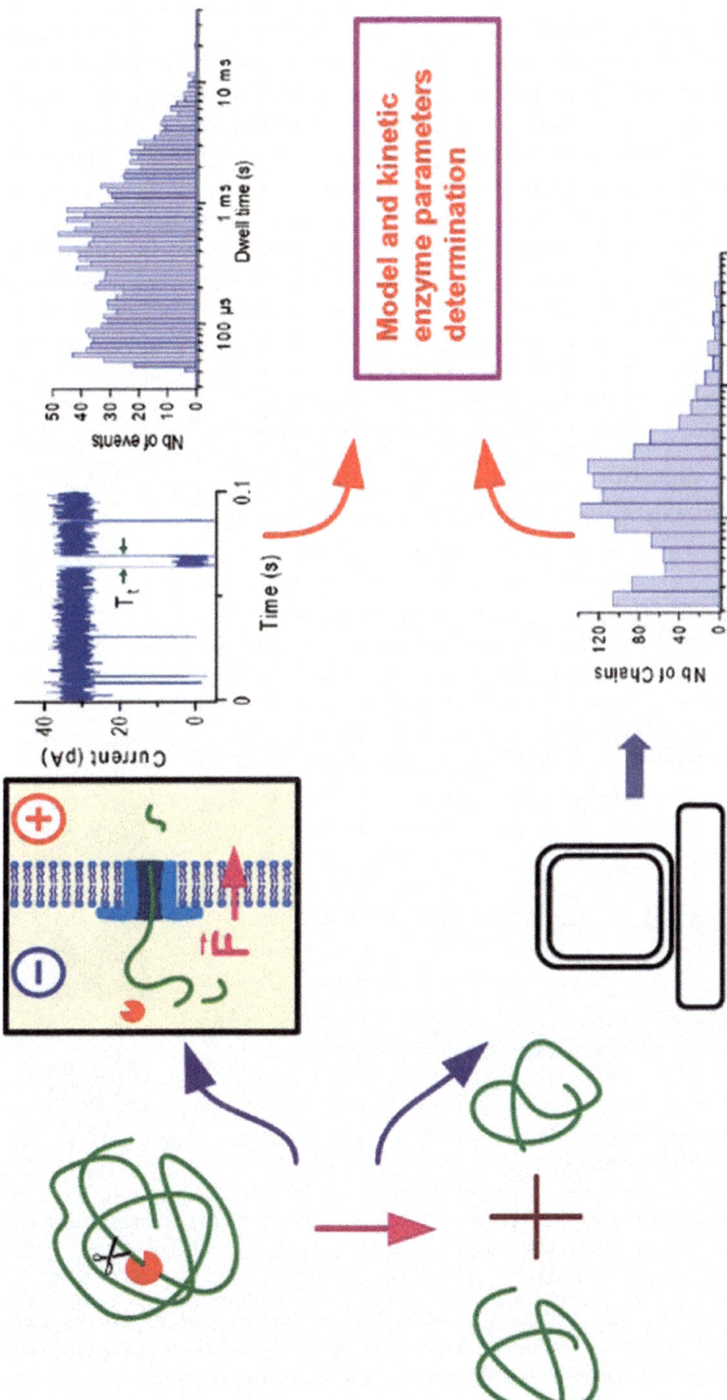

Figure 3.35 Monitoring of depolymerization of hyaluronic acid measure the corresponding kinetic constants at the single molecule level. Reproduced from ref. 71 with permission from American Chemical Society, Copyright 2013.

(−4.4 ± 0.3 and −12.2 ± 0.6 pA, respectively). The blockades with shorter duration are assigned to the cleavage products of Aβ (10-20), YEVHHQK (YK7) and LVFF fragments. Furthermore, to analyze the enzyme kinetics, quantitative analysis of the peptide Aβ (10-20) digestion process by trypsin was done. The Aβ (10-20) digestion rate decreased rapidly with reaction time. For instance, 93.5 (3.8% of the Aβ (10-20)) was cleaved in the first 20 min, but only an extra 6.1 (0.7% of the Aβ (10-20)) was cleaved in the second 20 min. The K_m of the reaction is 59.2 μM, while k_{cat} (= V_{max}/[trypsin]) is 4.43 s^{-1}). These experimental values are agreement with the value of other studies relating cleavage of Aβ peptide performed under experimental conditions similar to those involved in the present research.

Similar to α-HL, some researchers have described a nanopore-based method for the detection of BoNT-B based on aerolysin,[72] utilizing the developing aerolysin nanopore to detect the BoNT-B digestion of its substrate, a derivative of the synaptic protein Sb2 (Figure 3.37). The dynamic variation of the specific digestion product described the existence of the toxin at subnanomolar concentration in minutes. Further study proved the specificity of

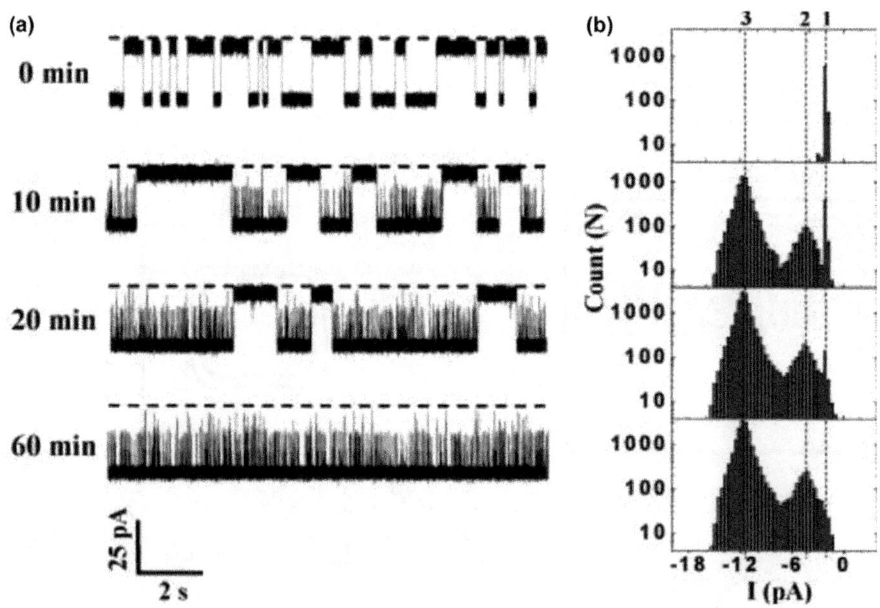

Figure 3.36 Monitoring of Aβ (10-20) cleavage by trypsin. (a) Representative segments of a single-nanopore recording trace at various times. Dashed lines represent the levels of zero current. (b) Corresponding time-dependent event amplitude histograms. Dashed lines 1, 2 and 3 represent the mean residual current levels for peptides YEVHHQKLVFF, YEVHHQK and LVFF, respectively. The experiment was performed at −40 mV with 10 μM Aβ (10-20) and 0.025 μM trypsin.
Reproduced from ref. 69 with permission from American Chemical Society, Copyright 2009.

detection in serum. In practical toxin detection, the presence of BoNT-B can be spotted by the observation of the blockades for the specific digest peptide.

3.3 Photo-responsive Molecule Analysis

3.3.1 Case Study I: Monitoring the Photo-responsive Molecule Induced Molecular Processes

The opening and closing of an ion channel in biosystems could be modulated by various stimuli, such as pH, light, potential and ligand. The natural biological nanopore functionalized by biochemical techniques would gain new features and extend their applications. For instance, a mutant monomer ofα-HL, α-HL-T117C-D8, was designed by derivatizing a single cysteine

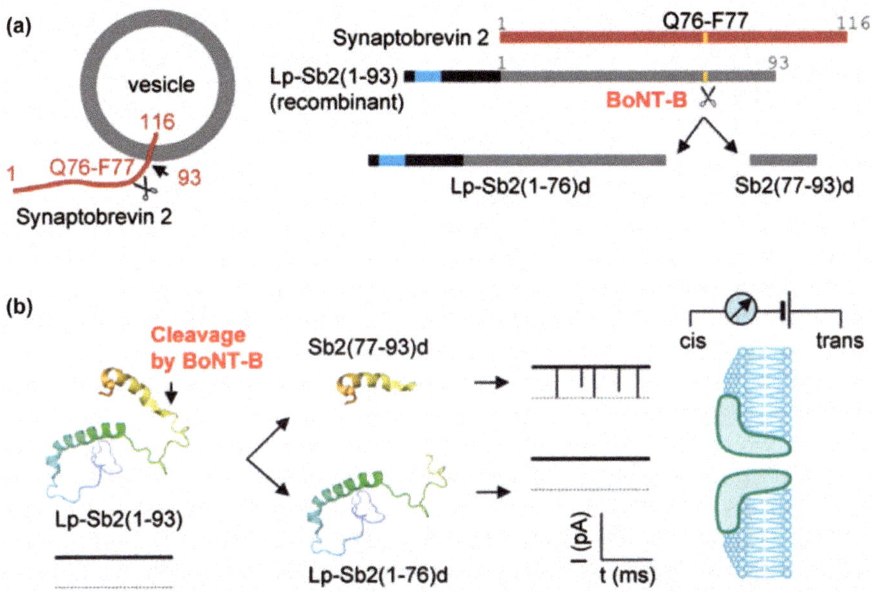

Figure 3.37 Schematized principle of BoNT-B detection using aerolysin nanopore. (a) BoNT-B substrate design and its cleavage products. (b) Illustration of the principle for nanopore BoNT-B detection. Lp-Sb2(1-93) (based on structure PDB ID code 2KOG in the Protein Data Bank (www.pdb.org)) would not affect the pore current, as shown by the hypothetical trace below the structure. After the cleavage of Lp-Sb2(1-93) by BoNT-B/Zn^{2+} (arrow), the short digested product Sb2(77-93) would generate a distinct modulation of pore currents, as shown by downward transient blocks in the hypothetical trace, while the long product Lp-Sb2(1-76)d would not affect the nanopore current. Dotted lines are levels of zero current. Analytes would be applied from the *cis* (pore vestibule) side of the bilayer that partitions the recording chamber.
Reproduced from ref. 72 with permission from American Chemical Society, Copyright 2015.

residue at position 117 with a water-soluble azobenzene derivative to construct a hetero-heptamer $WT_6(T117C\text{-}1\text{-}D8)_1$ (P_{AZO}) with other six wild-type ones.[74–76] Using this mutant P_{AZO}, the reversible photo-isomerization of individual azobenzene was observed for the first time in an aqueous environment.[74]

Different from the previous studies,[75,77] a light-regulated α-HL nanopore based on non-covalent interactions of a supramolecule was constructed, rather than the covalent modification of a photo-responsive molecule inside the nanopore.[78] In this work, sulfonato-calix 4 arene (SC4), seven times smaller than the constriction of an α-HL nanopore (~1.4 nm), could be employed to functionalize the nanopore. Our results revealed that SC4 caused a substantially deep and even full current blockages of the α-HL pore. The current inhibition through nanopore showed an orientation as well as voltage dependence. After adding SC4 to the *trans* chamber, a close-state of nanopore appeared and lasted for a long time at the potential less than −70 mV. Here, τ_{on} is defined as the time interval between two successive signals and exhibits inverse proportion to the potentials in the range varying from −70 to −140 mV (Figure 3.38a). This indicated that the full open-state of the pore is difficult to sustain substantially at the more negative potentials. Moreover, SC4 is repelled from the binding site using a repulsive potential across the membrane. The results above demonstrated that the close-state of the pore mainly originates from the strong host–guest interactions between the negatively charged SC4 and the positive residues (likely Lys^{131} and Lys^{147}) of the α-HL nanopore. As SC4 could induce blockages, the SC4:α-HL interaction could be modulated by either the competitive guest molecules or light stimuli. The photo-responsive 4,4′-dipyridinium-azobenzene (V^{2+}-Az) was introduced as a functional guest molecule (Figure 3.38b). The *trans* form of V^{2+}-Az (V^{2+}-*trans*-Az) and SC4 shows a two orders of magnitude larger binding constant of $\sim 10^5\ M^{-1}$ than the value of SC4:lysine ($\sim 753\ M^{-1}$) obtained at pH = 8.[79,80] With V^{2+}-*trans*-Az, the SC4:α-HL inhibitions are rarely observed in the potential range of −70 to −130 mV. When changing the potential to −140 mV, τ_{on} increased significantly to 2.41 ± 0.14 ms in comparison to that of SC4 (0.09 ± 0.04 ms). Thus, the SC4-functionalized α-HL can be applied to investigate the competition between V^{2+}-*trans*-Az and positive residues of the pore (likely Lys-131 or Lys-147). Because the association/dissociation of the V^{2+}-*trans*/*cis* -Az photo-isomers could be stimulated by light to and from SC4, V^{2+}-*cis*-Az inhibitions showed a larger frequency than those of V^{2+}-*trans*-Az (Figure 3.38c). The photo-isomerization efficiency of V^{2+}-Az in the SC4:α-HL system is 28%, which is comparable to 33% calculated from ^1H NMR. Further assays demonstrated that this SC4:α-HL system could be used to study the photo-isomerization dynamics of SC4:V^{2+}-Az at the single-molecule level in real time (Figure 3.38d). This work presented a novel strategy to construct a host-functionalized nanopore sensing system, providing a general tool for dynamics detection of the molecular machines using receptor-functionalized biomolecular nanopore.

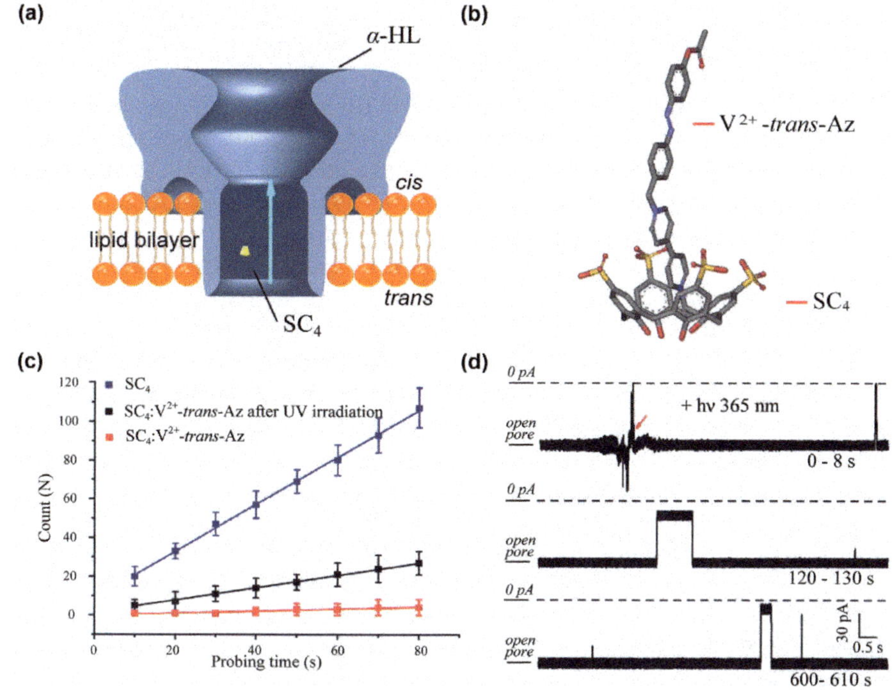

Figure 3.38 (a) The effect of voltage on the inter-event time intervals of α-HL (τ_{on}) induced by SC4 from the *trans* side. A large value of τ_{on} suggests that the inhibitions occur at a low frequency and *vice versa*. Inset: the model shows that SC4 is driven into the stem of the α-HL nanopore from the *trans* side. (b) Representation of the SC4:V^{2+}-*trans*-Az complex. (c) The cumulative number of blockage events *versus* the probing time for SC4 (blue), SC4:V^{2+}-*trans*-Az after UV irradiation (black) and SC4:V^{2+}-*trans*-Az (red) at the potential of −100 mV. (d) Real-time monitoring of the current traces induced by the photo-isomerization of SC4:V^{2+}-*trans*-Az by α-HL. The UV light induced V^{2+}-*trans*-Az to isomerize into V^{2+}-*cis*-Az to release SC4, which could bind with pore resulting in the current blockage of the nanopore.
Reproduced from ref. 77 with permission from Springer Nature, Copyright 2013.

Photo-chromism, a revisable transformation between two photo-isomers, has attracted widespread attention in artificial photo-regulated biosystems.[81–83] Our group adopted α-HL to study the changes of an oligonucleotide regulated by photochromic compounds with single-molecule resolution.[83] Spiropyran (SP),[92] one of the most commonly used photochromic compounds,[84] contributes to a great increase in geometric variations and dipole moments when the ring-closed SP photo-isomerized to the ring-opened merocyanine (MR) (Figure 3.39a). In this study, a SP:RNA complex was selected to investigate the SP-induced conformational variations of the oligonucleotide regulated by an external light stimulus. The selected RNA, adopted as an aptamer *in vitro*, owns a specific tertiary structure that is only able to bind with the ring-closed SP. Therefore, when alternating the ultraviolet/visible (UV/vis) irradiation, the translocation of RNA aptamer could be regulated by light because of the different binding affinity between RNA aptamer and SP/MR isomers (Figure 3.39b). The RNA induced typical three-level blockages of the α-HL nanopore, which contains levels 1, 2 and 3 (Figure 3.39c). Since the different current level could represent different translocation behaviors of target molecule demonstrated by previous work (ref. 85 and 86), level 1 shows current blockage with $I_1/I_0 = 0.56$ and duration time with $t_1 = 50.02 \pm 1.95$ ms, suggesting the captured RNA experiences partial breaks in the pore vestibule. Then, the current blockage increased to level 2 with $I_2/I_0 = 0.51$ and $t_2 = 5.96 \pm 0.25$ ms. The short duration of level 2 may represent the unstable intermediate form of partially unfolded RNA. Finally, the current blockage decreased to level 3, which almost fully blocked the nanopore. The statistical results demonstrate that level 3 owns current blockage of $I_3/I_0 = 0.87$ and duration time of $t_3 = 0.60 \pm 0.01$ ms. In line with the previous reports,[93,94] level 3 assigns to RNA translocation through the α-HL nanopore. When the premixed SP and RNA were added into the *cis* side of the nanopore and irradiated by vis light, the three-level blockades were observed with almost the same I_1/I_0, I_2/I_0 and I_3/I_0 values as those of RNA alone, implying that SP:RNA binding undergoes similarly the two-successive conformational changes inside the α-HL vestibule. The addition of SP has little influence on the alteration of RNA conformation. However, the duration times show a big difference for levels 1 and 2 when compared to those of RNA alone: t_1 increases from 50.02 ± 1.95 to 56.49 ± 2.13 ms, while t_2 decreases from 5.96 ± 0.25 to 4.92 ± 0.32 ms

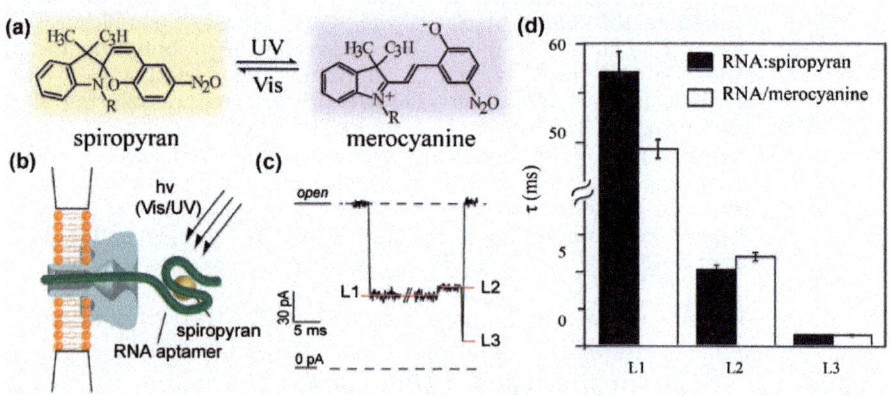

Figure 3.39 (a) Photochromic spiropyran (left) and its merocyanine form (right). R = C_2H_4COOH. (b) Illustration of an RNA:spiropyran translocating through the α-HL nanopore. (c) A typical three-level event for the translocation of RNA through α-HL nanopore. (d) Duration of each level of the typical current blockages induced by the RNA:spiropyran mixture after irradiation with vis light ($\lambda > 490$ nm) and UV light ($\lambda = 365$ nm).
Reproduced from ref. 83 with permission from the Royal Society of Chemistry.

(Figure 3.39d). Therefore, SP does interact with the bases of an oligonucleotide strand and affects RNA unfolding in the vestibule of the α-HL nanopore. Since level 1 likely represents the partial breakage of RNA inside the pore vestibule, the introduction of SP stabilizes RNA structure and slows down the process of RNA unfolding, contributing to an increased t_1 for level 1. Due to the SP:RNA interaction, an intermediate of the partially unfolded RNA may exist for level 2 of SP:RNA, which is related only to that of RNA. Level 3 represents the entrance of the full unfolded RNA aptamer into the pore stem and its translocation, following by the release of SP. In this translocation, an identical duration time and blockage current were observed with those of RNA only. After UV irradiation, the ring-closed SP was forced to switch to the ring-open MR, a form that cannot interact with RNA. The mixed MR and RNA generated more consistent t_1 and t_2 with RNA alone rather than the SP:RNA complex, while t_3 is still unchanged (Figure 3.39d). These results together confirmed that binding affinity is lacking between MR and RNA. Compared to RNA only, the SP:RNA interactions induce a different intermediate form residing in the pore cavity. This present work offers a general analytic model for understanding the photoregulation mechanism and some new insights for designing highly efficient biochemical systems at a single molecule level.

The structural stability of photo-controlling system is mostly imparted by the photo-responsive molecule for the ultimate association/dissociation of DNA duplex.[87,88] Although melting curves are generally used to describe the structural stability of the light-regulated DNA duplex,[89,90] such measurements are derived from the ensemble performance in bulk solution, obscuring information on the heterogeneous behavior of individual molecules in an exact space and time. The nanopore has been chosen as a high-sensitive single molecule tool for the identification of various molecules including nucleic acids, peptides, proteins and small molecules.[91–98] In this work, α-HL was employed to explore the structural stability of the *trans*/*cis* forms of an azobenzene-modified DNA duplex (azo-dsDNA) at the single-molecule level.[98] The azo-dsDNA consists of an azobenzene-modified single-strand DNA (azo-ssDNA) and its complementary DNA without modification (c-ssDNA). The azo-ssDNA was designed by intercalating an azobenzene group into the ssDNA strand through a D-threoninol linker, while its c-ssDNA contains a poly(dC)$_{30}$ overhang at its 3′ end. Since azo-dsDNA was trapped in the nanopore, α-HL recorded the current variation induced by the molecule capturing, unzipping and traversing across the nanopore, interpreting the difference of individual azo-dsDNA in structural stability into visualized information about the unzipping kinetics (Figure 3.40a). The unzipping of azo-dsDNA produce an obvious distribution with long duration times (>0.2 ms) and large current blockage ($I/I_0 < 0.23$). The duration times of azo-dsDNA are well described by a double exponential function and shows as two different distributions P1 and P2 (Figure 3.40b), suggesting the existence of two unzipping time constants resulting from the *trans*/*cis* form of azo-dsDNA. The UV light contributes to event frequencies of P1 and P2 with 6.0 ± 1.2 and 6.5 ± 0.9 min^{-1}, while P1 frequency decreased to

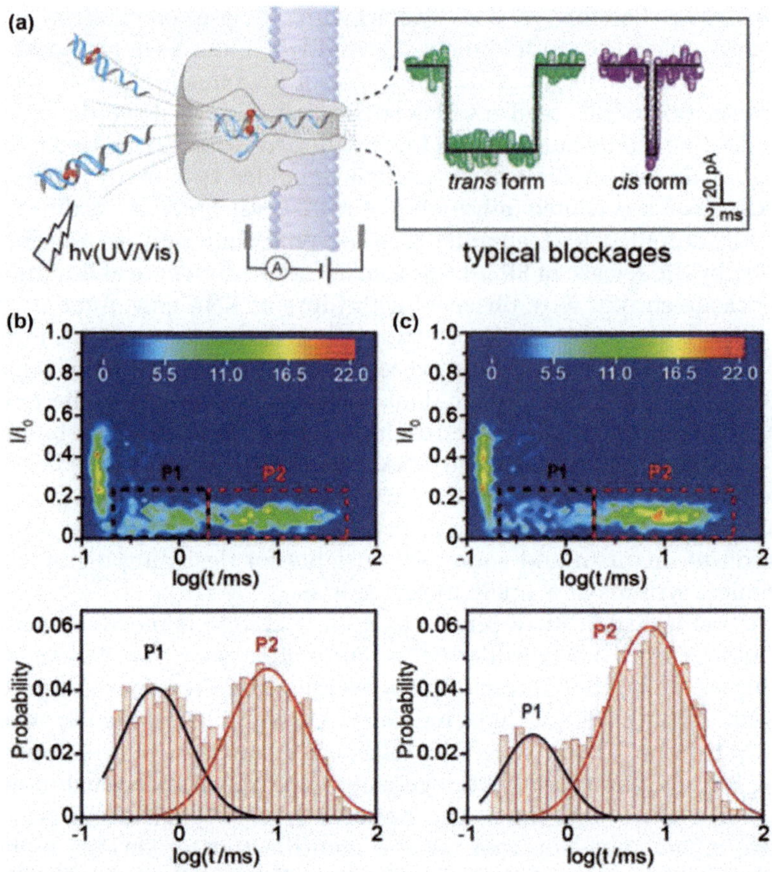

Figure 3.40 (a) Discrimination of the *trans* and *cis* forms of azo-dsDNA from the mixture by an α-HL nanopore. The *trans* form of azo-dsDNA generated the blockages with longer unzipping time than the *cis* form, which showed the high structural stability of the *trans* form. The illustration is not to scale. The 2D contour plots (upper) and the duration (t) histograms (bottom) of the unzipping evens for the annealed DNA solution containing azo-dsDNA after irradiation by (b) UV and (c) vis light, respectively. The duration distributions were fitted by a double exponential function. The experiments were conducted in solutions containing 1.0 M KCl buffered with 10 mM Tris and 1 mM EDTA (pH = 8.0) at + 80 mV.
Reproduced from ref. 99 with permission from the Royal Society of Chemistry.

2.8 ± 0.5 min^{-1} and P2 frequency increased to 9.6 ± 1.6 min^{-1} when altering to vis light. As the UV light facilitates *trans*-to-*cis* conversion and vis light contributes to *cis*-to-*trans* conversion of azo-dsDNA, P1 is assigned to the *cis* form and P2 is ascribed to the *trans* form of azo-dsDNA. Herein, the fitted values of duration distribution are used to represent the unzipping time constants (τ) to determine *cis*/*trans* form of azo-dsDNA. At + 80 mV, the τ

values are 0.52 ± 0.01 and 8.07 ± 0.36 ms for P1 and P2, respectively (Figure 3.40b). As the duration times of the two photo-isomers are well fitted into single exponential distributions, generating dissociation rate constants of 1920 ± 400 and 124 ± 15 s^{-1} for the *cis* and *trans* forms of the DNA duplex, respectively. Similarly, we obtained an unzipping rate constant of 287 ± 32 s^{-1} for the native DNA duplex, which is about two times more stable than the *cis* form and seven times less stable than the *trans* form of modified DNA duplex. These results together reveal that the *trans* form improves the structural stability and the *cis* form decreases the structural stability of DNA duplex in comparison to the native DNA duplex. The possible explanation could be that the non-planar *cis*-azobenzene breaks hydrogen bonding and disturbs base stacking to destabilize the DNA duplex, while the planar *trans*-azobenzene increases base stacking to stabilize the DNA duplex. This work provides new standards to design highly efficient photo-sensitive systems, which are the narrow peak widths and large peak gap for unzipping time distributions of the photo-isomers.

3.3.2 Case Study II: Directly Determination of Configuration of a Single Photo-responsive Oligonucleotide

The photo-responsive oligonucleotide has caused widespread concern over the excellent design of light-regulated biosystems and light-driven nanodevices.[100,101] However, the accurate identification of the configuration of each photo-responsive oligonucleotide molecule is still a big challenge because of the limited sensitivity and resolution of the present techniques,[90,102] let alone a precise presentation of the photo-conversion dynamics. As the small size of aerolysin (~1 nm) is comparable to that of an oligonucleotide, it offers a highly stereo-confined space for the photo-responsive oligonucleotide, which assures a high temporal-spatial resolution.[91,103] In this work, aerolysin was applied to unravel the subtle structural changes of an azobenzene-modified oligonucleotide (Azo-ODN) induced by an azobenzene moiety.[104] The Azo-ODN is a 5-mer oligonucleotide with the sequence of 5′-AA-Azo-AA-3′, where an azobenzene group was attached to the middle site of the oligonucleotide strand as a pseudo base *via* a deoxyribose liker (Figure 3.41a). When an Azo-ODN molecule stays inside an aerolysin, it causes a concentration redistribution of the cations and anions inside the nanopore, and the real-time visual ion current variation reflects the photo-conversion dynamics of Azo-ODN (Figure 3.41b). In this study, a transient current blockade represents a single molecule captured into an aerolysin nanopore, where I is the residual current of an occupied pore, I_0 is the ionic current of a vacant pore, and I/I_0 is the residual current blockade. At the photo-stationary state, two distinctly separated populations were obtained under both ultraviolet (UV) and visible (vis) irradiation. The current histogram shows a dominative distribution at $I/I_0 = 0.40 \pm 0.01$ under UV light, while a major population appears at $I/I_0 = 0.35 \pm 0.01$ under vis light

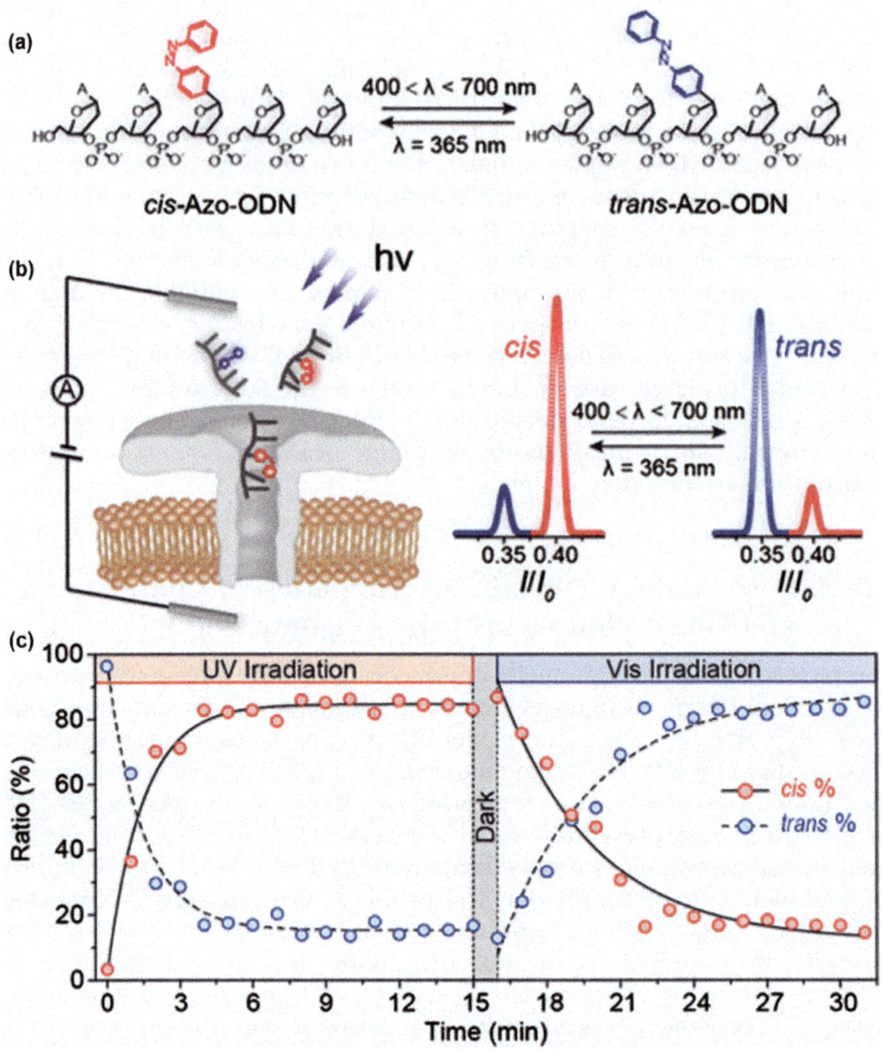

Figure 3.41 (a) Reversible conversion of *cis*/*trans*-Azo-ODN under UV (λ = 365 nm, 12 mW)–vis (λ = 400–700 nm, 12 mW) irradiation. 'A' in the molecular structural model represents the base 'adenine'. (b) Illustration of Azo-ODN detected *via* an aerolysin nanopore. Left: Azo-ODN modulation of the ionic current inside the aerolysin-based confined space. Right: the current distribution of *cis*/*trans*-Azo-ODN controlled by an external light stimulus. (c) Real-time evaluation of Azo-ODN photo-isomerization efficiency at a single-molecule level. As a surrogate of photo-isomerization efficiency, the photo-conversion ratio (%) was estimated by integrating cumulative change of *cis*/*trans*-Azo-ODN translocations in every minute. Data were obtained at 140 mV in 1 M KCl, 1 mM EDTA, and 10 mM Tris buffer, pH 8.0, in the presence of 2.0 μM Azo-ODN in the cap side of aerolysin.
Reproduced from ref. 104 with permission from American Chemical Society, Copyright 2018.

(Figure 3.41b). As previous works reported that the *trans* form of azobenzene is more stable than its *cis* form under vis light,[88,90] we assigned the blockades with $I/I_0 = 0.35 \pm 0.01$ and $I/I_0 = 0.40 \pm 0.01$ to *trans*-Azo-ODN and *cis*-Azo-ODN, respectively. Because the configuration of each Azo-ODN molecule could be well defined using the aerolysin nanopore technique, we performed the determination of *cis*/*trans*-Azo-ODN isomers in real time and the quantification of their photo-conversion dynamics. According to previous works,[105,106] the photo-conversion process of azobenzene could be described by the simple first order model. As the photo-conversion kinetics of *cis*/*trans*-Azo-ODN is well fitted with the simple first order model in this work, the calculations revealed photo response rate constants of 0.20 and 0.43 min^{-1} for *cis*-to-*trans* and *trans*-to-*cis*, respectively (Figure 3.41c). Through counting the number of *cis*/*trans*-Azo-ODN blockades in one minute, the dynamic photo-conversion efficiency was carried out in real time. These results unraveled photo-conversion values of 82.7% and 82.5% for *trans*-to-*cis* and *cis*-to-*trans* at each photo-stationary state, respectively (Figure 3.41c). Since the configuration of individual conformer is absolutely determined at single-molecule level, the kinetics and dynamics values obtained in this study are more convincing and precise when compared to those obtained in the bulk solution. Moreover, these values of Azo-ODN, including the photo-conversion rate and efficiency, were determined directly without complicated pre-treatment such as HPLC separation before spectroscopic analysis. Therefore, the kinetics and dynamics features of photo-responsive oligonucleotide could be described in a more simple and accurate manner, offering new insight for understanding the photo-regulated mechanism, activity and efficiency in photo-controllable biological systems and photon-fueled molecular machines. This study benefits the further rational design of photo-responsive oligonucleotides for diverse applications in nanodevices and biosystems by providing information of on-line molecular geometry, kinetics and dynamics at single molecule resolution.

3.3.3 Case Study III: Regulating the Translocation Speed of Single-molecule Train Using Photo-responsive Molecule

Aerolysin, extracted from Aeromonas sp., can spontaneously form a rivet-shaped heptameric biological nanopore with constriction of ~1 nm.[95,107] Due to its suitable size, aerolysin provide a stereo-confined nano-interface to significantly reduce the motion randomness of individual molecules and improve the spatio-temporal resolution for kinetics and dynamics study of the heterogeneous molecular behaviors. The aerolysin nanopore interface was recently adopted as a molecular machine to directly read out the current variation in real time induced by the photo-controlled structural and functional conversion of individual molecules.[108] By employing this aerolysin-based nano-interface, our group reported the first investigation of the photo-regulated dynamics of a 'photo-geared' oligonucleotide 'molecular train' with a single-molecule resolution.[109] The photo-geared molecular train is a single-strand oligonucleotide modified by an azobenzene (Azo-ssDNA), which was designed by attaching an azobenzene moiety to poly(dA)$_4$ *via* a D-threoninol linker (Figure 3.42a and b). When an Azo-ssDNA train is captured by aerolysin, the ionic current through the nanopore is interrupted, resulting in a blockade signal. Here, we defined I/I_0 as the residual current blockade, where I is the residual current of a blocked nanopore and I_0 is the ionic current of an open pore. The translocations of individual Azo-ssDNA trains fall into Gaussian distributions with two characteristic peak values of $I/I_0 = 0.25$ and 0.31 under vis light in the I/I_0-duration two-dimensional (2D) scatter plots (Figure 3.42c). Interestingly, the blockades gathering around $I/I_0 = 0.25$ have more than a six-fold population of the distribution at $I/I_0 = 0.31$. As Azo-ssDNA prefers to be in the *trans* form (*trans*-train) under vis irradiation, the blockades in the population of $I/I_0 = 0.25$ should belong to the *trans*-train and the blockades in the population of $I/I_0 = 0.31$ likely represent the *cis*-train. By changing to ultraviolet (UV) irradiation ($\lambda = 365$ nm), an increased population of $I/I_0 = 0.31$ was obtained with collection reduction around $I/I_0 = 0.25$ (Figure 3.42c). As expected, the blockades in the population of $I/I_0 = 0.31$ refer to a *cis*-train resulting from the *trans*-to-*cis* conversion of azobenzene under UV irradiation. According to previous studies, the *trans*-azobenzene prefers to protrude from the ssDNA backbone, while the *cis*-azobenzene is inclined to accommodate itself in the gap between two adjacent bases.[89,110] As a result, the Azo-ssDNA *trans*-train causes a deeper current blockade through aerolysin than the *cis*-train, mainly attributed to a greater steric hindrance. The high current-resolved *cis*/*trans* blockades support the direct discrimination of the conformation of individual Azo-ssDNA trains in real time using the aerolysin nanopore. The real-time current traces revealed each *trans*-train with an ultra-long duration time ($t_D = 2630.3$ ms) (Figure 3.42), resulting in a 500 times slower speed (1.9 bases/s) than that of the previously reported poly(dA)$_5$.[93] The azobenzene moiety performed as a molecular gear to decelerate the translocation speed of the DNA train. The speed of the *trans*-train is almost 3.3 times slower than the *cis*-train with $t_D = 794.3$ ms, leading to a relatively high speed of 6.3 bases/s (Figure 3.42c). Therefore, the gear effect could be regulated by photo-controlling interconversion of the *cis*/*trans*-train. By counting the accurately defined blockades of the *cis*/*trans*-train, the reversible photo-conversion process was uncovered in real time by aerolysin: the proportion of *cis*-train increased progressively with reduction in that of *trans*-train under continuous UV illumination; while the conversion process was inversed subsequently when altering to vis irradiation (Figure 3.42a). Compared with traditional strategies that fixed molecular machines onto silica nanoparticles or gold surfaces, the nanopore interface provides potential opportunities for applications in a single-molecule machine device, offering guidance for design of the more elaborate molecular machine systems.

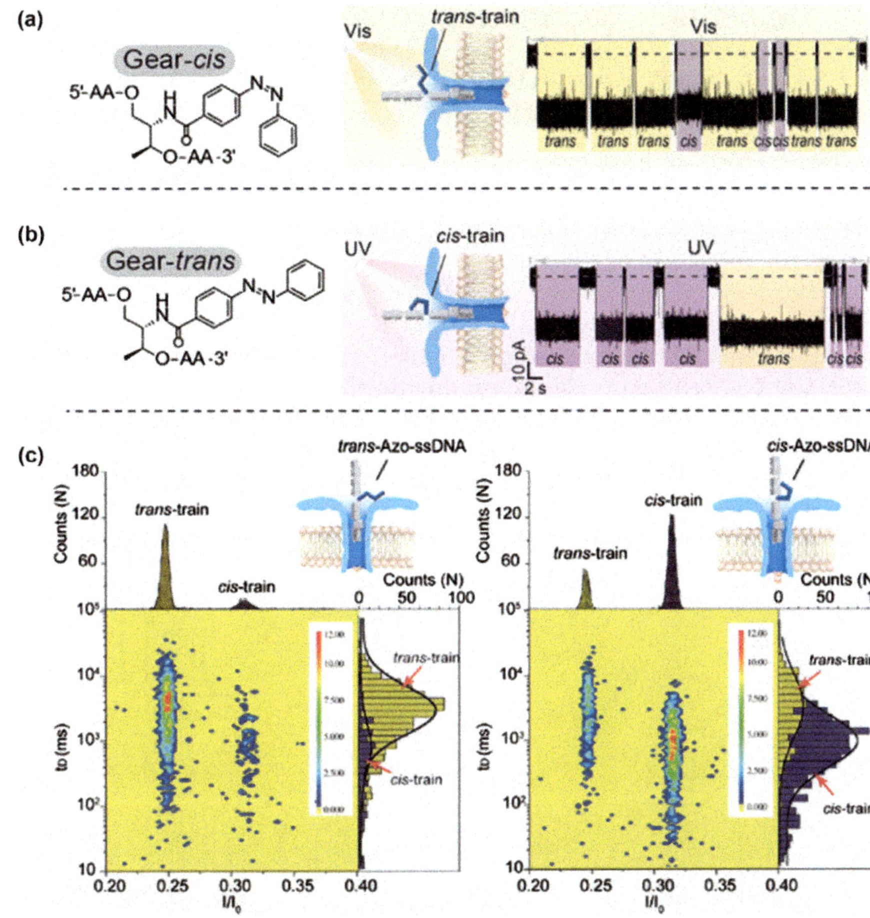

Figure 3.42 Representation of time-resolved electric current readouts of individual photo-reversible Azo-ssDNA trains based on an aerolysin nano-interface under (a) vis and (b) UV irradiation. The synthesized azobenzene moiety is introduced into poly(dA)$_4$ through the D-threoninol linker. The speed-regulated molecule train of 5′-AA-azobenzene-AA-3′ traverses through an aerolysin nanopore from its cap side. The UV light induces the high speed of designed *cis*-Azo-ssDNA, whereas the vis irradiation prolongs the translocation of *trans*-Azo-ssDNA. The illustration is not to scale. The current blockage of *trans*-Azo-ssDNA has a longer duration and deeper blockage amplitude than that of *cis*-Azo-ssDNA. (c) Analysis of the translocation events under vis (left) and UV (right) irradiation. The two populations in the 2D contour plots correspond to two Gaussian peaks in the current histogram (top), and their duration histogram on a log scale is fitted into Gaussian functions (right). In the current and duration histograms, the population for the *trans*-train is highlighted in yellow, and that for the *cis*-train is shown in purple. Data were acquired in 1 M KCl, 10 mM Tris, and 1 mM EDTA (pH 8.0) at +100 mV in the presence of 2.0 μM Azo-ssDNA.
Reproduced from ref. 109 with permission from Elsevier, Copyright 2018.

3.4 Host-Guest Interaction Analysis

3.4.1 Case Study I: Detection of Self-assembly Process of Host and Guest Molecules

The investigation of host–guest interaction permits the characterization of the self-assembly process by facilitating the construction of host to guest molecules with properties that are massively controlled and cooperative. In this regard, the study of the self-assemble recognition of the 1 : 1 and 1 : 2 complexes was achieved using an α-HL nanopore at the single molecule level, where para-sulfonatocalix 6 arenes (SC6) was the host and methyl viologen (MV^{2+}) as the guest (Figure 3.43a).[111] The SC6 was first examined *via* α-HL at a potential of −60 mV and it was found that it induced a short-lived duration time $\tau_{off} = 0.15$ ms. Approximately 90% of blockages fall in the distribution I, the plot I/I_0 *versus* τ_{off}, where I represents the blockage current

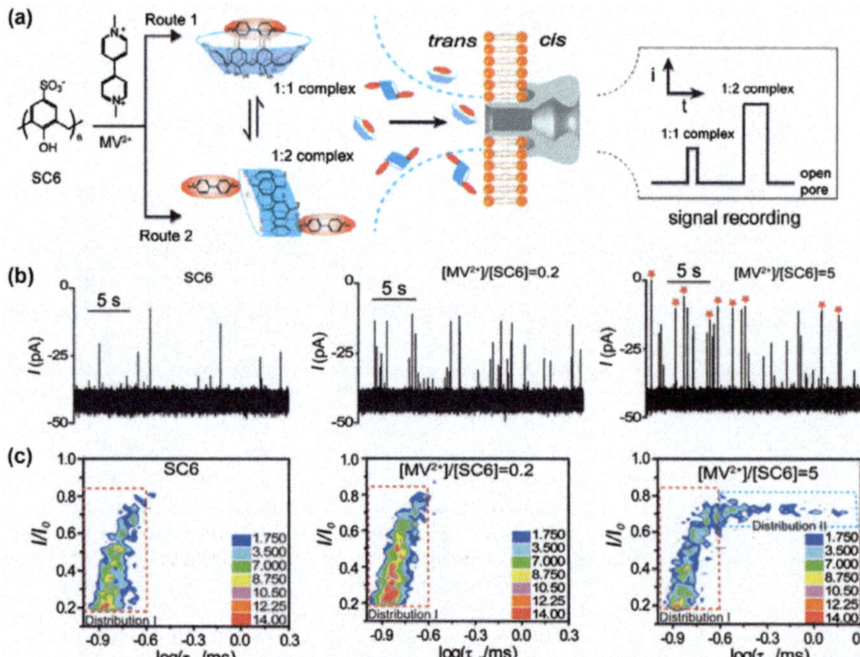

Figure 3.43 (a) Detection of the self-assembly process induced by SC6 and MV^{2+} at the single-molecule level using an α-HL nanopore. 1 : 1 and 1 : 2 complexes are self-assembled by SC6 and MV^{2+} at the ratio of 1 : 1 and 1 : 2, respectively. These two complexes generate distinguishable blockages. (b) Current traces were recorded in solutions containing 1.0 M KCl buffered with 10 mM Tris–HCl (pH = 8.0) at −60 mV. (c) 2D counter plots were obtained by continuously recording for 14 min during the nanopore experiments.
Reproduced from ref. 111 with permission from the Royal Society of Chemistry.

related to the analyte and I_0 represents the open pore current. Under similar experimental conditions, SC6 and MV^{2+} were mixed at a ratio of $[MV^{2+}]/[SC6] = 0.2$, and the mixture also showed short-lived blockages in distribution I. However, a strong increase in the frequency of the blockages current traces was observed compared to the SC6 one. Moreover, a linear increase in the cumulative number of blockages in the distribution I per min was seen with a slope of 94.1 events per minute for the mixture. This observation was ascribed to the 1 : 1 complex assembled in the mixture because it was found that MV^{2+} was too weak to generate any blockages. Consequently, the 1 : 1 complex was inherent to the translocation process *via* α-HL and the interaction with the *trans*-side opening of the pore than SC6. Furthermore, when the $[MV^{2+}]/[SC6]$ was in the range from 0.2 to 5, the event frequencies in the new region distribution II increased from 5.3 to 24.6 events per min, while those in distribution I decreased from 94 to 54.8 events per min, illustrating the formation of 1 : 2 complex because it was self-assembled *via* SC6 and MV^{2+} at the ratio 1 : 2 (Figure 3.43b and c). It was noted that distribution II had long-lived blockages estimated at 0.28 ms more than distribution I with 0.10 ms as duration time. Hence, two 1 : 1 and 1 : 2 complexes self-assembled were discriminated in the mixture $[MV^{2+}]/[SC6]$ through two blockage distributions in real time, indicating that α-HL has a great potential to probe self-assembled mechanisms of host–guest supramolecules at single-molecule conditions.

3.4.2 Case Study II: Measurement of Assembly Behaviors of Supramolecular Polymers

Exploring supramolecular oligomers is of keen interest due to their key role in the formation of supramolecular polymers for the characterization of the assembly behaviors of single functional supramolecular polymers. Using an α-HL nanopore, the analysis of individual supramolecular oligomers in aqueous solution has been investigated and the tracking of self-assembly of the oligomers at the single molecular level as well, as described in Figure 3.44.[112] In this work, bis(p-sulfonatocalix [4]-arene) (bisSC4) and bis(methylviologen) ($bisMV^{4+}$) were studied in the *cis* side of α-HL to overcome the interplay between bisSC4 and the stem of the pore (Figure 3.44a and b). Thereby, the study of 1 μM $bisMV^{4+}$ at 100 mV *via* α-HL indicated a population *P* of the blockages with dwell time of 0.13 ms and $I/I_0 = 0.88$. A large value of $\tau_P = 0.32$ ms was found and attributed to the interaction with the pore. The combination of $bisMV^{4+}$ and bisSC4 led to the formation of supramolecular oligomers. The mixture of $[bisSC4]/[bisMV^{4+}] = 0.2$ showed an increase in the event frequencies upon comparing with $bisMV^{4+}$ alone and generated long-lived blockages (Figure 3.44c). In addition, the plot cumulative number of blockages *versus* time displayed a linear growth. The increase of the blockages in the population *P* was assigned to the self-assembly oligomers in the mixture. The voltage-dependent study denoted a single exponential decay distribution, while the applied potential was raised

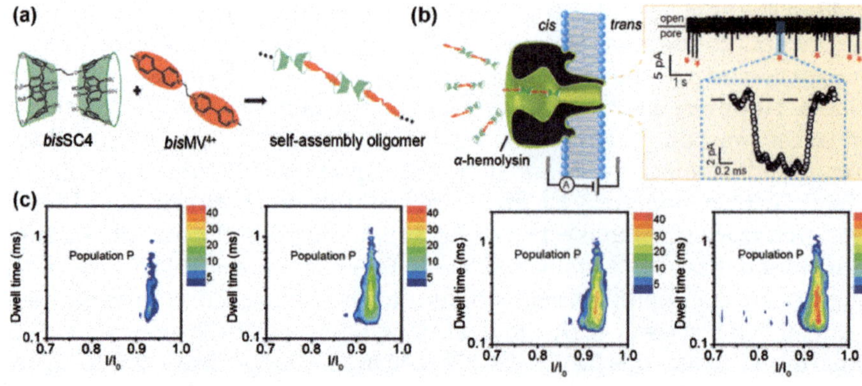

Figure 3.44 (a) Schematic representation of a self-assembled oligomer formed by the iterative host – guest complexation based on bisSC4 and bisMV^{4+}. (b) Illustration of supramolecular oligomers determined *via* an α-HL nanopore and the raw data of current blockages in the presence of 1 μM bisSC4 and 1 μM bisMV^{4+} at +100 mV within the *cis* chamber containing 1.0 M KCl, 10 mM Tris, and 1 mM EDTA (pH = 7.8). (c) 2D contour plots for the mixtures with [bisSC4]/[bisMV^{4+}] = 0, 0.2, 1 and 10 (left to right). The blockages used in the plots were obtained by continuously recording for 30 min during the nanopore experiments. Reproduced from ref. 112 with permission from American Chemical Society, Copyright 2016.

with a decrease in the dwell time for the mixture at the ratio 0.2. These results have illustrated that the single supramolecular oligomers have traversed the α-HL nanopore. To further increase the mixture at the ratio 20, a large increase in the population *P* was found, which was caused by the growth of the self-assembled oligomers of bisMV^{4+} and bisSC4. This strategy constitutes an ideal platform for better understanding the structure–function relationship in the preorganized precise structure of supramolecular oligomers.

3.5 Measurement of Covalent Bonding at the Single-molecule Level

Monitoring single-molecule reactions provides the detailed information about the microscopic reaction mechanisms such as the bond strengths and conformational changes of the synthesized molecules. A mutant aerolysin K238C with a dimeter ~1 nm was used to monitor a single-molecule reaction when 5,5′-dithiobis-(2-nitrobenzoic acid) (DTNB) comes into collisions at the interface of the pore, illustrated by Figure 3.45.[113] To fulfil this aim, a +100 mV was applied across the bilayer membrane, leading DTNB to traverse from the *cis* to *trans* sides of the pore with the onset of a characteristic blockade current. This process has shown the formation of disulfide bonds DTNB and cysteine residues of the pore. No blockade currents were produced when an unfunctionalized wild-type aerolysin was employed to follow the

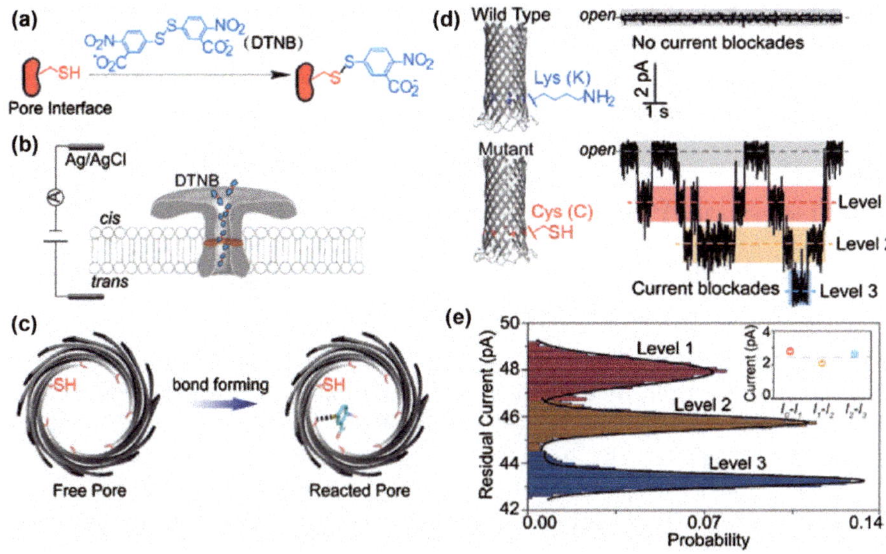

Figure 3.45 (a) Single-molecule reaction and structure of mutant aerolysin in the presence of DTNB. A proposed reaction Cys-238 reacting with DTNB in a reversible reaction to form the mixed disulfide bonds. (b) A cross section of the aerolysin K238C with Cys at position 238. (c) Top view of the β-barrel of mutant aerolysin K238C and seven Cys residues. (d) The β-barrel of wild-type aerolysin with Lys colored blue at the position 238 and the current trace of wild-type aerolysin. The β-barrel of mutant aerolysin with Cys colored red at the position 238 and the current trace of mutant aerolysin (middle right). (e) The histograms of level 1, level 2 and level 3 fitted to Gaussian distribution. The values of current for $I_0 - I_1$, $I_1 - I_2$, and $I_2 - I_3$ were very close (insert in (e)). All experiments were performed in the buffer of 1.0 M KCl, 1.0 mM EDTA and 10 mM Tris, at 24 ± 1 °C, pH 8.0.
Reproduced from ref. 113 with permission from Springer Nature, Copyright 2018.

translocation process of DTNB molecules (Figure 3.45d, top). Likewise, it was not possible to distinguish the blockage current traces whenever DTNB was simultaneously introduced into both *cis* and *trans* sides owing to the small volume of DTNB molecules to induce current drops during the translocation process or the duration time was very short for the interactions between the analytes molecules and the sites of the pore.

Otherwise, by adding DTNB molecules in the *cis* side only, well-defined current blockades were observed and showed three levels, indicating the formation of chemical bonds forming and breaking (Figure 3.45d bottom). Moreover, the residual current *I* was further used to characterize the reaction events and it was found that each level corresponded to one step of the residual current when a single-molecule DTNB passed through the pore K238C and reacted with the cysteine residue. The histograms of the residual current indicated the existence of three individual molecule reactions (Figure 3.45e).

This was corroborated by the values of current for $I_n - I_{n+1}$ ($n = 0, 1, 2$), which revealed that just one reaction occurred at the interface of the pore. Therefore, the engineered aerolysin K238C has been proven to be an ultrasensitive tool for tracking chemical bonds being made and broken in single molecule reactions.

3.6 Detection of Small Biomolecule

3.6.1 Case Study I: Transient Conformation Measurement of Flavin Adenine Dinucleotide (FAD)

Moreover, the non-covalent conformations differences of a single molecule can be also monitored by biological nanopore. Since the super confinement construction of aerolysin nanopore as described above, the researchers applied the aerolysin in the analysis of conformation of flavin adenine dinucleotide (FAD) at single molecule level (Figure 3.46). As FAD contains flavin and adenine simultaneously, it is considered as a cofactor for electron transportation and photoreceptors in living systems, exhibiting conformational dynamics in performing its biological functions. The FAD in 'open' and 'stack' conformations were observed by aerolysin, and the population of intermediates were also detected at single-molecule level. The results of temperature dependent experiments and mutant aerolysin nanopores provide the direct evidences for the correlation of distinct current blockages and FAD conformations. Moreover, both experimental results and MD simulation indicated the FADs were stabilized in quasi-stack conformations in various polarity solvents. Also, a direct correlation between the residual current ratio (I/I_0) and FAD backbone angle was further established, providing new insights for studying the transient conformations of a single FAD molecule, which may enhance the understanding of the structure–activity relationship.

3.6.2 Case Study II: Antibiotic Molecule Analysis by Biological Nanopores

Nanopores can also identify small antibiotic molecules (<600 Da) by passing the membrane protein pore. Taking OmpF from *E. coli* as an example, such a water-filled membrane pore is composed of 16 beta-sheets. There are 8 long loops (L1–L8) connecting each pair of strands, typically L3 loop folds back the nanopore lumen forming the narrowest constriction region (0.7×1.1 nm). The well defined confined space is suitable for the detection of small molecules *via* nanopore electrochemistry technology. An innovative approach for the translocation/binding mechanism of small molecules was designed (Figure 3.47). In this work, since OmpF is overall cysteine free, a cysteine is introduced at the exit of the OmpF nanopore *via* site-direct mutagenesis (E181C), far away from the original constriction region. Thanks to the sulfate side chain, cysteine chemistry is performed by covalently binding to MTSES or GLT (blocker). Any molecule translocation process in

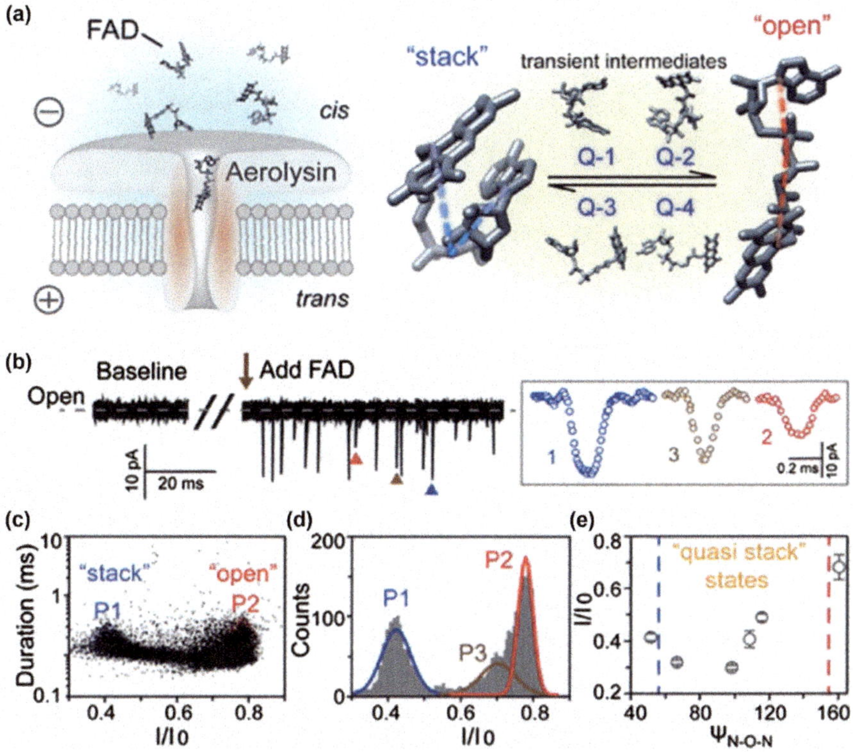

Figure 3.46 Identification of FAD conformations by an aerolysin nanopore. (a) Left: schematic illustration of a nanopore sensor. Right: the 'stack' and 'open' conformation of the FAD. (b) Left: continuous current recording of the baseline and FAD translocating through an aerolysin nanopore. Right: typical events of three distributions. (c) Scatter plots and (d) I/I_0 histograms of FAD blockages at +80 mV. (e) The ψ_{N-O-N} *versus* I/I_0 curve of the FAD molecule. The data were acquired in 1.0 M KCl, 10 mM Tris, and 1.0 mM EDTA at pH 8.0 in the presence of 2.0 mM FAD. The amplified signal was low-pass filtered at 5 kHz and sampled at 100 kHz. Reproduce from ref. 117, copyright 2020 Royal Chemical Society.

the wild-type case would behave differently due to the presence of the blocker, in another case, molecules of binding process would not 'see' the blocker, consequently no electrochemical fluctuation would be detected. Such work merges the mutagenesis as well as chemical modification, creating a well-designed nanopore for small molecules detection. Moreover, the resolution could reach only two amino acid difference (mono-Arginine *vs.* tri-Arginine).[115]

Multi-drug resistance bacteria are some of the notorious issues among the WHO concerns. One can hardly find a broad-spectrum antibiotic, mainly due to the high susceptibility of bacteria species. Porins are regarded as the potential pathways of the antimicrobial agents. One of the strategies of antimicrobial resistance is to decrease the permeability of such by decrease the

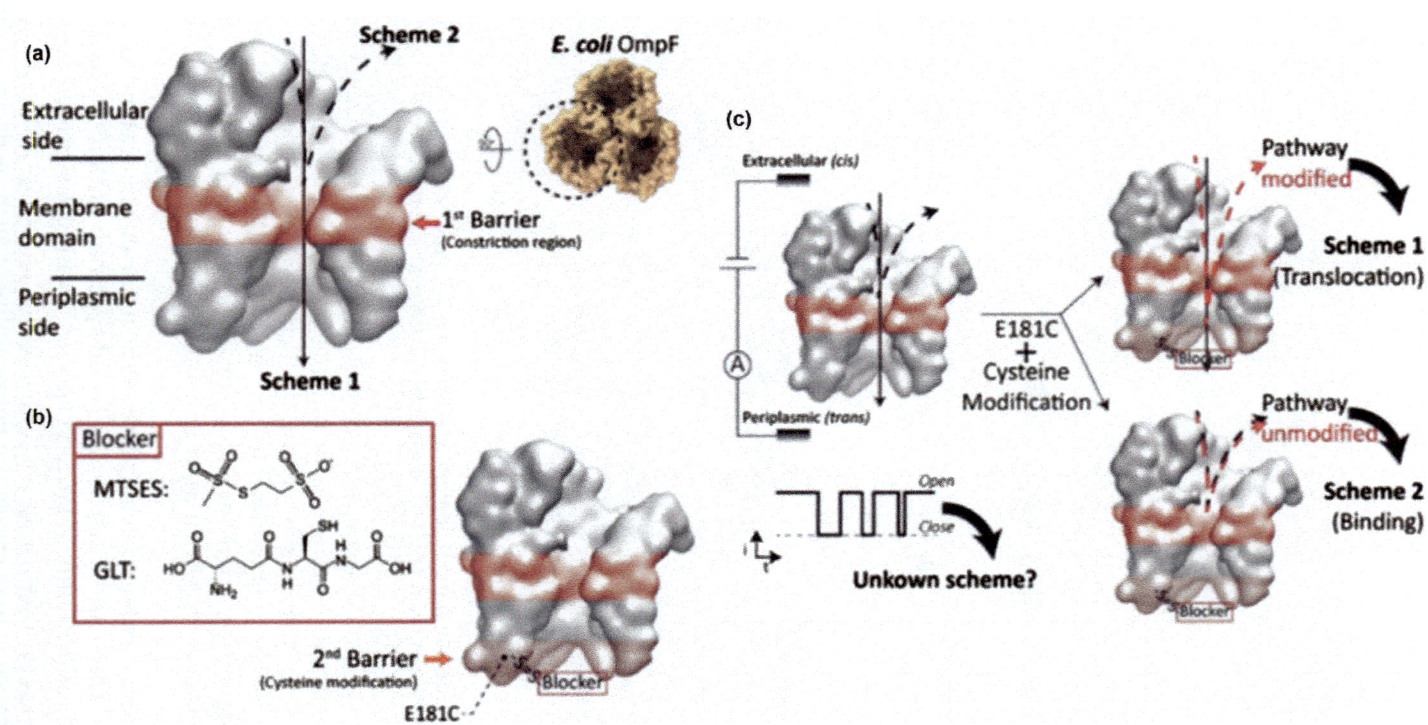

Figure 3.47 (a) Schematic diagram of detecting molecule translocation (scheme 1) or binding and bouncing back (scheme 2) with OmpF (only one monomer is shown). (b) Introducing a molecule blocker at the exit (far beyond the constriction region) *via* a single-site mutation (E181C) along with cysteine chemistry modification. Blockers with different spatial occupancy (*i.e.* MTSES and GLT) are chosen. (c) Schematic demonstrating single-molecule analysis. By adding a blocker at the exit, molecules fitting into scheme 1 (translocation) would then alter their pathway and the dwell time. Other molecules unable to translocate (scheme 2) will not encounter the blocker so the dwell time should not be modified.
(Permission from ref. 115 © 2019 Wiley-VCH Verlag GmbH & Co. KGaA, Weinheim).

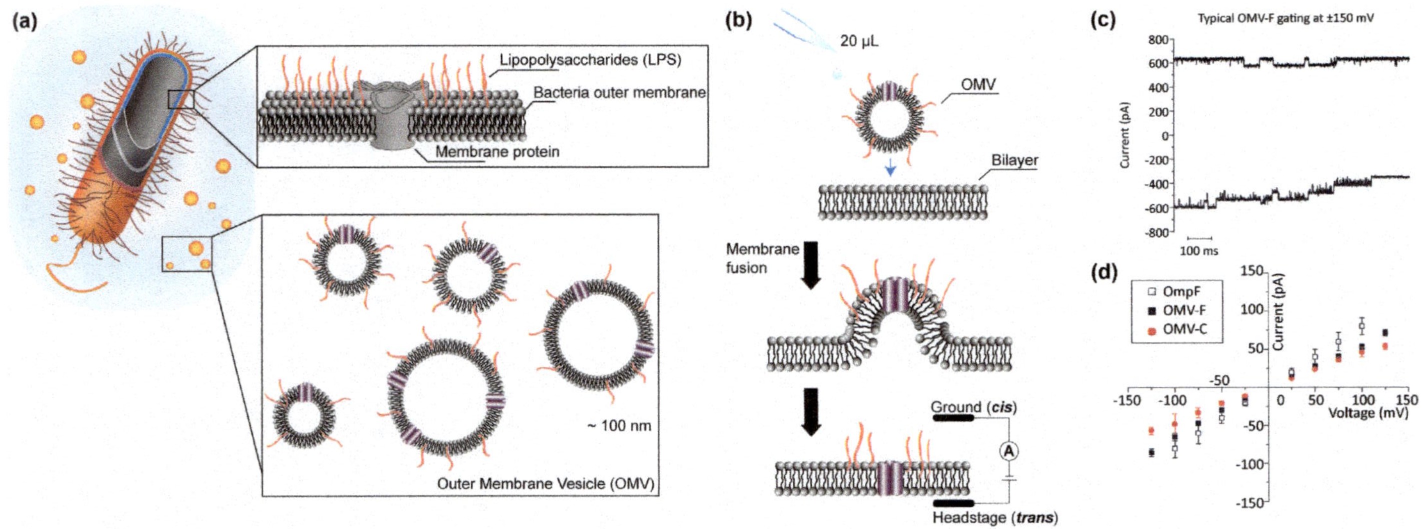

Figure 3.48 (a) Schematic outer membrane structure of Gram-negative bacteria, composed of a phospholipids inner leaflet, lipopolysaccharides (LPS) outer leaflet as well as of membrane proteins. The bacilli secrete outer membrane vesicles (OMV) whose diameters are around 100 nm. It is expected that their composition is similar to the outer membrane. (b) OMVs are added directly to the planar lipid bilayer, forming BLM-OMV. Fusion and protein activity is followed by applying external voltages. The ion current steps correspond to open porin insertion. (c) Two typical ion current traces after OMV fusion containing likely three trimeric OmpF porins. Conclusions on single-pore conductance are taken from the current steps caused under high applied voltage ±150 mV inducing gating. (d) I–V response of OmpF and OmpC from OMV fusion. 200 mM KCl, 20 mM MES, pH 6.0 was used throughout. At least three individual experiments were performed.
(Permission from ref. 116 © 2020 Wiley-VCH Verlag GmbH & Co. KGaA, Weinheim).

expression of critical porins or mutagenize into restriction orthologues. Making use of nanopore electrochemistry, OmpF and OmpC are the major porins from *E. coli*, by reconstituting the purified porins into artificial bilayer, detection of the passage of a single antibiotic molecule without labelling could be achieved. However, another barrier at the bacterial cellular membrane surface is that the lipopolysaccharides (LPS) are not feasible in such an artificial system. Such a dilemma was tackled by collecting the outer membrane vesicle from the *E. coli* host,[116] then direct fusing the secrets to the artificial bilayer to mimic the native environment. Surprisingly, low pore activity with overexpressed membrane protein was recorded at high reproducibility (Figure 3.48). Specifically, by overexpressing OmpF, its orthologue OmpC is relatively suppressed but not extinct. From the nanopore electrochemistry recordings, OmpF is the only responsive pore; same case was performed on OmpC overexpressed strains. This OMV-fusion approach is fairly easy for membrane proteins de-novo design as better biosensors for small molecules detection.

Acknowledgements

We would like to thank Yong-Xu Hu, Dong-Fang Liao and Bo Yuan for contributing to the section on nucleic acid and peptide sensing. Yao Lu is acknowledged for her help with the section on photo-responsive molecule analysis. We thank Fu-Na Meng for her assistance with the section on host-guest interaction.

References

1. J. J. Kasianowicz, E. Brandin, D. Branton and D. W. Deamer, *Proc. Natl. Acad. Sci. U. S. A.*, 1996, **93**, 13770–13773.
2. M. Akeson, D. Branton, J. J. Kasianowicz, E. Brandin and D. W. Deamer, *Biophys. J.*, 1999, **77**, 3227–3233.
3. G. D. Fasman, C. Lindblow and L. Grossman, *Biochemistry*, 1964, **3**, 1015–1021.
4. P. K. Sarkar and J. T. Tang, *Biochemistry*, 1965, **4**, 1238–1244.
5. S. Arnott, R. Chandrasekaran and A. G. W. Leslie, *J. Mol. Biol.*, 1976, **106**, 735–748.
6. A. Meller, L. Nivon, E. Brandin, J. Golovchenko and D. Branton, *Proc. Natl. Acad. Sci. U. S. A.*, 2000, **97**, 1079–1084.
7. Y.-L. Ying, D.-W. Li, Y. Liu, S. K. Dey, H.-B. Kraatz and Y.-T. Long, *Chem. Commun.*, 2012, **48**, 8784–8786.
8. Y.-L. Ying, H.-Y. Wang, T. C. Sutherland and Y.-T. Long, *Small*, 2011, **7**, 87–94.
9. Y.-L. Ying, D.-W. Li, Y. Li, J. S. Lee and Y.-T. Long, *Chem. Commun.*, 2011, **47d**, 5690–5692.
10. K. R. Lieberman, G. M. Cherf, M. J. Doody, F. Olasagasti, Y. Kolodji and M. Akeson, *J. Am. Chem. Soc.*, 2010, **132**, 17961–17972.

11. T. Lindahl, *Nature*, 1993, **362**, 709–715.
12. B. K. Duncan and J. H. Miller, *Nature*, 1980, **287**, 560–561.
13. K. Imai, G. Slupphaug, W.-I. Lee, P. Revy, S. Nonoyama, N. Catalan, L. Yel, M. Forveille, B. Kavli, H. E. Krokan, H. D. Ochs, A. Fischer and A. Durandy, *Nat. Immunol.*, 2003, **4**, 1023–1028.
14. R. Savva, K. Mcauleyhecht, T. Brown and L. Pearl, *Nature*, 1995, **373**, 487–493.
15. N. An, A. M. Fleming and C. J. Burrows, *J. Am. Chem. Soc.*, 2013, **135**, 8562–8570.
16. Q. Jin, A. M. Fleming, R. P. Johnson, Y. Ding, C. J. Burrows and H. S. White, *J. Am. Chem. Soc.*, 2013, **135**, 19347–19353.
17. M. S. Cooke, R. Olinski and M. D. Evans, *Clin. Chim. Acta*, 2006, **365**, 30–49.
18. J. Cadet, J. L. Douki and T. Ravanat, *Free Radic. Biol. Med.*, 2010, **49**, 9–21.
19. J. G. Muller, V. Duarte, R. P. Hickerson and C. J. Burrows, *Nucleic Acids Res.*, 1998, **26**, 2247–2249.
20. M. Hada and B. M. Sutherland, *Radiat. Res.*, 2006, **165**, 223–230.
21. D. Fologea, M. Gershow, B. Ledden, D. S. Mcnabb, J. A. Golovchenko and J. Li, *Nano Lett.*, 2005, **5**, 1905–1909.
22. D. Stoddart, A. J. Heron, E. Mikhailova, G. Maglia and H. Bayley, *Proc. Natl. Acad. Sci. U. S. A.*, 2009, **106**, 7702–7707.
23. A. E. P. Schibel, N. An, Q. Jin, A. M. Fleming, C. J. Burrows and H. S. White, *J. Am. Chem. Soc.*, 2010, **132**, 17992–17995.
24. R. Lister, M. Pelizzola, R. H. Dowen, R. D. Hawkins, G. Hon, J. Tonti-Filippini, J. R. Nery, L. Lee, Z. Ye, Q.-M. Ngo, L. Edsall, J. Antosiewicz-Bourget, R. Stewart, V. Ruotti, A. H. Millar, J. A. Thomson, B. Ren and J. R. Ecker, *Nature*, 2009, **462**, 315–322.
25. M. G. Goll and T. H. Bestor, *Annu. Rev. Biochem.*, 2005, **74**, 481–514.
26. Y.-F. He, B.-Z. Li, Z. Li, P. Liu, Y. Wang, Q. Tang, J. Ding, Y. Jia, Z. Chen, L. Li, Y. Sun, X. Li, Q. Dai, C.-X. Song, K. Zhang, C. He and G.-L. Xu, *Science*, 2011, **333**, 1303–1307.
27. S. Schiesser, T. Pfaffeneder, K. Sadeghian, B. Hackner, B. Steigenberger, A. S. Schröder, J. Steinbacher, G. Kashiwazaki, G. Höfner, K. T. Wanner, C. Ochsenfeld and T. Carell, *J. Am. Chem. Soc.*, 2013, **135**, 14593–14599.
28. R. P. Johnson, A. M. Fleming, R. T. Perera, C. J. Burrows and H. S. White, *J. Am. Chem. Soc.*, 2017, **139**, 2750–2756.
29. C. Cao, Y.-L. Ying, Z.-L. Hu, D.-F. Liao, H. Tian and Y.-T. Long, *Nat. Nanotechnol.*, 2016, **11**, 713–718.
30. C. Cao, J. Yu, M.-Y. Li, Y.-Q. Wang, H. Tian and Y.-T. Long, *Small*, 2017, **13**, 1702011.
31. J. Yu, C. Cao and Y.-T. Long, *Anal. Chem.*, 2017, **89**, 11685–11689.
32. D.-F. Liao, C. Cao, Y.-L. Ying and Y.-T. Long, *Small*, 2018, **14**, 1704520.
33. C. Cao, M.-Y. Li, N. Cirauqui, Y.-Q. Wang, M. Dal Peraro, H. Tian and Y.-T. Long, *Nat. Commun.*, 2018, **9**, 2823.
34. Y.-Q. Wang, C. Cao, Y.-L. Ying, S. Li, M.-B. Wang, J. Huang and Y.-T. Long, *ACS Sens.*, 2018, **3**, 779–783.

35. Y.-Q. Wang, M.-Y. Li, H. Qiu, C. Cao, M.-B. Wang, X.-Y. Wu, J. Huang, Y.-L. Ying and Y.-T. Long, *Anal. Chem.*, 2018, **90**, 7790–7794.
36. C. S. Tan, A. M. Fleming, H. Ren, C. J. Burrows and H. S. White, *J. Am. Chem. Soc.*, 2018, **140**, 14224–14234.
37. Y. Wang, D. Zheng, Q. Tan, M. X. Wang and L. Q. Gu, *Nat. Nanotechnol.*, 2011, **6**, 668–674.
38. M. Wanunu, T. Dadosh, V. Ray, J. Jin, L. McReynolds and M. Drndic, *Nat. Nanotechnol.*, 2010, **5**, 807–814.
39. R. W. Carthew and E. J. Sontheimer, *Cell*, 2009, **136**, 642–655.
40. R. C. Lee, R. L. Feinbaum and V. Ambros, *Cell*, 1993, **75**, 843–854.
41. V. N. Kim, J. Han and M. C. Siomi, *Nat. Rev. Mol. Cell Biol.*, 2009, **10**, 126–139.
42. B. Wightman, I. Ha and G. Ruvkun, *Cell*, 1993, **75**, 855–862.
43. K. Tian, Z. He, Y. Wang, S.-J. Chen and L.-Q. Gu, *ACS Nano*, 2013, **7**, 6–10.
44. M. T. Landi, Y. Zhao, M. Rotunno, J. Koshiol, H. Liu, A. W. Bergen, M. Rubagotti, A. M. Goldstein, I. Linnoila, F. M. Marincola, M. A. Tucker, P. A. Bertazzi, A. C. Pesatori, N. E. Caporaso, L. M. McShane and E. Wang, *Clin. Cancer Res.*, 2010, **16**, 430–441.
45. R. Garzon, G. A. Calin and C. M. Croce, *Annu. Rev. Med.*, 2009, **60**, 167–179.
46. C. Ortholan, M. P. Puissegur, M. Ilie, P. Barbry, B. Mari and P. Hofman, *Curr. Med. Chem.*, 2009, **16**, 1047–1061.
47. J. Yang, Y.-Q. Wang, M.-Y. Li, Y.-L. Ying and Y.-T. Long, *Langmuir*, 2018, **34**, 14940–14945.
48. S. Sakakibara, Y. Kishida, K. Okuyama, N. Tanaka, T. Ashida and M. Kakudo, *J. Mol. Biol.*, 1972, **65**, 371–373.
49. G. Oukhaled, J. Mathé, A.-L. Biance, L. Bacri, J.-M. Betton, D. Lairez, J. Pelta and L. Auvray, *Phys. Rev. Lett.*, 2007, **98**, 158101.
50. M. Pastoriza-Gallego, L. Rabah, G. Gibrat, B. Thiebot, F. G. can der Goot, L. Auvray, J.-M. Betton and J. Pelta, *J. Am. Chem. Soc.*, 2011, **133**, 2923–2931.
51. L. Payet, M. Martinho, M. Pastoriza-Gallego, J.-M. Betton, L. Auvray, J. Pelta and J. Mathé, *Anal. Chem.*, 2012, **84**, 4071–4076.
52. K. J. Freedman, M. Jürgens, A. Prabhu, C. W. Ahn, P. Jemth, J. B. Edel and M. J. Kim, *Anal. Chem.*, 2011, **83**, 5137–5144.
53. Y.-X. Hu, Y.-L. Ying, Z. Gu, C. Cao, B.-Y. Yan, H.-F. Wang and Y.-T. Long, *Chem. Commun.*, 2016, **52**, 5542–5545.
54. C. M. Dobson, *Nature*, 2003, **426**, 884–890.
55. G. Schatz and B. Dobberstein, *Science*, 1996, **271**, 1519–1526.
56. W. Wickner and R. Schekman, *Science*, 2005, **310**, 1452–1456.
57. J. J. Kasianowicz, J. W. F. Robertson, E. R. Chan, J. E. Reiner and V. M. Stanford, *Annu. Rev. Anal. Chem.*, 2008, 1737–1766.
58. Y.-L. Ying, J. Zhang, R. Gao and Y.-T. Long, *Angew. Chem., Int. Ed.*, 2013, **52**, 13154–13161.

59. L. Mereuta, A. Asandei, C.-H. Seo, Y. Park and T. Luchian, *ACS Appl. Mater. Interfaces*, 2014, **6**, 13242–13256.
60. A. Asandei, M. Chinappi, H.-K. Kang, C.-H. Seo, L. Mereuta, Y. Park and T. Luchian, *ACS Appl. Mater. Interfaces*, 2015, **7**, 16706–16714.
61. J. Sanchez-Quesada, M. R. Ghadiri, H. Bayley and O. Braha, *J. Am. Chem. Soc.*, 2000, **122**, 11757–11766.
62. H.-Y. Wang, Y.-L. Ying, Y. Li, H.-B. Kraatz and Y.-T. Long, *Anal. Chem.*, 2011, **83**, 1746–1752.
63. H.-Y. Wang, Z. Gu, C. Cao, J. Wang and Y.-T. Long, *Anal. Chem.*, 2013, **85**, 8254–8261.
64. M. S. Prakash, A. Matouschek and L. Movileanu, *J. Am. Chem. Soc.*, 2008, **130**, 4081–4088.
65. Y.-L. Ying, D.-W. Li, Y. Li, J. S. Lee and Y.-T. Long, *Chem. Commun.*, 2011, **47**, 5690–5692.
66. T. C. Sutherland, Y.-T. Long, R.-I. Stefureac, I. Bediako-Amoa, H.-B. Kraatz and J. S. Lee, *Nano Lett.*, 2004, **4**, 1273–1277.
67. F. Piguet, H. Ouldali, M. Pastoriza-Gallego, P. Manivet, J. Pelta and A. Oukhaled, *Nat. Commun.*, 2018, **9**, 966.
68. L. Mereuta, M. Roy, A. Asandei, J. K. Lee, Y. Park, I. Andricioaei and T. Luchian, *Sci. Rep.*, 2014, **4**, 3885.
69. Q. Zhao, R. S. de Zoysa, D. Wang, D. A. Jayawardhana and W. Guan, *J. Am. Chem. Soc.*, 2009, **131**, 6324–6325.
70. A. Asandei, I. Schiopu, S. Iftemi, L. Mereuta and T. Luchian, *Langmuir*, 2013, **29**, 15634–15642.
71. A. Fennouri, R. Daniel, M. Pastoriza-Gallego, L. Auvray, J. Pelta and L. Bacri, *Anal. Chem.*, 2013, **85**, 8488–8492.
72. Y. Wang, V. Montana, V. Grubišić, R. F. Stout, V. Parpura and L.-Q. Gu, *ACS Appl. Mater. Interfaces*, 2015, **7**, 184–192.
73. L. Movileanu, J. P. Schmittschmitt, J. M. Scholtz and H. Bayley, *Biophys. J.*, 2005, **89**, 1030–1045.
74. S. Loudwig and H. Bayley, *J. Am. Chem. Soc.*, 2006, **128**, 12404–12405.
75. L. Song, M. R. Hobaugh, C. Shustak, S. Cheley, H. Bayley and J. E. Gouaux, *Science*, 1996, **274**, 1859–1866.
76. S. Loudwig and H. Bayley, *J. Am. Chem. Soc.*, 2006, **128**, 12404–12405.
77. Y.-L. Ying, J. Zhang, F.-N. Meng, C. Cao, X. Yao, I. Willner, H. Tian and Y.-T. Long, *Sci. Rep.*, 2013, **3**, 1622.
78. N. Douteau-Guével, A. W. Coleman, J.-P. Morel and N. Morel-Desrosiers, *J. Chem. Soc., Perkin Trans. 2*, 1999, **2**, 629–634.
79. D. Guo, L. Wang and Y. Liu, *J. Org. Chem.*, 2007, **72**, 7775–7778.
80. A. A. Beharry and G. A. Woolley, *Chem. Soc. Rev.*, 2011, **40**, 4422–4437.
81. G. C. Ellis-Davies, *Nat. Methods*, 2007, **4**, 619–628.
82. J. Zhang, Q. Zou and H. Tian, *Adv. Mater.*, 2013, **25**, 378–399.
83. X. Zhang, J. Zhang, Y.-L. Ying, H. Tian and Y.-T. Long, *Chem. Sci.*, 2014, **5**, 2642–2646.
84. V. I. Minkin, *Chem. Rev.*, 2004, **104**, 2751–2776.

85. T. Z. Butler, J. H. Gundlach and M. a. Troll, *Biophys. J.*, 2006, **90**, 190–199.
86. J. A. Cracknell, D. Japrung and H. Bayley, *Nano Lett.*, 2013, **13**, 2500–2505.
87. Y. Nakasone, H. Ooi, Y. Kamiya, H. Asanuma and M. Terazima, *J. Am. Chem. Soc.*, 2016, **138**, 9001–9004.
88. H. Asanuma, T. Ito, T. Yoshida, X. Liang and M. Komiyama, *Angew. Chem., Int. Ed.*, 1999, **38**, 2393–2395.
89. T. Goldau, K. Murayama, C. Brieke and H. Asanuma, *Chem. – Eur. J.*, 2015, **21**, 17870–17876.
90. Y.-L. Ying, D.-W. Li, L. Yang, J. S. Lee and Y.-T. Long, *Chem. Commun.*, 2011, **47**, 4–7.
91. H.-Y. Wang, Y. Li, L.-X. Qin, A. Heyman, O. Shoseyov, I. Willner, Y.-T. Long and H. Tian, *Chem. Commun.*, 2013, **49**, 1741–1743.
92. C. Cao, Y.-L. Ying, Z.-L. Hu, D.-F. Liao, H. Tian and Y.-T. Long, *Nat. Nanotechnol.*, 2016, **11**, 1–7.
93. Y. Wang, D. Zheng, Q. Tan, M.-X. Wang and L.-Q. Gu, *Nat. Nanotechnol.*, 2011, **6**, 668–674.
94. R. Stefureac, Y.-T. Long, H. B. Kraatz, P. Howard and J. S. Lee, *Biochemistry*, 2006, **45**, 9172–9179.
95. Y.-X. Hu, Y.-L. Ying, Z. Gu, C. Cao and B. Yan, *Chem. Commun.*, 2016, **52**, 5542–5545.
96. H. Ouldali, K. Sarthak, T. Ensslen, F. Piguet, P. Manivet, J. Pelta, J. C. Behrends, A. Aksimentiev and A. Oukhaled, *Nat. Biotechnol.*, 2020, **38**, 176–181.
97. M. Pastoriza-Gallego, M. Breton and F. Discala, *ACS Nano*, 2014, **8**, 11350–11360.
98. G. S. Pulcu, E. Mikhailova, L.-S. Choi and H. Bayley, *Nat. Nanotechnol.*, 2014, **10**, 76–83.
99. F.-N. Meng, Z.-Y. Li, Y.-L. Ying, S.-C. Liu, J.-J. Zhang and Y.-L. Long, *Chem. Commun.*, 2017, **53**, 9462–9465.
100. Y. Kamiya and H. Asanuma, *Acc. Chem. Res.*, 2014, **47**, 1663–1672.
101. R. Wang, C. Jin, X. Zhu, L. Zhou, W. Xuan, Y. Liu, Q. Liu and W. Tan, *J. Am. Chem. Soc.*, 2017, **139**, 9104–9107.
102. E. Kazuma, M. Han, J. Jung, J. Oh, T. Seki and Y. Kim, *J. Phys. Chem. Lett.*, 2015, **6**, 4239–4243.
103. Y.-L. Ying and Y.-T. Long, *Sci. China: Chem.*, 2017, **60**, 1187–1190.
104. Z.-L. Hu, Z.-Y. Li, Y.-L. Ying, J.-J. Zhang, C. Cao, Y.-T. Long and H. Tian, *Anal. Chem.*, 2018, **90**, 4268–4272.
105. C.-C. Chu, Y.-C. Chang, B. K. Tsai, T.-C. Lin, J.-H. Lin and V. K. S. Hsiao, *Chem. – Asian J.*, 2014, **9**, 3390–3396.
106. M. Moniruzzaman, J. D. R. Talbot, C. J. Sabey and G. F. Fernando, *J. Appl. Polym. Sci.*, 2006, **100**, 1103–1112.
107. I. Iacovache, S. De Carlo, N. Cirauqui, M. Dal Peraro, F. G. van der Goot and B. Zuber, *Nat. Commun.*, 2016, **7**, 12062.

108. C. Soc, M. A. Watson and S. L. Cockroft, *Chem. Soc. Rev.*, 2016, **45**, 6118–6129.
109. Y.-L. Ying, Z.-Y. Li, Z.-L. Hu, J.-J. Zhang, F.-N. Meng, C. Cao, Y.-T. Long and H. Tian, *Chem*, 2018, **4**, 1893–1901.
110. E. Merino and M. Ribagorda, *Beilstein J. Org. Chem.*, 2012, **8**, 1071–1090.
111. F.-N. Meng, Z.-Y. Li, Y.-L. Ying, S.-C. Liu, J.-J. Zhang and Y.-T. Long, *Chem. Commun.*, 2017, **53**, 9462–9465.
112. F.-N. Meng, X. Yao, J. Zhang, Y. Ying and H. Tian, *ACS. Sens.*, 2016, **1**, 1398–1402.
113. B. Zhou, Y. Wang, C. Cao, D. Li and Y. Long, *Sci. China: Chem.*, 2018, **61**, 1385–1388.
114. J.-J. Wang, M.-Y. Li, J. Yang, Y.-Q. Wang, X.-Y. Wu, J. Huang, Y.-L. Ying and Y.-T. Long, *ACS Cent. Sci.*, 2020, **6**, 76–82.
115. J.-J. Wang, J. A. Bafna, S. P. Bhamidimarri and M. Winterhalter, *Angew. Chem., Int. Ed.*, 2019, **58**, 4737–4741.
116. J.-J. Wang, R. Terrasse, J.-A. Bafna, L. Benier and M. Winterhalter, *Angew. Chem., Int. Ed.*, 2020, DOI: 10.1002/anie.201913618.
117. M.-Y. Li, Y.-Q. Wang, Y.-L. Ying and Y.-T. Long, *Chem. Sci.*, 2019, **10**, 10400–10404.
118. Y. Wang, D. Zheng, Q. Tan, M.-X. Wang and L.-Q. Gu, *Nat. Nanotechnol.*, 2011, **6**, 668–674.
119. V. V. Meervelt, M. Soskine, S. Singh, G. K. Schuurman-Wolters, H. J. Wijma, B. Poolman and G. Maglia, *J. Am. Chem. Soc.*, 2017, **139**, 18640–18646.
120. N. S. Galenkamp, A. Biesemans and G. Maglia, *Nat. Chem.*, 2020, **12**, 481–488.
121. E. A. Manrao, I. M. Derrington, A. H. Laszlo, K. W. Langford, M. K. Hopper, N. Gillgren, M. Pavlenok, M. Niederweis and J. H. Gundlach, *Nat. Biotechnol.*, 2012, **30**, 349–353.
122. C. Cao, J. Yu, Y. Q. Wang, Y. L. Ying and Y. T. Long, *Anal. Chem.*, 2016, **88**, 5046–5049.

CHAPTER 4

Solid-state Nanopore Confinement for Single Molecule Detection

YAO LIN,[a] QIAO LI,[a,b] SHAO-CHUANG LIU,*[b] SI-MIN LU,[b] YI-LUN YING[b] AND YI-TAO LONG[b]

[a] East China University of Science and Technology, Shanghai, China;
[b] Nanjing University, Nanjing, China
*Email: shaochuangliu@nju.edu.cn

4.1 Fabrication Methods

Biological nanopores have proved to be accessible to a broad range of single molecule sensing, as discussed in Chapter 3. Nonetheless, there are two conundrums that biological nanopores cannot overcome, *i.e.* the relative fixed pore size limits the application of confinement effect, and the weak robustness hinders the application under harsh conditions. The advantages of solid-state nanopores are tunable pore size, high thermal and chemical stability, adjustable surface properties and their facile integration into devices.[1] To date, solid-state nanopores have been widely used for detecting a variety of single entities including DNA, peptides, proteins, biomolecular complexes, nanoparticles, *etc.*[2–4] To fabricate a solid-state nanopore with desirable diameter for various sensing applications, an increasing number of methods have been developed. For example, solid-state nanopores with diameters comparable to biological nanopores could be successfully fabricated with focused ion beams, focused electron beams and controlled

Detection Science Series No. 17
Confining Electrochemistry to Nanopores
Edited by Yi-Tao Long

Published by the Royal Society of Chemistry, www.rsc.org

dielectric breakdown. By modification of the inner surface of solid-state nanopores, the interface properties such as charge and specific receptors are tuned for the selective sensing of a wide range of analytes. Increasing efforts have been put into the development of novel solid-state nanopore platforms combined with other promising detection techniques, and thus have further expanded the application of nanopore sensing and provide a number of exciting opportunities in the fields of biotechnology, cell biology and material sciences.[5]

The goal of this chapter is to introduce the current techniques for fabricating solid-state nanopores, and the promising applications using the nanopore confinement in terms of both electrochemistry and optical measurement. Firstly, we fabricate solid-state nanopores *via* focused ion beam, which is the widely used nanopore fabrication method. Then, we describe the applications of the solid-state nanopore confinement for characterization of DNA duplex unzipping process, detection of target proteins, and the study of the dynamics of single DNA–protein complexes. Finally, we introduce the nanopore-based optical detection of single entities at nano-scale by confining an electromagnetic field inside solid-state nanopores.

Researchers have utilized different approaches to fabricate a variety of solid-state nanopores on different kinds of supporting membrane materials. Examples include porous polymer membranes fabricated by the track–etch technique,[6] porous anodic alumina fabricated by the anodic oxidation method,[7] and nanopores on silicon-based materials or 2D materials fabricated by focused ion beam,[8] electron beam[9] or the newly emerging method controlled dielectric breakdown.[10]

As for the solid-state membranes, the most commonly used materials include SiN_x, SiO_2, SiC, Al_2O_3, and graphene, which feature superior chemical and mechanical robustness and precisely tunable pore size.[5] Low-stress silicon nitride is the most widely used membrane material because it is an excellent insulator with high resistivity and dielectric strength, and has high stability over a range of temperatures, pH and electrolyte concentrations. The fabrication of low-stress free-standing SiN_x can be briefly described[5,11] in the following steps (Figure 4.1): (1) SiN_x films are deposited on both sides of the silicon wafer by using low-pressure chemical vapor deposition (LPCVD). (2) Photolithographic techniques are used to define areas for the SiN_x etch. (3) Reactive ion etching (RIE) is subsequently performed to remove the SiN_x on the bottom side. (4) The exposed silicon substrate is removed using anisotropic wet etching (KOH etching solution). Typically, with this method, the side length of the membrane could range from 7 to 50 mm, while the thickness of the membrane could range from 10 to 200 nm.

After finishing the membrane fabrication, the final stage is to fabricate the pore on the membrane. This process can be achieved in a variety of ways and we will discuss three most commonly used approaches for silicon nitride nanopore fabrication: focused ion beam, high-power electron beam and the emerging method named controlled dielectric breakdown (CBD).

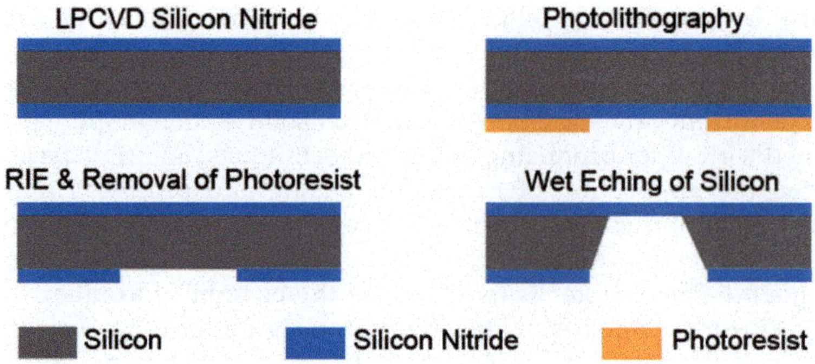

Figure 4.1 The process for the fabrication of a freestanding silicon nitride membrane.

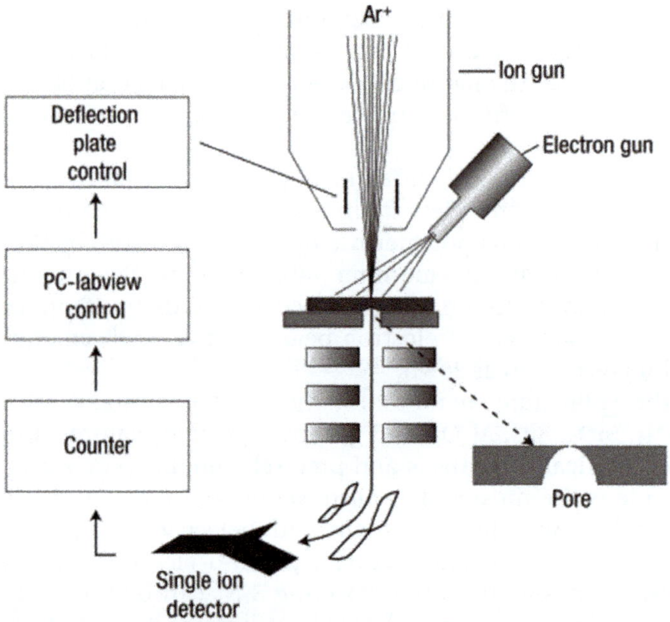

Figure 4.2 Fabrication of solid-state nanopores using the focused ion beam technique. The Ar^+ beam is used to sputter away the membrane and form a nanopore. The electron gun is used to neutralize surface charging effects during ion beam exposure. When the ions drilling the hole penetrate through the membrane, a feedback system provides signals to stop the beam milling.
Reproduced from ref. 8 with permission from Springer Nature, Copyright 2001.

The first fabrication of solid-state nanopores with nanometer resolution using ion beam sculpting was reported in 2001.[8] In this experiment, focused ion beams (FIBs) were employed to mill a nanopore in the membrane (Figure 4.2). The fabrication of a molecular-sized pole requires precise control of the sputtering process, thus a feedback system below the membrane that detects the ions transmitted through the pore stops the milling when the desired pore diameter is reached. By modulating the ion rate and temperature, pores can be enlarged and shrunken, thereby allowing the fine-tuning of pores with a precision of ~1 nm. This method has been successfully used to fabricate 1–5 nm solid-state nanopores for DNA translocation measurements. FIBs are predominantly utilized to fabricate small nanopores with diameters of 1–10 nm. In addition, FIBs attract much interest because they provide high flexibility with the pore dimensions and membrane materials. Basically, any ions with sufficient energy to sputter away material from the membrane can be employed for nanopore fabrication. For instance, ions such as Ga^+, Ar^+, He^+, Ne^+, and Xe^+ have successfully been employed to mill or sculpt solid-state nanopores.[12] Ga^+ is the most frequently used for nanopore milling and it typically mills pores with diameters greater than 20 nm. Recently, a sub-5 nm pore was fabricated on a SiC membrane using a Ga^+ beam.[13] The FIB approach for nanopore fabrication was further developed using a scanning helium ion microscope (HIM).[14] The HIM utilizes an ion source with an atomically small tip that enables producing a highly focused ion beam and achieves the fabrication of pores as small as 4 nm in diameter. Compared to traditional Ga^+ beam milling, this technique has advantages such as higher resolution, no ion implantation and direct thinning of the membrane.

Apart from the ion beam, another commonly used method for nanopore fabrication is high-energy electron beams. It enables the fabrication of solid-state nanopores with nanometer precision with a transmission electron microscope (TEM). The electron beam-based fabrication of solid-state nanopores with sub-10 nm diameter was first performed in 2003 (Figure 4.3).[9] They combined electron-beam lithography with anisotropic etching to fabricate nanopores in a silicon membrane. Then nanopore shrinkage could occur with the introduction of an electron beam at an appropriate intensity. The shrinking process could be terminated within seconds by reducing the beam intensity, or shutting it off. This approach provides a way to fabricate solid-state nanopores with sub-nanometer resolution, and gives the unique possibility of imaging the pore size *in situ* during the shrinking process with TEM. Later, the electron beam is also employed to fabricate nanopores in a SiN membrane.[15] By maximizing the electron beam intensity, it achieves a nearly five-fold decrease in the fabrication time of a single nanopore. In addition, high energy electron beams have been used for nanopore fabrication on numerous membrane materials including metal oxides,[16] graphene,[17] *etc.*

The use of ion and electron beams has shown great success in solid-state nanopore fabrication. The solid-state nanopores obtained using these methods have been applied for a wide range of studies. However, these methods suffer several drawbacks such as low throughput, expensive instrumentation and complicated fabrication procedures. Recently, a promising alternative for solid-state nanopore fabrication on a silicon nitride

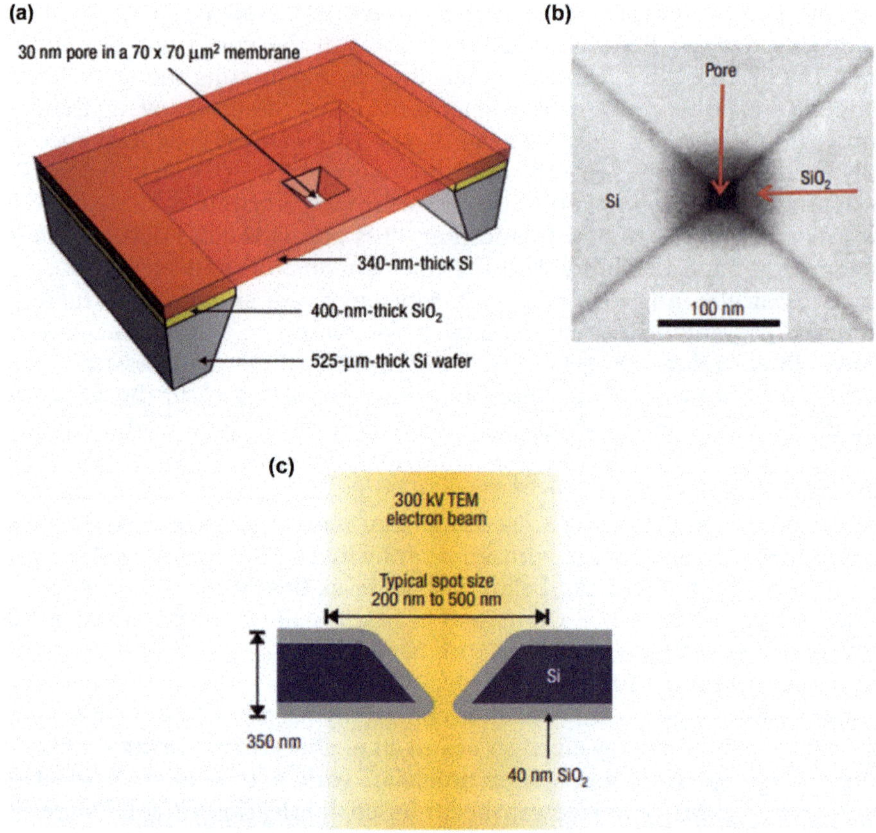

Figure 4.3 Fabrication of silicon oxide nanopores using an electron beam. (a) Cross-section view of the device. It consists of a free-standing single crystalline silicon membrane, supported by a KOH-etched Si wafer. (b) Top-view scanning electron micrograph of a nanofabricated pore after thermal oxidation. The pore is about 20 nm, and is surrounded by a SiO_2 layer of about 40 nm thickness. (c) Cross-sectional view of the pore inside the electron microscope. An electron beam drills a hole in the membrane. The pore size can be directly imaged in the electron microscope and the electron beam can also be used to enlarge or shrink the nanopore. Reproduced from ref. 9 with permission from Springer Nature, Copyright 2003.

membrane has been proposed that involves controlled dielectric breakdown (CDB).[10] This fabrication method utilizes high electric fields to form a single pore on the membrane and it is automated, simple and low-cost. More importantly, the CDB allows nanopores to be generated directly in aqueous solution, thus the fabricated pore could be directly used for single molecule experiments without further procedures to increase the pore hydrophilicity. In this approach, the SiN_x membrane is positioned between two chambers containing electrolyte solution. A pair of Ag/AgCl electrodes are put into the electrolyte solution and connected to a feedback current amplifier. When a sufficient voltage is applied across the SiN_x membrane, the pore is generated under the high field strength (Figure 4.4a–d). The voltage applied to fabricate small nanopores in thin membranes is in the range of 5–10 V, which can reduce the damage to the membrane after the dielectric breakdown event. The nanopore fabrication process could be monitored in real time through the value of the leakage current, $I_{leakage}$ (Figure 4.4e and f). A sudden irreversible increase in $I_{leakage}$, which is due to the conduction of ionic current, indicates the creation of a single nanopore. The nanopore further enlarges as the current continues to increase. A LabVIEW program has been designed to rapidly stop the potential when the $I_{leakage}$ exceeds a preset threshold current. This method could successfully be used to fabricate individual SiN_x nanopores in the range of 1 to 25 nm in diameter.

The experiments as a function of applied voltage, membrane thickness, electrolyte composition, pH and concentration were performed to further investigate the mechanism of the fabrication process.[10] Figure 4.5 shows

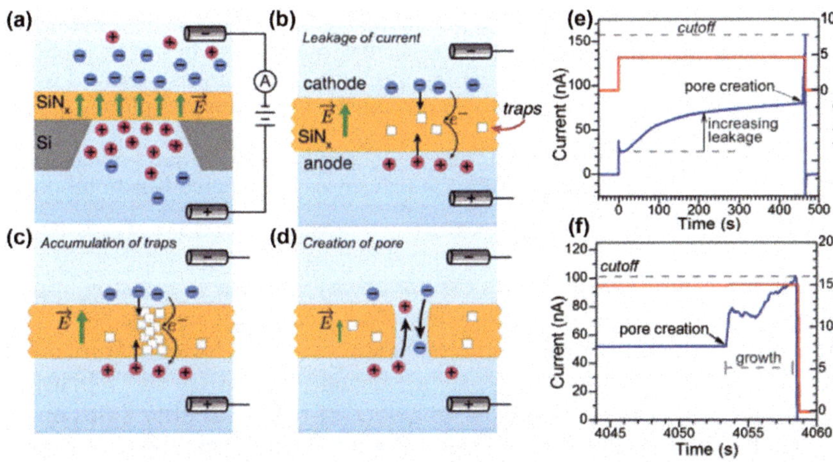

Figure 4.4 Nanopore formation using controlled dielectric breakdown. (a) Application of a trans-membrane potential generates an electric field inside the SiN_x and charges the interfaces with opposite ions. (b) Leakage current through the membrane follows a trap-assisted tunneling mechanism. Free charges (electrons or holes) can be produced by redox reactions at the surface or by field ionization of incorporated ions. The number of available charged traps (structural defects) sets the magnitude of the observed leakage current. (c) Accumulation of charge traps produced by electric field induced bond breakage or energetic charges carries leads to a highly localized conductive path, and a discrete dielectric breakdown event. (d) A nanopore is formed following removal of the defects. (e) Leakage current at 5 V on a 10 nm thick SiN_x membrane in 1 M KCl. Pore created is 5 nm. (f) Leakage current at 15 V on a 30 nm thick SiN_x membrane. Pore created is 3 nm.

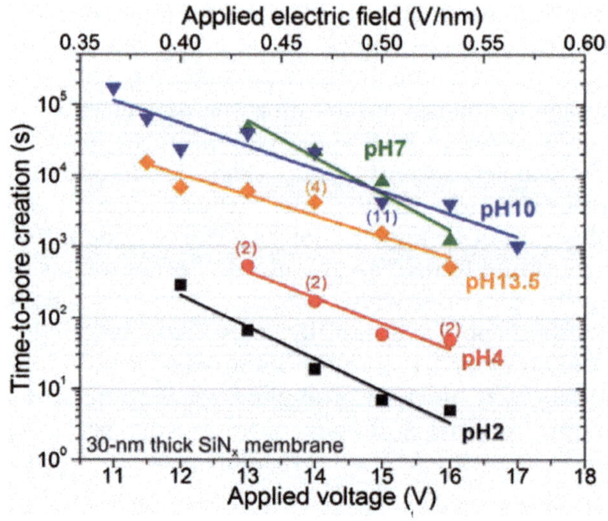

Figure 4.5 Time to pore creation as a function of experimental conditions. The pores are fabricated in 30 nm thick SiN_x membranes in 1 M KCl solution. The number of separate nanopores each data point is averaged over is indicated in parentheses. The vast majority of nanopores plotted are sub-5 nm in size.
Reproduced from ref. 10, https://doi.org/10.1371/journal.pone.0092880, under the terms of a CC BY 4.0 license, https://creativecommons.org/licenses/by/4.0/.

that the fabrication time decreases exponentially with the increasing applied voltage irrespective of other conditions. For a constant voltage, pH has a strong effect on the fabrication process. When the pH is lowered from 7 to 2, the time for pore creation can be reduced 1000-fold. Also, the fabrication time at lower salt concentration is longer than that of higher salt concentration. As for the membrane thickness, the fabrication time for 10 nm thick SiN_x membranes is decreased by approximately one third compared with 30 nm thick membranes. These observed results suggest the primary driving force for initiating the fabrication of the nanopore is the electric field applied in the membrane. The electric fields are in the range of 0.4 to 1 V nm^{-1}, which are close to the dielectric breakdown strength of SiN_x membranes,[18] thus being considered as the main cause leading to the breakdown in low-stress SiN_x membranes.

The fabrication of sub 2 nm SiN_x nanopores was achieved automatically and reproducibly by using the CDB method.[19] There are two key conditions for reproducibly creating sub-2 nm SiN_x nanopores. Firstly, the applied voltage must be terminated rapidly upon detection of the breakdown event, thus avoiding significant nanopore growth. Secondly, the use of thinner membranes should be chosen that can minimize the variability in the damage size as a result of fabricating smaller nanopores. The results show that the fabrication of sub-2 nm nanopores in 10 nm thick SiN_x is easier than that in 30 nm thick SiN_x membranes. In addition to these two conditions mentioned above, the electrolyte solution used for nanopore fabrication plays a secondary role in nanopore fabrication. It shows that the use of 1 M KCl pH 10 has a clearer indication of a nanopore generation event and less initial ionic current rectification than other electrolyte solutions. Based on the dielectric breakdown strategy, a technique called multilevel pulse-voltage injection (MPVI) was later proposed for small nanopore fabrications (Figure 4.6).[20] This method uses pulse voltages for creating the nanopores as well as enlarging the nanopores, and the generation of the nanopores could be verified at the same time by measuring the current through a membrane. With this method, the generation of nanopores with diameters smaller than 1 nm could be achieved with a probability of 90%. The diameters of the generated nanopores could be further enlarged to the desired diameters with sub-nm precision.

The CBD method is further extended to fabricate solid-state nanopore arrays within specific microfluidic devices.[21] The use of the CBD method enables integrating nanopores into lab-on-chip devices, which significantly simplifies the fabrication and assembly process. In addition, the integration of microchannels with nanopore arrays could significantly increase the application in nanopore sensing. For instance, multiple samples can be added to different microfluidic channels for multiplexed analysis on a single device. Recently, this technique has also been employed for fabrication of individual nanopores in metallized dielectric membranes.[22] Further, the CDB method could be integrated into an electric circuit that could be

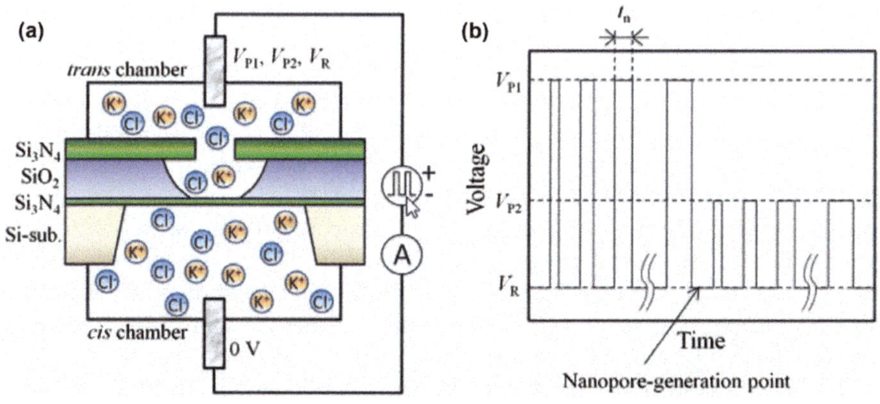

Figure 4.6 Schematic diagram of multilevel pulse-voltage injection (MPVI). (a) Setup for MPVI. *Cis* and *trans* electrodes are immersed in both chambers and are connected to a voltage-pulse generator and an ammeter. (b) Pulse chart of MPVI, which uses three different voltages (V_{P1}, V_{P2}, and V_R). A high-voltage pulse (V_{P1}) is used to create a nanopore. A mid-voltage pulse (V_{P2}) is used to widen the nanopore to an intended size. A low-voltage (V_R) is used to measure the current between the electrodes. Reproduced from ref. 20, https://doi.org/10.1038/srep05000, under the terms of a CC BY 3.0 license, https://creativecommons.org/licenses/by/3.0/.

employed to simultaneously achieve nanopore fabrication and single molecule detection.[23]

Despite the simple fabrication procedure, the unpredictability of the pore position remains the major limitation of this approach. The position of the pore is generated randomly on the membrane, thus it is very difficult for researchers to locate and image the pore. Recently, a method was demonstrated to localize pore formation by selectively thinning silicon nitride membrane.[24] The increase in local electric field strength in the thinned region yields a higher probability of nanopore formation compared to the surrounding region. Another promising method is using plasmonic assisted CBD to fabricate nanopores at directed locations.[25] By introducing a large electric field across a silicon nitride membrane, the generation of a nanopore is induced at the optical field hot spot of the bowtie nanoantenna.

4.2 Real-time Monitoring of the DNA Unzipping Process

Real-time probing and monitoring of single DNA molecules dynamics give significant insights into the fundamental studies of mechanisms in biological processes. When the narrowest constriction of the nanopore is smaller than that of DNA duplexes, the DNA duplexes can translocate through the pore until it experiences an unzipping process. To date, α-hemolysin nanopores have been widely applied to study the unzipping dynamics of DNA duplexes since they have vestibules with similar internal dimensions to DNA duplexes.[26–29] Solid-state nanopores with diameters as small as sub-2 nm are fabricated with the newly rising methods such as CDB, which have been employed to study the DNA unzipping process. For example, the unzipping kinetics of individual DNA duplex has been reported through sub-2 nm solid-state nanopores.[30] By analyzing the dwell-time distributions, the unzipping kinetics can be probed with a high-bandwidth electrical measurement. The translocation of hybrid DNA structures containing duplex regions of different lengths and sequences shows significant differences in the dwell time. In addition, the subtle mismatches in the duplex region can be resolved due to its relatively shorter dwell time than the perfect complement. The temperature-dependent studies of the unzipping kinetics demonstrate that the unzipping process follows first-order kinetics. In addition, molecular dynamics simulations have been utilized for revealing the unzipping or stretching pathway of hairpin DNA translocation through solid-state nanopores.[73–75]

Recently, the unzipping process of DNA duplexes through sub-2 nm SiN_x solid-state nanopores has been demonstrated by characteristic current blockages.[31] In this work, the sub-2 nm silicon nitride nanopores were fabricated in 10 nm thick membrane *via* CDB. *I–V* curves were used to estimate the pore diameter, and single-strand DNA (ssDNA) was employed as a molecule ruler to further verify the pore diameter. The translocation of DNA duplexes generated long-lived blockades with deep current amplitudes, which could be basically assigned as type I and type II. Characteristic blockades with three levels of type I events was ascribed to the unzipping process of the DNA duplex, while the type II events with only one blockade were suggested to be the DNA duplex retracting back to the solution without translocation. Since the DNA duplex molecule has a 40 bp duplex region and a 10 nt single-strand overhang, the amount of tail-first orientation should be slightly higher than that of the head-first orientation. Thus, the ratio between type I and type II (56:44) was consistent with the proportion of tail-first and head-first orientated DNA duplexes arriving at the pore mouth, further confirming the explanations for these two types of current blockades. To have a detailed investigation of the unzipping process of the DNA duplexes, the current blockades of each level for type I events were carefully analyzed (Figure 4.7a). Level 1 corresponded to the unzipping of the DNA duplex and its duration time was comparable with the reported timescales for DNA unzipping by nanopores (Figure 4.7b). Level 2 with a smaller current blockade was attributed to the state that the short-strand ssDNA was unzipped from the duplex and temporarily resided in the interior of the pore. The short-stand ssDNA leaved the pore as a result of the short-lived level 3 state (Figure 4.7c). The blockade of level 3 was consistent with that of the ssDNA translocation. The duration times of level 1 and level 3 showed exponential decrease with increasing applied voltage (Figure 4.7d). In order to further examine the observations, a simple model was employed to calculate the current conductance corresponding to each level. The calculated results from the presented model matched closely with the experimental results. The characteristic current signatures have great potential in the determination and quantification of the target analytes and further applications in real-sample detection.

More recently, solid-state nanopores with small diameter fabricated by CDB were also used for the detection of human telomere repeat sequences (Figure 4.8a).[23] The ultra-low current detection system in this study employed an electric circuit, which could achieve solid-state nanopore fabrication and single molecule detection, simultaneously (Figure 4.8b). The telomere repeat sequence was used as a probe to determinate the confinement effect in the nanopore analysis. The results showed that the nanopore with comparable volume of analytes exhibited an excellent capability for single molecule sensing. For example, the translocation of telomeric DNA through the confined nanopore with a diameter of 1.7 nm generated distinct current blockades with long duration time, which reflected the conformational changes of the DNA molecules (Figure 4.8c). However, the translocation events for the human telomeric DNA through a larger nanopore with a dimeter of 6.2 nm generated short duration time (Figure 4.8d). It showed that the remarkable electrochemical confinement effect would enhance the sensing performance of nanopore platform, and the unique nanopore confinement could be further used for the accurate analysis of short-lived metastable conformations of DNAs or peptides.

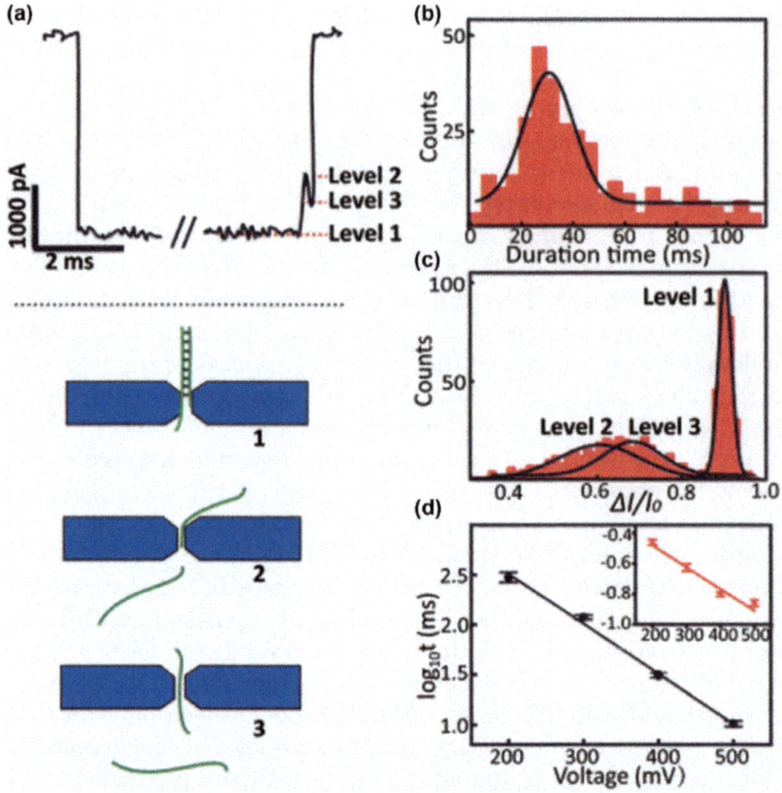

Figure 4.7 Real-time monitoring the DNA duplex unzipping process with the characteristic current signals. (a) Top: a current signature of a type I event. Bottom: diagrams for the molecular mechanism of the DNA duplex unzipping and translocation processes. (b) Histogram of the duration time of level 1, which refers to the unzipping time. The histogram is fitted to an exponential function. (c) Histograms of current blockade ($\Delta I/I_0$) of level 1, level 2 and level 3, respectively. The histograms of $\Delta I/I_0$ are fitted to the Gaussian function. (d) Relationship between the duration time of level 1 and the applied voltage for the unzipping of DNA duplexes. Inset: relationship between the duration time of level 3 and the applied voltage.
Reproduced from ref. 31 with permission from the Royal Society of Chemistry.

4.3 DNA-based Detection of Target Proteins

The protein translocation through solid-state nanopores typically occurs too quickly to be resolved by a conventional electrochemical sensing system. A previous study demonstrated that the mean protein translocation time (~1 μs) was faster than the detection of limits, resulting in a substantial fraction of protein translocation events being undetected.[32] Enhancing the time resolution of the current detection system is a good way to resolve more protein translocation events that are unresolved by conventional systems.[33]

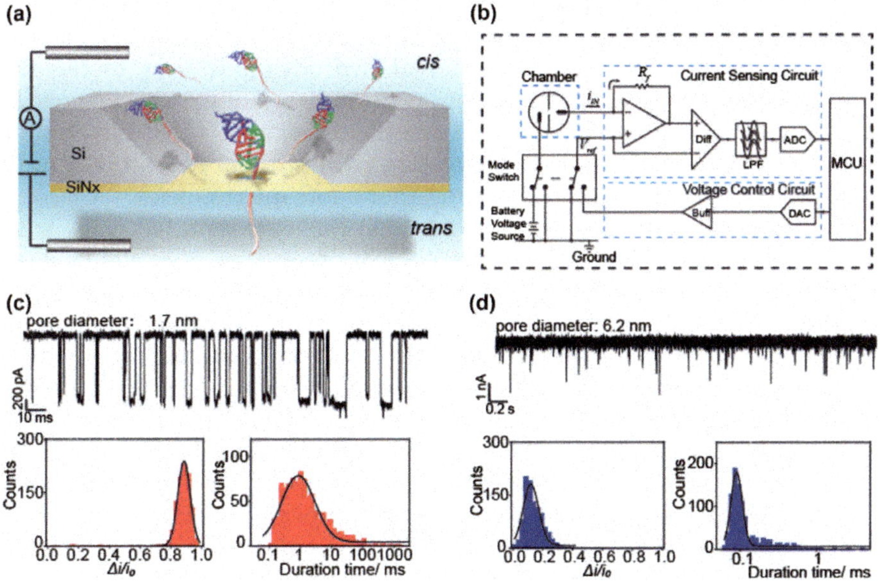

Figure 4.8 Detection of telomere repeat sequences using solid-state nanopores. (a) Schematic illustration of human telomeric DNA detection using a solid-state nanopore. The DNA molecule is driven into the pore under an applied potential. (b) The diagram of electric circuit for integrating the nanopore fabrication with the electrical detection device. (c) Current trace and the corresponding histograms of current blockade ($\Delta i/i_0$) and duration time for human telomeric DNA using a 1.7 nm solid-state nanopore. The applied potential is at 300 mV. (d) Current trace and the corresponding histograms of current blockade ($\Delta i/i_0$) and duration time for human telomeric DNA using a 6.2 nm solid-state nanopore. The applied potential is 300 mV.
Reproduced from ref. 23 with permission from Springer Nature, Copyright 2017.

Meanwhile, the DNA probe that binds with the proteins can prolong the translocation time of proteins, making it resolvable with standard electronics. The protein bound to a designed DNA molecule can further generate a characteristic blockade current, thus providing a novel approach for selective protein detection. Recently, several research groups have made progress toward the use of DNA–protein complexes, allowing the detection and identification of the target proteins.

Modified oligonucleotides containing a single biotin moiety was used to bind with streptavidin protein and achieved specific detection and quantification of a DNA–protein complex using solid-state nanopores.[34] In this study, the monovalent streptavidin (mSA) and mono-biotinylated 90 bp dsDNA (bio90) did not yield significant translocation signals over a range of 50–200 mV. However, significant increases in the number of events per unit time were observed when MS and bio90 were incubated together before measurements. The significant negative charge of MS contributed to a strong electrophoretic driving force, making its translocation time beyond the bandwidth limitations of the apparatus. When the mSA bonded with bio90, the bio90 supplied an additional hydrodynamic drag to the fast-moving protein, thus slowing the translocation speed of mSA to a resolvable level. The solid-state nanopore approach based on a DNA–protein complex could be applied to not only the detection of specific proteins but also the monitoring of specific microRNA and other short nucleic acids.

The detection of local proteins along DNA was first reported using solid-state nanopores.[35] The authors used RecA protein as a model of DNA-binding protein to bind to the dsDNA molecule. When the bare dsDNA and the dsDNA fully coated with RecA translocated through the pore, events with low-blockade level and high-blockade level were observed, respectively. These two different current blockades could be used as indicators to identify the absence and presence and position of proteins along a DNA molecule. Interestingly, when random patchiness of RecA were attached along dsDNA, characteristic current signatures could clearly resolve each RecA patch on the DNA. This work provided a promising approach for directly reading of the position of proteins along a single DNA molecule with high resolution by solid-state nanopores. More recently, the same group introduced a new system of anti-DNA antibodies binding to lambda phage DNA.[36] The DNA–antibody complex showed characteristic ionic current signatures, which could be associated with the presence of the antibodies. This work provides a proof-of-concept for target protein detection based on DNA probes. This strategy was extended to a rapid identification of the position of zinc finger protein (ZFP) along a double-stranded DNA using solid-state nanopores.[37] The translocation of a bare DNA through pores generated single-step ionic current blockades, while the movement of ZFP-labelled DNA induced an additional spike on top of the DNA blockade. The characteristic current signatures caused by the ZFP/DNA complexes enabled us to locate the position of the ZFP binding site. Two DNA molecules with ZFP binding at different positions were investigated and the results showed that the position of the additional spike originating from the DNA–ZFP complex was consistent with the theoretical values.

A new method was demonstrated for specific protein detection by using designed DNA carriers with a nanopipette. In this method, target proteins bound to DNA carrier generate characteristic current signatures, which allows for the positioning of protein binding sites at nanometer precision.[38] The characteristic current spike of the translocation event corresponds to the passage of the protein. The amplitude of the spike increases with increasing protein binding sites and could be used for counting the number of the binding sites on DNA carriers. In addition, DNA carriers were employed to determine the concentration of target proteins by measuring the proportion of DNA carriers that have a secondary current spike, indicating the presence of a protein.[39] Very recently, this approach has been used for multiplexing sensing of proteins by using digitally encoded DNA nanostructures.[40] In this work, barcode regions are constructed along a long double-strand DNA by using dumbbell hairpin motifs as individual digital bits. The authors demonstrated that the 3 bit barcode exhibits an assignment accuracy of 94% while the designed DNA structures translocate through a solid-state nanopore. Binding sites for specific analytes were further functionalized to detect a target antibody by the introduction of an antigen at designed positions on the DNA. This system can achieve simultaneous detection of different antibodies of the same isotype at nanomolar concentration.

DNA aptamer is a promising probe for protein detections because it can recognize target protein with high affinity and specificity. A solid-state nanopore approach with RNA aptamers for single molecule detection of a biomarker of the human immunodeficiency virus was demonstrated, the nucleocapsid protein (NCp7).[41] Recently, a rapid and direct method for detection of prostate specific antigen (PSA) based on DNA aptamer using solid-state nanopores has been demonstrated (Figure 4.9).[42] In this work, the PSA–DNA complex generated a characteristic deep current blockade, which could be easily distinguished from the signals of DNA probe. The frequency of the characteristic blockade showed a linear relationship with the PSA concentration ranging from 1 pM to 25 nM, which could be applied for a direct quantification of PSA. This strategy potentially provides a fast, convenient, and low-cost method for quantification of cancer biomarkers in real samples.

4.4 The Dynamics of Individual Protein–DNA Complexes Docked in a Confined Space

Currently, single molecule translocation is the most widely used and extensively developed strategy for nanopore sensing. However, the fast translocation of molecules has limited the amount of information that can be obtained from the measurements. A number of methods have been proposed to extend the residence time of the biomolecules in the nanopore confined space. One strategy is to slow down the molecule transportation speed to increase the resolution through adjusting the electrolyte,[43,44] electric field,[45] *etc.* Another strategy is to tune the nanopore surface properties to enhance the interaction between the molecule and the pore, such as genetic engineering of protein pores[46,47] or surface modification of solid-state nanopores.[48,49] In addition, some subtle techniques have been developed to precisely localize and confine the target molecules within the nanopore sensing regions. Trapping individual molecules in a confined space not only increases the duration time of the signals but also enables the direct observation of intrinsic dynamics of the biomolecule in real time. The successful confinement and trapping of individual molecules have been achieved by using optical tweezers,[50] magnetic tweezers,[51] and the atomic force microscope,[52] but these techniques suffer complicated operations, and show low throughput.

Using a molecule plug in larger size that could hardly translocate through the pore is a promising method to persistently confine the biomolecules

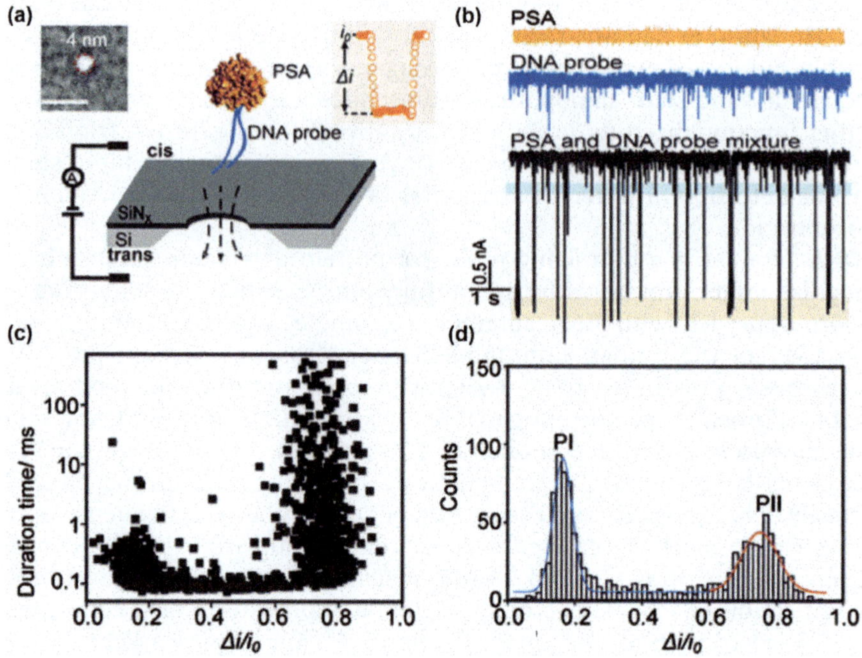

Figure 4.9 Analysis of a PSA–DNA probe complex using solid-state nanopores. (a) Schematic for detection of PSA binding with a DNA probe using solid-state nanopores. Upper left: A TEM image of a typical 4 nm solid-state nanopore fabricated by controlled dielectric breakdown. Scale bar is 10 nm. Upper right: a current blockade generated by the PSA–DNA probe complex. (b) Current trace after addition of PSA (orange), DNA probe (blue), and PSA incubated with DNA probe at a molar ratio (PSA–DNA probe) of 1:5 (black). (c) Scatter plot of the current blockade *versus* duration time of PSA and DNA probe mixture. (d) Histogram of current blockade for PSA and DNA probe mixture.
Reproduced from ref. 42 with permission from the Royal Society of Chemistry.

within the nanopore. The target analyte binding with the molecule plug, typically a larger protein, could be trapped inside the sensing region of the nanopore. Whereas, this technique has been widely applied in the biological nanopores platform to probe the biomolecules dynamics such as the monitoring of transient conformations of DNA or proteins,[53,54] this unique technique has been barely used in solid-state nanopore technology. Recently, this plug-docking method has been extended to the investigation of protein–DNA interactions.[55]

Recently, the dynamics of a single protein–DNA complex trapped in a confined solid-state nanopore was reported by real-time reading of the ionic current (Figure 4.10a).[56] The monovalent streptavidin, as a molecule plug, can bind with biotinylated DNA. Solid-state nanopores with a diameter of 3–5 nm fabricated by the CDB method were employed for the analysis pf

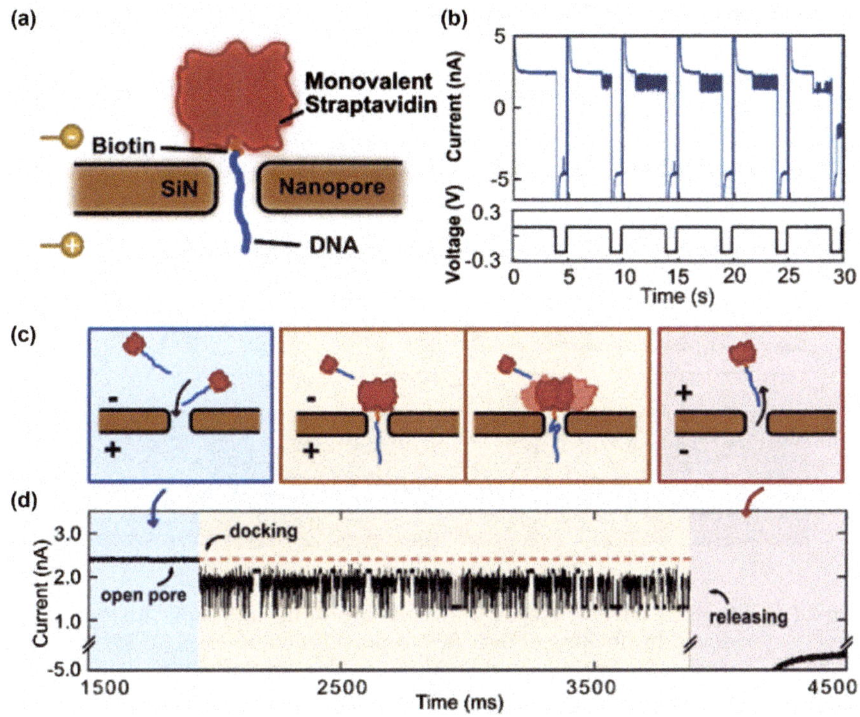

Figure 4.10 Schematic of the experimental setup. (a) Schematic of the protein–DNA complex trapped inside a solid-state nanopore. The biotinylated DNA molecule is tethered on a monovalent streptavidin. (b) Example of a current trace showing several individual capture events of the DNA–protein complexes. (c) Schematic illustrations of the complex capture, docking and release. At a positive bias voltage, the streptavidin–DNA complex is docked on the nanopore. At a negative bias, the complex is released from the pore. (d) A representative ionic current trace corresponding to the docking and release of a streptavidin–ssDNA complex. The blockade current fluctuates between several discrete levels. Reproduced from ref. 56 with permission from American Chemical Society, Copyright 2018.

protein–DNA complex. The confined structure of nanopore prevented the translocation of the protein plug but allowed the DNA molecules to enter into the pore. Upon the application of an electrical potential, the negatively charged DNA–protein complex was driven into and captured inside the pore, generating sustained ionic current blockades (Figure 4.10b–d). Then the DNA–protein complexs were released from the pore and back to the bulk solution by applying reverse potential.

The ionic current traces produced by the streptavidin–DNA complexes docking inside the pore were further investigated. Figure 4.11 shows the typical ionic current traces and the corresponding all-points histograms of the streptavidin–dsDNA complexes. The ionic current blockades were fluctuating with high amplitude and between several distinct levels.

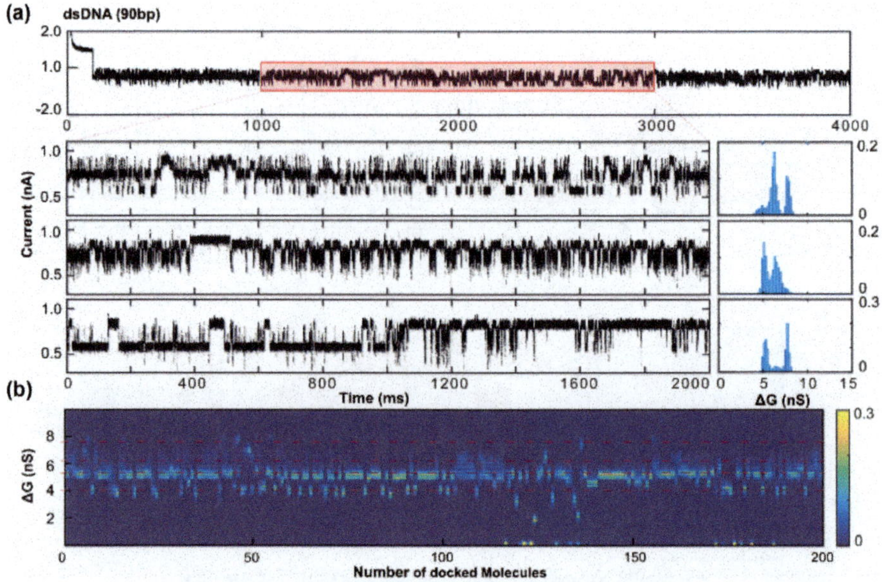

Figure 4.11 Typical current traces and corresponding all-points histograms produced by docking of dsDNA–streptavidin complexes. (a) Typical current traces for dsDNA–streptavidin docking experiments. The top panel shows a four-second current trace from one docking event, followed by three zoomed-in views of representative blockade current traces (left) and the corresponding normalized current histograms (right). (b) Collection of all-points current histograms from 200 dsDNA–streptavidin docking events.
Reproduced from ref. 56 with permission from American Chemical Society, Copyright 2018.

Interestingly, the similar current blockade patterns were also observed for the streptavidin–ssDNA complexes. The large amplitude fluctuation of the current blockade was ascribed to the motion of the protein plug. To further confirm it, the current blockades for the streptavidin protein with different orientations and positions at the entrance of the pore were theoretically calculated. The ionic current was calculated during the rotation of streptavidin molecule along the x, y and z axes (Figure 4.12). The protein rotation angle randomly ranged from 0° to 360° with the fixed center of the mass of the streptavidin. The change caused a conductance variation with a ~0.5 nA (3.3 nS) of the blockade current. Moreover, it caused another ~0.5 nA of current variation when moving the center of the mass of the protein along the pore axis. This clearly indicates that the protein motion inside the nanopore access region resulted in ~7 nS conductance variation. In the experiments, the order of the variation of the blockade current was as same as that in the simulation.

As for the dynamics of the docked molecule, one explanation would be the Brownian motion of the biomolecules. However, a protein experiencing

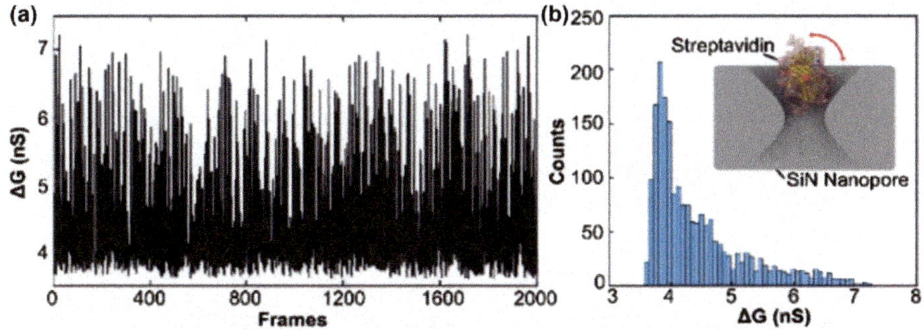

Figure 4.12 Theoretical current blockades produced by streptavidin. (a) Current blockade fluctuation caused by the stochastic rotation of the streptavidin in the access region. The ionic current was obtained under an applied potential of 150 mV while the streptavidin molecule was randomly rotated about the *x*, *y* and *z* axes, with its center of mass [red dot in inset of (b)] fixed. (b) Histogram of the current shown in (a). Reproduced from ref. 56 with permission from American Chemical Society, Copyright 2018.

stochastic rotational diffusion is uncorrelated in the microseconds time scale, and it would lead to high-frequency noise in the ionic current instead of discrete transitions between the current levels. Beyond that, another possible explanation is the viscous drag of the electroosmotic flow. For dsDNA, the viscous drag force compensates three quarters of the pull force in the electric field. For the weakly charged proteins, the viscous drag leads to electrostatics and determines the direction of molecule transport. Although there is a reduction in electroosmotic flow in the pore, the weakly charged streptavidin still has a cross-section larger than the DNA tether and would see a much larger drag. Based on this, the net force that acts on the complex can be much reduced to slightly release the molecular plug in the pore and make a reorientation of it before being pressed onto the membrane once more. As a result, the docked complexes in long and sporadic trapping situations cannot be explained by either Brownian motion or electroosmotic forces. According to the previous simulations and experiments, the non-specific interactions between DNA/protein and the inner surface of the pore could transiently prevent the translocation events. This non-specific adhesion could stabilize a docked conformation, meanwhile, the effective force on the complex could be also reduced, which prompts the complex to adjust the conformation. Hence, stick-and-slip transportation in nanopores from surface interactions may induce this spontaneous complex rearrangement of docked protein–DNA molecules.

The dynamics of the docked molecule were also demonstrated at different applied potentials (Figure 4.13). Although the fluctuating patterns are different, the locations of the peaks barely changed with the voltage increasing from 100 mV to 200 mV. Moreover, a higher voltage did not lead to an elimination of the current fluctuations. The experiments were performed for

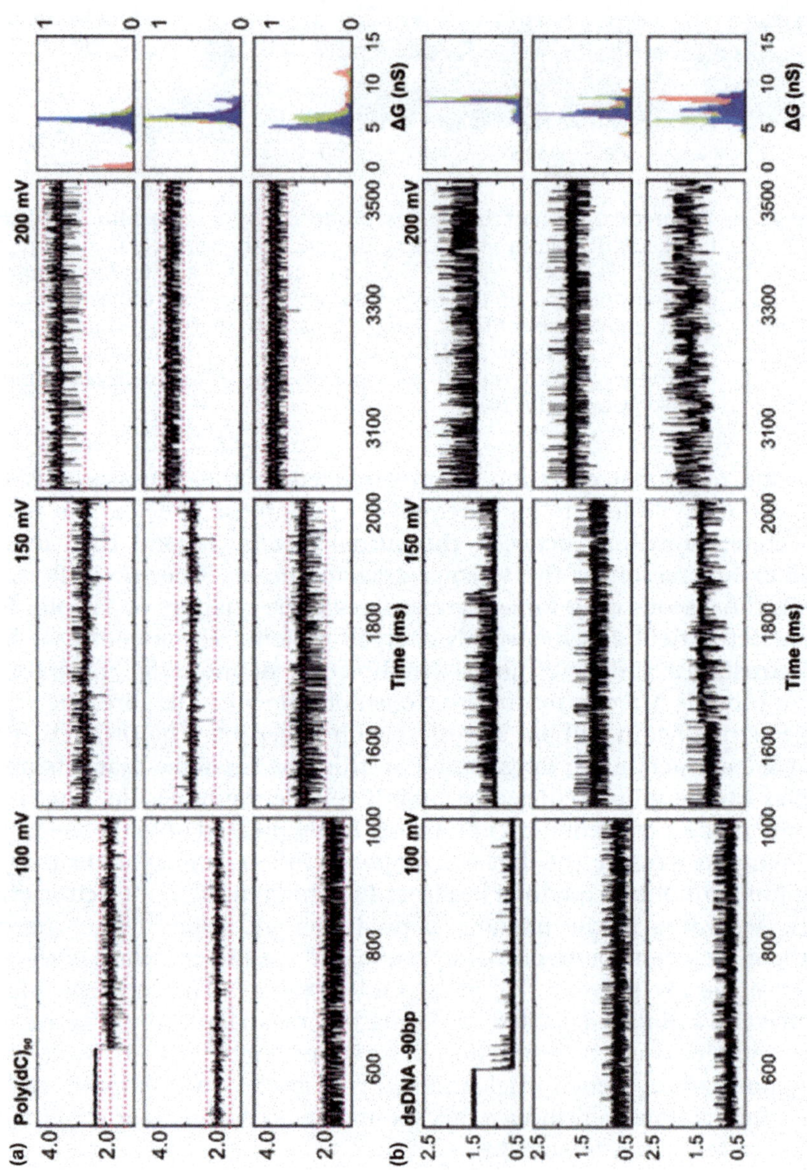
(a) Poly(dC)90
100 mV
150 mV
200 mV
Time (ms)
ΔG (nS)
(b) dsDNA -90bp
100 mV
150 mV
200 mV
Time (ms)
ΔG (nS)

both the streptavidin–ssDNA complexes and streptavidin–dsDNA complexes, and the outcome was consistent. To demonstrate the influence of the voltage on the lifetime of the current states, the lifetime for the most populated state of 100 dockings of streptavidin–poly(dC)$_{90}$ molecules at 100 mV, 150 mV and 200 mV were analyzed. The results showed that there are similar lifetime distributions at different voltages and increasing the voltage hardly prolongs or shortens the lifetime of certain states. Both the measurement and the molecular dynamics simulations indicate that the interaction between DNA and the nanopore surface has a major role in the stabilization of the docked molecule motion with compensation of an effective force on the complex as well as the production of a finite lifetime of voltage independence.

To further investigate the DNA tether's function in governing the docking dynamics, a C-rich ssDNA strand in a short length from human telomere replaced the DNA tether owing to its ability to fold into an intercalated motif (i-motif) structure (Figure 4.14). This kind of tetrameric structure can be formed at an acidic pH but not at an alkaline pH. As shown in Figure 4.14a, the current oscillations, which are very similar to that from the poly(dC)$_{90}$ and the dsDNA tethers experiment, could be observed under alkaline conditions (pH 8) after the docking happened. However, the current levels are less pronounced. When the electrolyte changed to acidic conditions (pH 5), the quantity of fluctuations reduced in the residual current, which appears much more stable. At the same time, the blockage amplitude increased from ~10 nS to ~20 nS when changing the pH from 8 to 5.

The differences in structure and the motion between ssDNA tether and protein–ssDNA complexes determined the variation of behaviors in two pH conditions. First, in the acidic solution, it is easier to form an i-motif structure for an ssDNA strand. For the i-motif structure, the cross section is larger than that of the unstructured ssDNA, which could be expected to see a larger current. Second, the i-motif makes the DNA stationary inside the pore resulting sufficient temporal resolution. In addition, all the isoelectric points of streptavidin, DNA and the SiN_x in this experiment are around 4–5, hence, at pH 5 the lower net charge of all components results in a reduced electrostatic interaction between the molecules and the solid surface and hence a reduced electroosmotic flow, further promoting a more stable docking of the complex. Thus, the significant variation in the blockade current

Figure 4.13 Voltage dependence of docking dynamics. (a) Representative current traces produced by docking of three individual streptavidin–poly(dC)$_{90}$ complexes (one in each row) under the bias voltages of 100 mV, 150 mV and 200 mV. Right panel shows the corresponding all-points histograms. Red: 100 mV, green: 150 mV and blue: 200 mV. (b) Representative current traces produced by docking of three individual streptavidin–dsDNA complexes (one in each row) under the bias voltages of 100 mV, 150 mV and 200 mV. Right panel shows the corresponding all-points histograms. Red: 100 mV, green: 150 mV, and blue: 200 mV.
Reproduced from ref. 56 with permission from American Chemical Society, Copyright 2018.

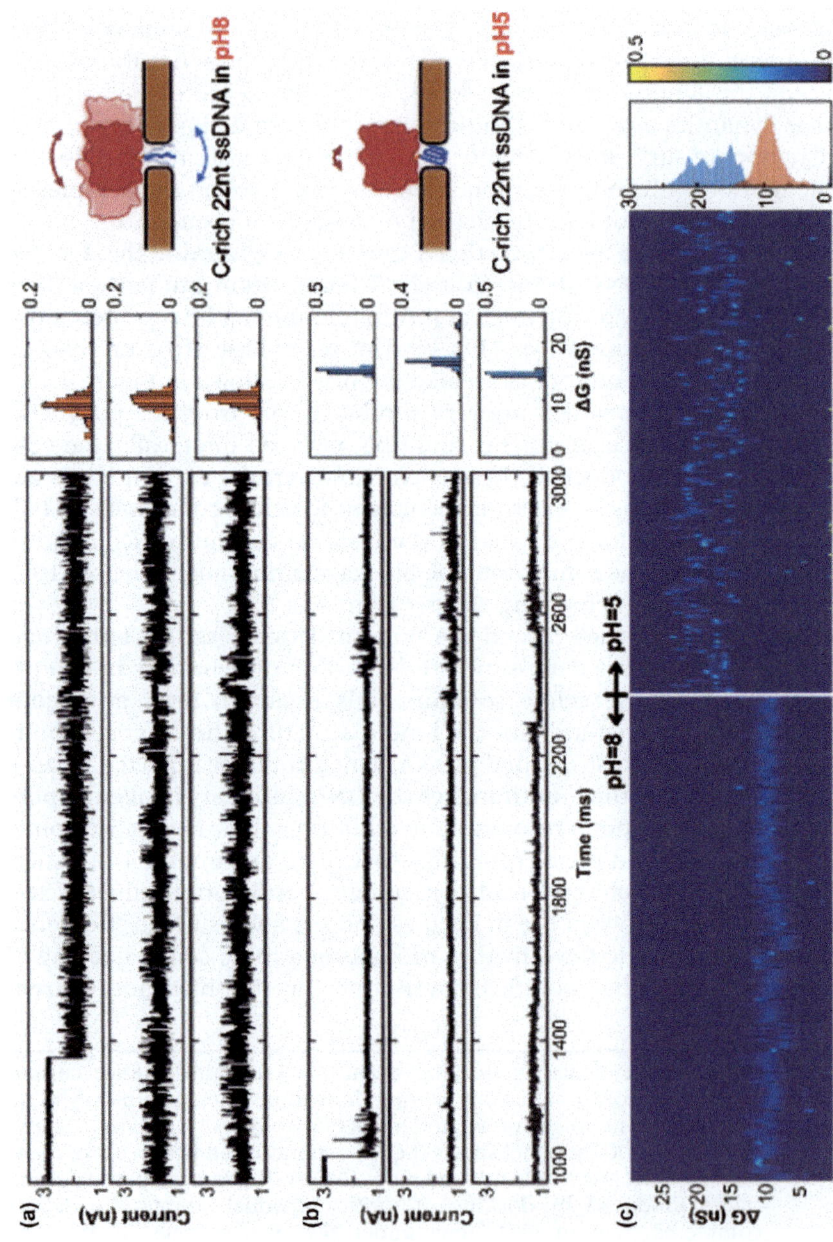
(a)
Current (nA)
C-rich 22nt ssDNA in pH8
(b)
Current (nA)
C-rich 22nt ssDNA in pH5
Time (ms)
ΔG (nS)
(c)
ΔG (nS)
pH=8
pH=5
Number of docked Molecules

demonstrates the potential advantages of utility of the docking experiments for direct observations of conformational transitions at the single molecule level. Reproduced from ref. 56 with permission from American Chemical Society, Copyright 2018.

4.5 Confining Optical Enhancement Inside Nanopores

To directly observe the heterogeneous of each molecule, optical measurements were introduced into nanopore-based detection. In 2007, a solid-state nanopore was intergrated with single molecule fluorescence to detect labeled DNA with 'two-channel optical readouts'.[57] Although the fluorescence label has provided more specific nanopore information, incorrect labeling and photobleaching can affect detection accuracy. Moreover, some samples are not suitable for fluorescence labeling. Calcium fluorescence as a label free optical method is employed to overcome these disadvantages. Fluorescence dyes and Ca^{2+} encounter each other at the nanopore region and emit fluorescence, whose intensity can be converted into ionic current intensity. Observing the fluorescence signal, some information about what happened in the pore can be obtained.[58] However, both label and label-free methods experience extensive use of fluorescence in optical nanopores. The introduction of fluorophores will complicate the experiment procedures and affect the intrinsic structure of the analyte. An alternative optical approach without any additional molecules would be more suitable for nanopore detection.

In the past decade, sub-wavelength apertures have inspired great interest due to their properties, such as greatly enhanced light transmission and wavelength filtering owing to the interaction between light and the surface of metal films. According to Bethe, light with longer wavelength would not propagate through a sub-wavelength hole that exists on a metal film with zero thickness.[59] Based on this theory, optical imaging methods similar to near-field optical microscopy (NFOM) have been developed to enable detection beyond the diffraction limit. However, much more light has been detected due to surface plasmons induced by the interaction between electromagnetic waves and the free electrons of the metal, which is beyond Bethe's prediction. Moreover, the direction of diffracted light and the

Figure 4.14 Effect of tether conformation on docking dynamics. (a), (b) Typical current traces and corresponding all-points histograms produced by docking of a 22 nt c-rich ssDNA–streptavidin complex at pH 8 (a) and pH 5 (b). Images to the right illustrate the proposed pH-dependent conformational transitions of the ssDNA tether. (c) All-points histograms of the current traces obtained from docking of the c-rich ssDNA–streptavidin complexes at pH 8 and 5 along with the corresponding all-points histogram of all 100 traces at pH 8 (orange) and100 traces at pH 5 (blue).
Reproduced from ref. 56 with permission from American Chemical Society, Copyright 2018.

spectrum of transmitted light could be regulated by adjusting the material and geometry of the sub-wavelength aperture. Optical signals beyond the diffraction limit can be excited and collected by the nanoscale aperture through the near-field optical microscope. And the zero mode waveguide based on a nanohole in a metal film has been used for single molecule sequencing. Moreover, metal nanopores holding surface plasmons property have been used as biosensors for single molecule sensing.

Vibrational spectroscopy methods have been used along with nanopore detection to provide more accurate nanopore readouts. Surface-enhanced resonance Raman scattering (SERRS), as one of the approaches in vibrational spectroscopy, has been widely used to integrate with the nanopore technique. When a dye-coated gold nanoparticle approaches a nanopore with metal film or metal nanostructure, the SERRS signal of the analyte could be extremely enhanced by the 'hot spot' between the nanoparticle and the nanopore edge. The translocation events can be supervised and the molecular structures can be obtained by monitoring the characteristic band in the SERRS spectrum.[60] The combination of SERRS and nanopores is a perfect choice to detect, and even identify the analyte at the single molecular level. Therefore, an idealized sub-wavelength hole with plasmonic effects is a high-efficiency combination to strongly enhanced transmission of light and highly improve the nicety of detection. The design of an ideal optical nanopore is of vital importance for optical-electric detection.

4.5.1 Fabrication of Optical Nanopores

Optical nanopores can be fabricated in a similar way to general solid-state nanopores by transmission electron microscope (TEM) drilling, focus ion beam (FIB) milling and dielectric breakdown. For optical nanopores, the metal nanostructures around the nanopores possess the 'hotspots' effect and are able to enhance the optical intensity (*e.g.* scattering light intensity, Raman intensity) when an incoming laser beam focuses on the pore.

4.5.1.1 Single Optical Nanopore

Figure 4.15 presents a typical fabrication process of a single optical nanopore:[61]

1. Coat silicon nitride (SiN_x) with tens of nanometers thickness on both sides of silica by low-pressure chemical vapor deposition (LPCVD).
2. Protect parts of SiN_x using photolithography, and etch the rest of the SiN_x using a reactive ion along with the protecting photoresist.
3. Expose to a high energy e-beam to make room for the desired nanostructures.
4. Coat the metal membrane with the desired thickness by electron beam evaporation or magnetron.
5. Drill a nanohole through the membrane by TEM or FIB.

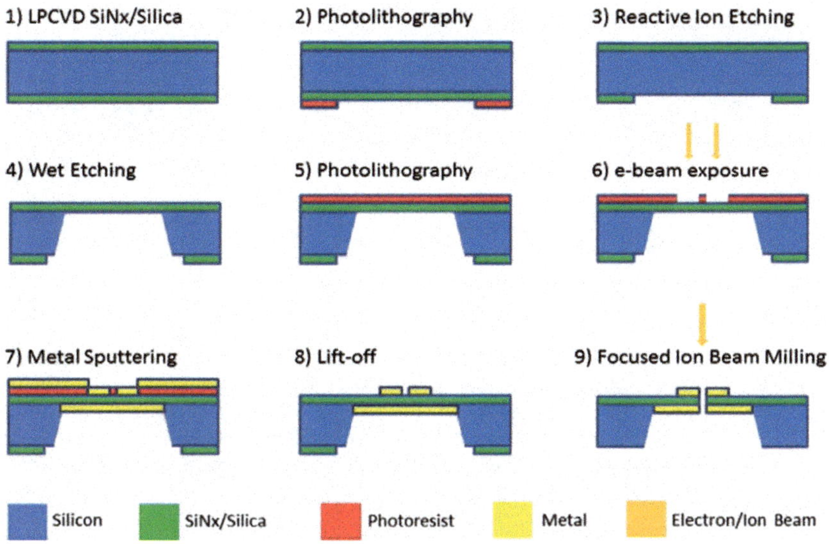

Figure 4.15 Typical fabrication process of optical nanopores. Reproduced from ref. 61 with permission from Elsevier, Copyright 2016.

A satisfactory optical nanopore can be fabricated by varying the energy, angle and exposure time of a high e-beam and ion-beam. The fabrication processes based on TEM and FIB are demanding and costly, so to simplify the fabrication process and lower the cost, the laser illumination and electrical bias was combined to fabricate an optical nanopore.[25]

As shown in Figure 4.16a, an optical nanopore chip is sealed to separate the electrolyte into two independent chambers and are exposed simultaneously to a DC electric field and an optical field that is locally enhanced by two gold nano-antennas. An applied voltage of 6 V is the highest possible voltage that does not induce a dielectric breakdown without laser illumination. Then 5 mW laser power is added at the center of the bowtie antenna to produce an optical field of 1.5 MV cm^{-1} in the plasmonic hot spot. The membrane current is shown in Figure 4.16b. For several minutes the chip is exposed to the two powers; a rapid increasing current across the membrane indicates the occurrence of a laser-assistant dielectric breakdown event and the creation of an optical nanopore. Both the laser illumination and the bias voltage are switched off when the current reaches the threshold. The pore is in a line with the two nano-antennas. The approach to confirm nanopore alignment within the plasmonic hot spot is applying a low-power laser beam to scan the membrane and monitor ionic current at the same time. The laser illumination on the bowtie structure heats the plasmonic structure and increases the electrolyte conductivity, resulting in an increase in the ionic current. The appearance of maximum current when the laser illuminates the center of the plasmonic bowtie will certify the alignment of the nanopore with the plasmonic hot spot.

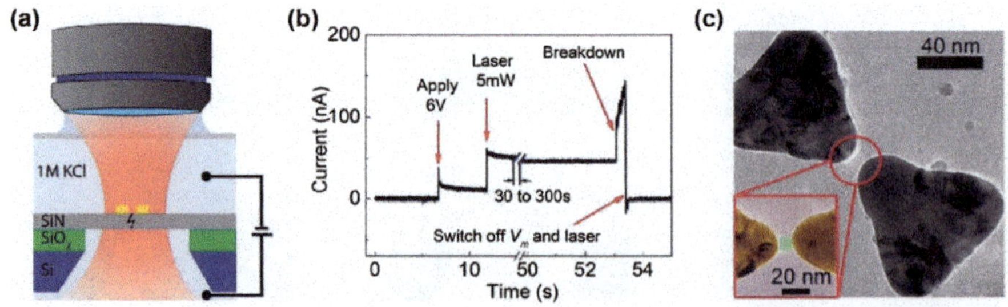

Figure 4.16 (a) Side view schematic of the self-aligned fabrication method. (b) Time trace of the membrane current during the formation of an optical nanopore by laser-assisted dielectric breakdown. (c) Typical TEM image of an optical nanopore fabricated by laser-assisted dielectric breakdown. The zoom in the inset shows a false-colored image of the gap of the gold bowtie structure that more clearly visualizes the nanopore. Reproduced from ref. 25 with permission from American Chemical Society, Copyright 2015.

The optical activities of nanopores benefit from the metal layer or metal nanostructures. If the diameter of a metal nanopore is far less than the wavelength of the incoming light, the surface electrons of the positive nuclei will resonate with the transmission of the light. When the frequency of the oscillating electrons overlaps with the photons from incoming light, the 'plasmon resonance' effect will occur and generate optical enhancement, coupling the incoming light on the surface of the metal nanopore. Optical nanopores thus become an excellent tool for a wide range of applications in single molecule sensing owing to the 'plasmon resonance' effect, which gives the pore its characteristic scattering and absorption bands.

The enhancement characteristics of optical nanopores, taking a nanobowtie pore with a diameter of 5 nm as an example, have been evaluated by simulating the field enhancement as a function of the bowtie gap, the bowtie angle, the pore's size and the space thickness separately. Figure 4.17 shows the results of each simulation. In the first simulation, when the gap size decreases, the electric field becomes stronger and the resonance wavelength becomes longer. The decreased gap gives stronger plasmonic coupling to enhance the electric field; charge attraction from closer dipoles with lower frequency causes the resonance wavelength to red-shift. In the second simulation, an increase in the bowtie angle causes less enhancement of the electric field and blue-shift of the resonance wavelength, meaning the enhancement of a 30° bowtie is double that of the 60° bowtie. However, the decreased angle also caused a dramatic shift of over 100 nm in resonance wavelength. Moreover, the 'lightning rod effect' occurs and generates a high local field at the tip of the bowtie, this caused the nano-plasmonic hot spots when the tip's axis and the exciting light's polarization were parallel to each other. The change of enhancement factor in Figure 4.17b can also be explained because the 'lightning rod effect' becomes even stronger as the tip becomes sharper. In the third simulation, the 20 nm gap size is compared with different smaller size

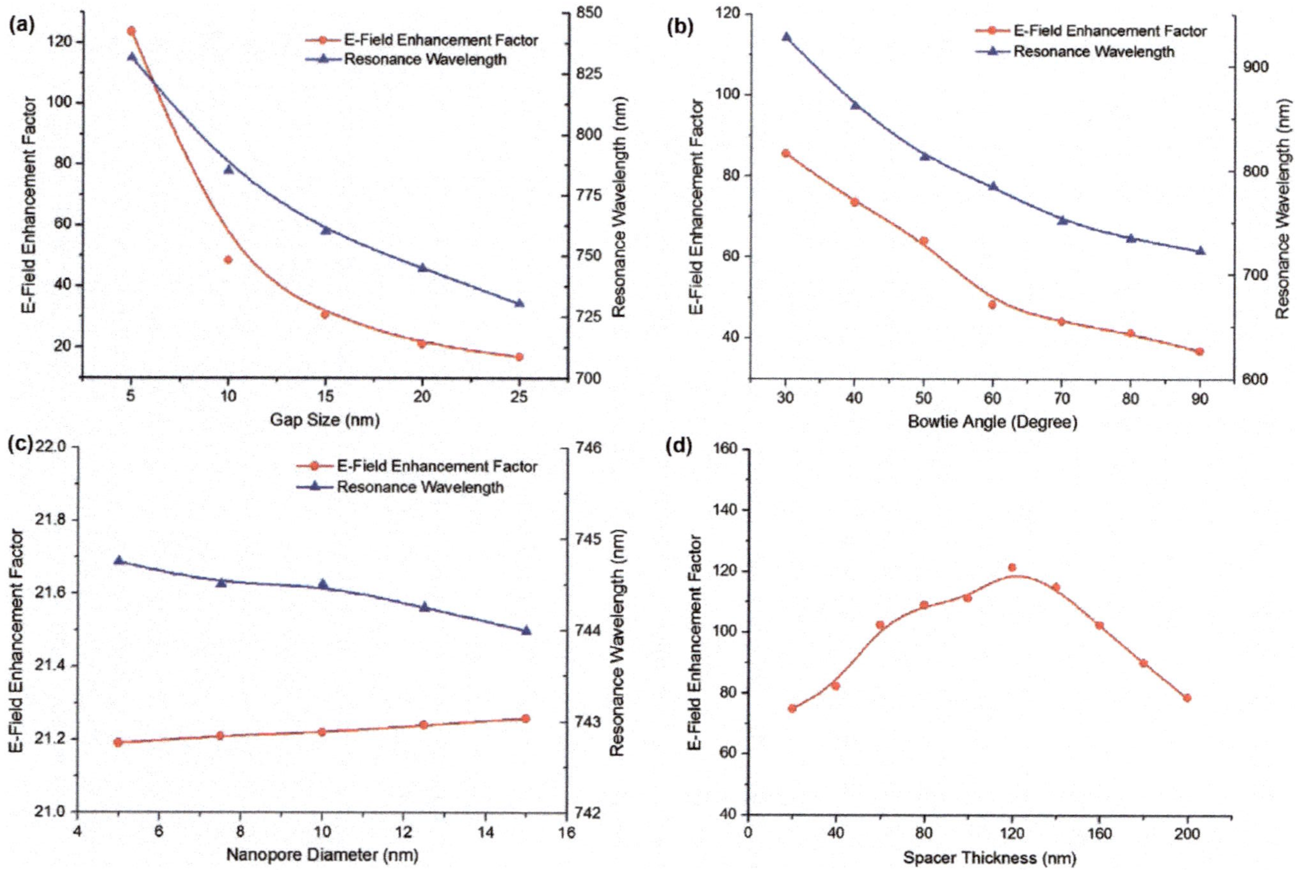

Figure 4.17 Electrical field enhancement as a function of several parameters, such as the bowtie gap size (a), the bowtie angle (b), the nanopore size (c) and the spacer thickness (d).
Reproduced from ref. 61 with permission from Elsevier, Copyright 2016.

pores to find their influence on the field and wavelength. It is apparent that pore size has a limited effect on the nano-antenna's field enhancement and resonance wavelength, as seen in the relatively stable values across all pore sizes. This is because the nano-antenna characteristics are determined mostly by the bowtie's characteristics, such as material, shape and size. In the last simulation, the effect of different thickness of silica substrate on field enhancement is shown in Figure 4.17d. The interface of the metal film and the dielectric material would cause the π phase shift during reflection. Therefore, the optimal thickness for the silica substrate in-phase reflection could be estimated as follows:[62]

$$2 \times h \times n_{\text{silica}} = (m + \frac{1}{2})\lambda \tag{4.1}$$

where n_{silica} is the silica index of 1.45, λ is the laser wavelength, which is 785 nm in this system, and m is a whole number. Therefore, the optimized thickness is approximately 135 nm which is slightly more than the actual optimized thickness of 120 nm (Figure 4.17d). Since gold is not a perfect electric conductor, the additional phase shift at the gold–silica interface caused the deviation. The four results discussed above indicate a narrower gap and a sharper bowtie angle are of most benefit to stronger plasmonic electric enhancement. Duc to limitations of current nanopore fabrication techniques, the size of the nanopore and the angle of the bowties cannot be reduced without limit. Therefore, a balance between performance and feasibility of the whole experiment system is needed.

4.5.1.2 Optical Nanopore Array

Single optical nanopores can be easily fabricated using FIB/TEM drilling and laser-assisted dielectric breakdown, but an array of optical nanopores are often needed in biomolecule sensing, which cannot be mass produced. One method for producing nanopore arrays is called 'applied colloidal lithography', by which two kinds of optical nanopores, *i.e.* the nanowells and the nanocaves, can be mass fabricated.[63] The fabrication procedure is shown in Figure 4.18a. Starting with a clean fused silica support, the surface is covered by a colloid as a short-range order monolayer of monodispersed polystyrene particles whose diameter can be reduced by oxygen plasma. Then physical vapor deposition coast gold and alumina is applied, the role of alumina being to protect the gold. To remove the colloids, the whole surface is kept in liquid and rubbed with soft materials. Nanowells and nanocaves are formed by dry etching with CF_4 mixed with O_2 and wet etching with HF, respectively. Figure 4.18b shows TEM images of the nanostructures made by this method. Compared to the spectral signature from the binding to the planar top surface, the spectral signature from binding to the interior of the nanopores has a better resolution. Both the enlarged the liquid volume and increased field strength improve the refractometric sensitivity. However, the size of the nanostructures made by this method are too large to detect and analyze determinants of interests at the single molecule level.

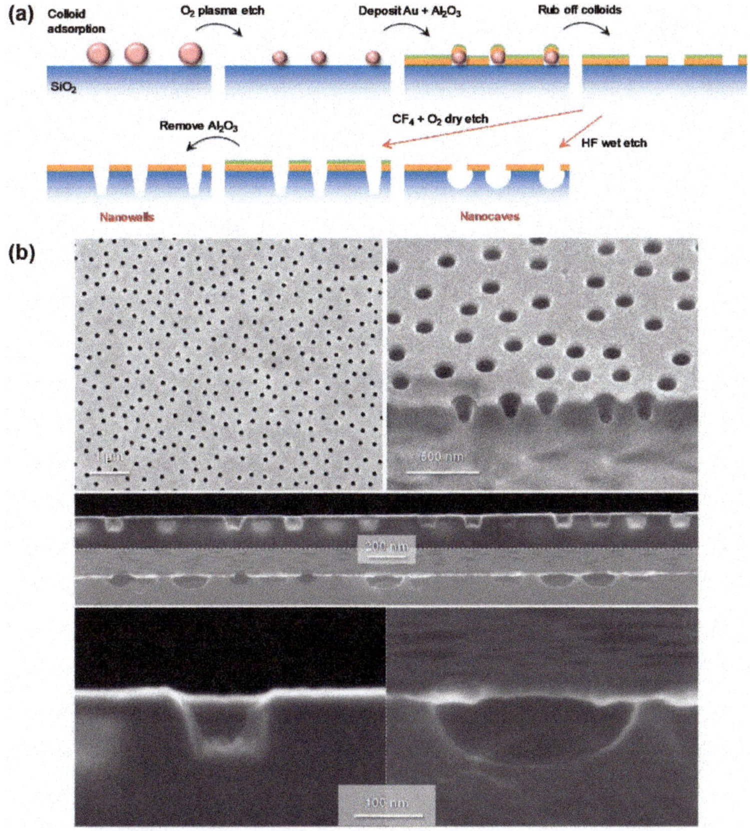

Figure 4.18 (a) Fabrication steps for preparing nanowells and nanocaves. (b) TEM images of the nanostructures made by the colloidal lithography method. Reproduced from ref. 63, https://doi.org/10.3390/s17061444, under the terms of a CC BY 4.0 license, https://creativecommons.org/licenses/by/4.0/.

Another method is chemically etching 2D plasmonic Ag nanoparticle (AgNP) supercrystals.[9] Sub-10 nm arrays of nanopores can be fabricated using this method. 2D AgNP supercrystals can be synthesized using a mesoporous silica template, EP-FDU-12, to be used in the nanopore array fabrication process. Common templates, such as KIT-6 or SBA 15, cannot give results of thin wall thickness as thin as 5 nm and pore size as large as 27 nm like EP-FDU-12 can. EP-FDU-12 possesses significant virtues not only in the nanocasting process but also in their applications in plasmonic nanooptics.

During the nanocasting process, $AgNO_3$ precursors were implanted into the SiO_2 EP-FDU-12 template without a functionalized surface. Then the SiO_2 template and the $AgNO_3$ precursors were both dried using a vacuum for 12 h at 323 K. Many 2D AgNP super crystals can be obtained from this high reduction temperature. In the experiment that follows, plasmon modes were created from the 2D AgNP super crystals using an etching process. Nanopores with sub-10 or even sub-5 nm sizes were found unexpectedly during

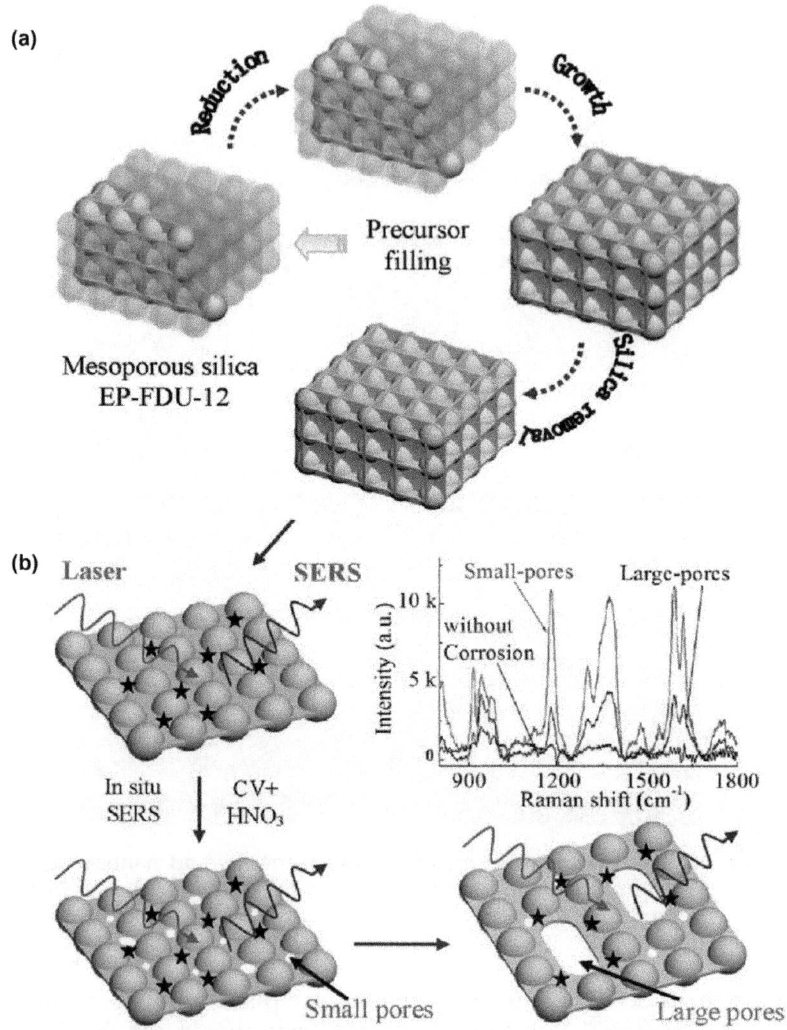

Figure 4.19 (a) A schematic fabrication process of 2D plasmonic Ag NP supercrystals *via* nanocasting using mesoporous silica, EP-FDU-12, as a template. (b) The chemical etching process using HNO_3 aqueous solution and *in situ* SERS detection.
Reproduced from ref. 64 with permission from John Wiley and Sons, Copyright © 2016 WILEY-VCH Verlag GmbH & Co. KGaA, Weinheim.

the early etching stages within the nanogap regions, as depicted in Figure 4.19b. An *in situ* SERS was used along with the etching process to determine how the electromagnetic field enhancement were influenced by nanopores being fabricated. The *in situ* SERS needs an etchant of crystal violet (CV) or 4-aminothiophenol (4-ATP) mixed with HNO_3 to be used to monitor the etching process. At about 25 min into the etching process, the characteristic peaks of CV molecules increased to a maximum intensity. The common facile etching method can promote nanopore fabrication with different compositions for various applications. However, it must be noted that nanoparticles may randomly attach to the surface of AgNP super crystals, and this may influence the signal uniformity SERS detection and the electromagnetic field enhancement.

4.5.2 An Integrated System for Optical and Electrical Detection for a Scattering Nanopore

When both optical and electrical methods are synchronously used in nanopore detection, the physical and chemical properties of targets are determined, such as structure, conformation, volume, composition, and concentration. Its advantages are less apparent with small molecules, but it becomes especially useful when analyzing complex samples.

A schematic of the integrated system is shown in Figure 4.20.[65] This integrated system can be divided into two parts: dark-field microscopy based

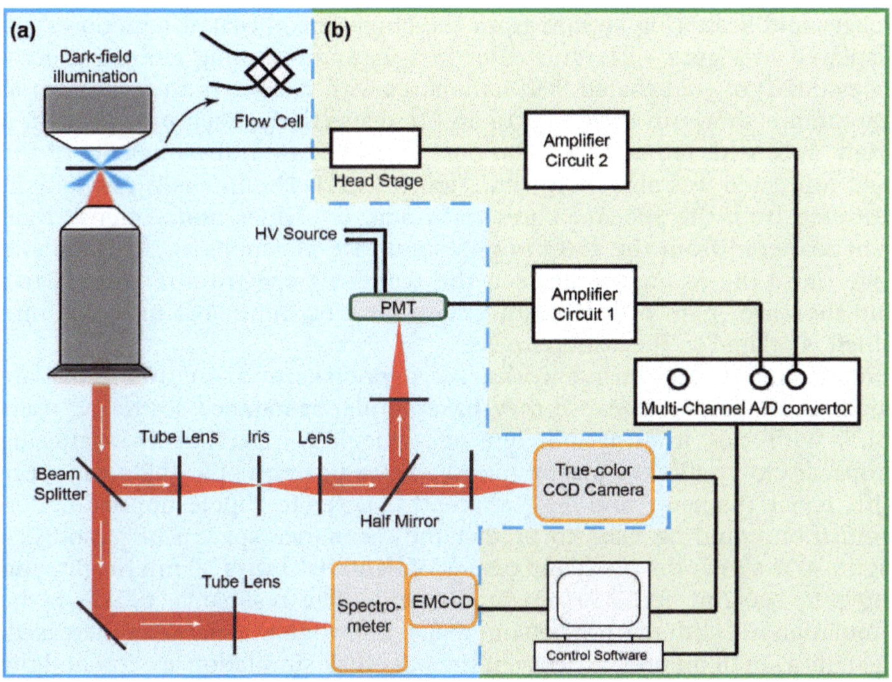

Figure 4.20 Schematic of the integrated system. Experimental configuration of the dark-field microscopy system (a) and the home-built ultralow ionic current acquisition system (b).
Reproduced from ref. 65 with permission from the Royal Society of Chemistry.

optical detection and current amplifier based electrical detection. The functions of inverted dark-field microscopy are illumination and monitoring the optical nanopore chip. The chip is exposed through a dark-field condenser under unpolarized white light emitted from 100 W halogen. A 40x object is used for collecting the scatted light from the nanopore. Dispersed scatted light, accomplished using a Czemy–Tumer spectrograph, is recorded using an electron multiplying charge coupled device (EMCCD) array camera. A true-color digital CCD camera can reconstruct the output image from the objective. And the scatted light intensity is displayed in real time by the integration of a home-designed optical path equipped with a photomultiplier tube (PMT), a home-designed current amplifier and a multichannel data acquisition system. A PMT can collect 40% of the scatted light through this optical path. Replacement of the 20:80 beam splitter with a mirror and dislodgement the half mirror can ensure that the PMT collects all of the scatted light.

4.5.2.1 Optical Detection

A gold-coated solid-state nanopore array with diameter ranging from 150 nm to 300 nm is used to test the optical detection part. The typical dark-field images and scattering spectra from the single metal-coated nanopores are displayed in Figure 4.21. True-color images and scanning electron microscopy (SEM) of gold-coated SiN_x nanopores with diameters from 150 nm to 300 nm are shown in Figure 4.21a and b, respectively. Each pore gave off a bright spot with red color; as the pore size increased, the intensity of the light diffracted was also increased (Figure 4.21c). The intensity of the light scattered from the 300 nm pores was about ten times more intense than light scattered from the 150 nm pores. Surface plasmons at the nanopore edge cause the resonance peak in the scattering spectrum (Figure 4.21d). And the wavelength of maximum intensity varies from 650 nm to 700 nm, which is related to the nanopore size.

According to pioneering works, Au nanodisks and Au thin films are qualitatively similar because they have similar resonance scattering spectra.[66] With this information, one can calculate a nanodisk's scattering properties to predict the plasmonic resonance position of a single nanopore with equal diameter and height (depth). Discrete dipole approximation simulation could be used to predict the resonance spectra of nanodisks. Figure 4.22 shows the predicted results of nanodisks with 90 nm height, and 150 nm, 200 nm and 250 nm in diameter. The resonance peaks in the simulation red shifted from 600 nm to 800 nm when the diameter increased. There is a small difference between the nanodisk simulation spectra and the gold coated nanopore experimental result, which may be resulted from several factors. First, the nanopores are fabricated by FIB and its geometry may have a slight difference to the simulation. Second, a thin gold layer could cover the silicon nitride wall from inside the pore because the gold atoms around the pore could redistribute due to the FIB. This can cause the

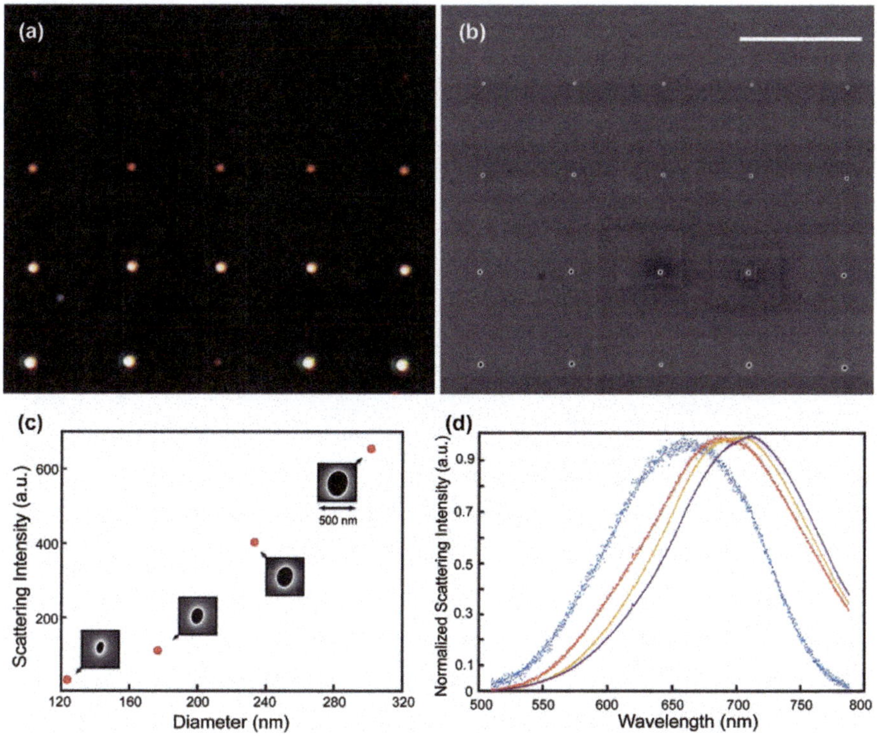

Figure 4.21 Characterization of single metal-coated nanopores under a dark-field. (a) True-color image of single metal-coated nanopores with diameters of ~150 nm, ~200 nm, ~250 nm and ~300 nm in ambient air. (b) Corresponding scanning electron micrograph for each nanopore array (scale bar is 10 mm). (c) Plot of the scattering intensity for each nanopore *versus* diameter, which shows the intensity increasing along with an increase in the pore diameter. (d) Corresponding scattering spectra for each single metal-coated nanopore with diameters of 150 nm (blue), 200 nm (red), 250 nm (yellow) and 300 nm (purple) (an average of four nanopores).
Reproduced from ref. 65 with permission from the Royal Society of Chemistry.

formation of a structure in the shape of a volcano inside the pore and then change the nanopore's dark-field scattering spectrum. Currently, the more suitable methods for predicting the scattering spectra of such metal-coated nanopores are being explored.

A board scattering spectrum is observed because of localized surface plasmons at the edge of the Au nanopore. The light diffraction will be changed as an analyte traverses the nanopore. Using the reflected dark-field illumination method, the sensitivity of an optical detection system and translocation signal variation could both be increased using laser beams as an exciting light source (Figure 4.23). The halogen light source is replaced by

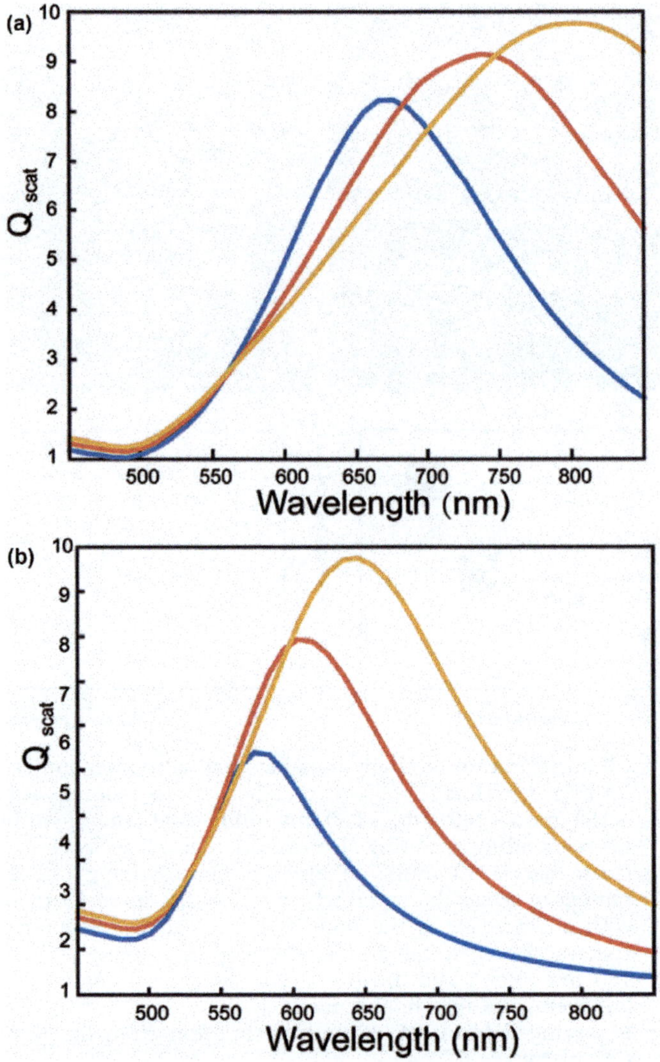

Figure 4.22 Simulated scattering spectra of single gold nanodisks with a height of 90 nm and diameters of 150 nm (blue), 200 nm (red) and 250 nm (yellow) in ambient water (a) and air (b).
Reproduced from ref. 65 with permission from the Royal Society of Chemistry.

a He–Ne laser beam. The gold layer will reflect the background light, and the metal nanopore will diffract a fraction of the monochromatic light. Thus, the CCD camera will capture bright red spots, which a photomultiplier will also be able to detect. Figure 4.23b shows the laser illuminated dark-field image; it is apparent that the size of the aperture will determine the intensity of diffracted light.

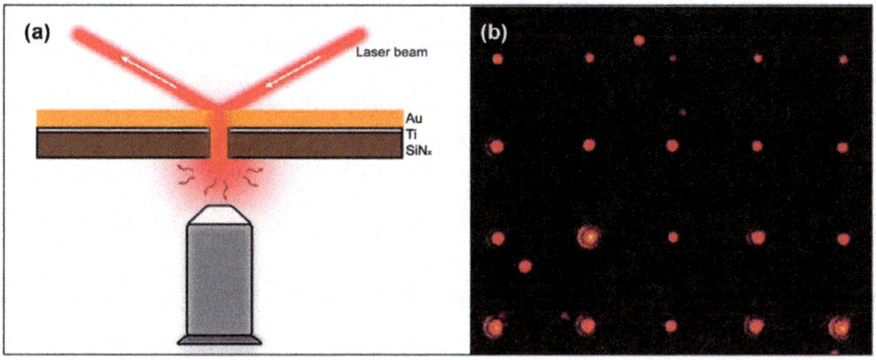

Figure 4.23 (a) Schematic of the reflected dark-field illumination method for detecting a single nanopore. (b) True-color image of the same nanopore array corresponding to Figure 4.21 under reflected dark-field illumination.
Reproduced from ref. 65 with permission from the Royal Society of Chemistry.

4.5.2.2 Electrical Detection

A self-designed integrated system used to collect an ultralow current signature includes the headstage, the current amplifier, the filter circuit and the analog–digital converter (ADC). Two function of the headstage are applying a voltage bias and amplifying the ionic current by a feedback transimpedance amplifier circuit. Further, the current amplifier with multistage amplifier circuits continue to regulate the subsequent signal. The noise introduced in the detection is eliminated by a three-pole Bessel low-pass filter. The analogue signals from the detection cell are converted into digital signals by the ADC. Then the digital signals are displayed on the computer in real time and recorded.

The ultralow current measurement system performed at the same level as the common commercial patch-clamp amplifiers used in the nanopore analyses, such as the Axopatch 200B and HEKA EPC 10 patch-clamp systems. Synchronized optical-electrical detection can be achieved based on this integrated system by amplifying the ionic current and collecting the scattered light signals. The ionic current is measured by the headstage and filtered at 5 kHz and 50 kHz sampling rate.

Under illumination of a 100 W halogen lamp, the scatter light of a 150 nm Au nanopore could be transferred by a PMT into a 50 nA photocurrent that is recorded with a 50 kHz sampling rate (Figure 4.24).

The noise conditions of the Au nanopore are assessed by analyzing the noise power spectrum density (PSD) (Figure 4.24a). The low region frequency below 1000 Hz is called flicker noise, also signed as $1/f$ noise, which is attributed to the electrolyte fluctuations in the chambers.[67] For an uncoated bare SiN_x solid-state nanopore, the power density of flicker noise is less than 1 $pA^2\,Hz^{-1}$ at 1 Hz (blue line in Figure 4.24a). Comparing the power density of flicker noise with (red line in Figure 4.24a) and without (green line in Figure 4.24a) light illumination, the heat effect and surface plasmons that are generated at the nanopore region increased the power density under light illumination. In addition, the frequency in range of 1000 Hz to 10 000 Hz is called white noise, which is suppressed by the gold coating compared to the bare SiN_x nanopore. The power density of the white noise decreases with the increasing of the frequency regardless of illumination of the Au nanopore, but keeps a similar level to 1/*f* noise power density for a bare SiN_x nanopore without Au coating.

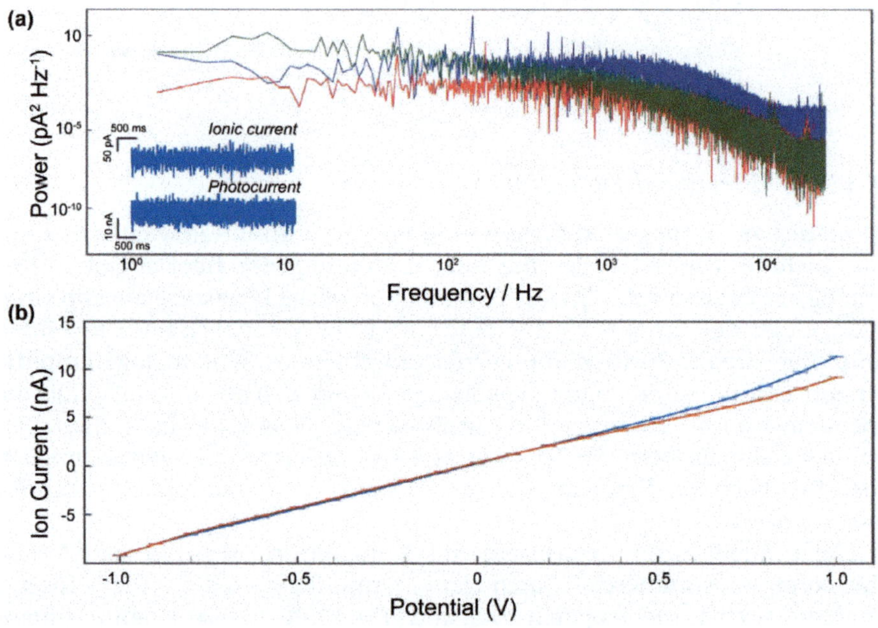

Figure 4.24 (a) Power spectral densities of the ionic current detected from a single gold-coated nanopore, before (red) and after (green) turning on the halogen light, and from a single nanopore without the gold layer in a conventional flow cell (blue). Inset: baseline ionic current at 0 mV, under white light illumination for a single 150 nm gold-coated nanopore, and baseline photocurrent from PMT at 1100 V for the same nanopore at the same time period. (b) Ionic current–voltage characteristics of a single gold-coated nanopore with (blue) and without (red) white light illumination. The error bars are standard deviations of 1 s sampling traces under different bias. The light-induced effects of the ionic current through the nanopore might mainly originate from the plasmonic enhanced heat effects and plasmoelectric potential. Reproduced from ref. 65 with permission from the Royal Society of Chemistry.

The current–voltage (*I–V*) curves of ionic current with and without illumination are shown in Figure 4.24b. The diameter of a large solid-state nanopore can be estimated based on the conductance model:[68]

$$G = \sigma \frac{\pi d^2}{4l} \tag{4.2}$$

where G, l and d are the conductance, length and diameter of the pore respectively, and σ is the conductivity of the electrolyte. The bias that the estimated diameter value of 130 nm is smaller than the value measured by SEM, which is due to the model being more suitable for a solid-state nanopore with a small diameter. A good linearity is shown in the *I–V* curve from −1000 mV to 1000 mV. Under the illumination of a halogen lamp, the heat effect caused by the surface plasmon and plasmoelectric potential induced by the incident light increases the ionic current.

4.5.3 Optical Nanopore-based Sensing Application

Extraordinary optical transmission (EOT) means the larger transmission of electromagnetic fields are generated when electromagnetic resonances occur through a sub-wavelength aperture in a metal film which could be flat or corrugated. And this EOT phenomenon that can enhance the optical transmission has been widely used to sensing, lenses, metamaterials, color filters, optical trapping, enhancement of nonlinear effects and more. Plasmonic nanopores have often used the EOT effect in the fields of biomolecule and nanoparticle detection. When a metal-coated nanopore is illuminated, the oscillation between the metal surface electrons and the photonics generates the surface plasmon which could significantly enhance the light transmission. Moreover, the refractive index inside the nanopore has a effect on the intensities and frequencies of scattering light. A novel label-free optical nanopore read-out method with this detection principle is recently demonstrated. A 100 nm thick gold film is layered onto a custom silicon nitride chip, and gold-coated nanopores are fabricated by TEM drilling a single sub-wavelength aperture. An optical detection system can easily monitor the scattered light through the Au nanopore significantly enhanced by a surface plasmon. A dark-field illumination system could eliminate the interference of the incident light to significantly suppress the background.

As shown in Figure 4.25a, when single analytes, such as polystyrene (PS) beads, enter the nanopore and occupy the inner space of the nanopore, the refractive index inside the pore will change along with the translocation event.[69] The refractive index of PS beads is 1.58, which is larger than water (1.33). Therefore, the replacement of PS beads with water locally increases the refractive index and light flux inside the nanopore. Applying a dark-field illumination system to provide the incident light, the objective will only collect the scattered light from the nanopore. PS beads with zeta-potential of −25 mV at the experiment condition can be electrically driven across the gold-coated nanopore at property bias voltage. The scattered light signals obtained by the photomultiplier and the ionic current signals obtained by the low-noise amplifier can be monitored simultaneously. Typical trace and representative of both optical and electrical signals are shown in Figure 4.25b. Electrical signals and optical signals both increase during the single particle traversing event. Compared with the 10 mM KCl electrolyte used in this experiment, PS beads are highly surface charged and have a

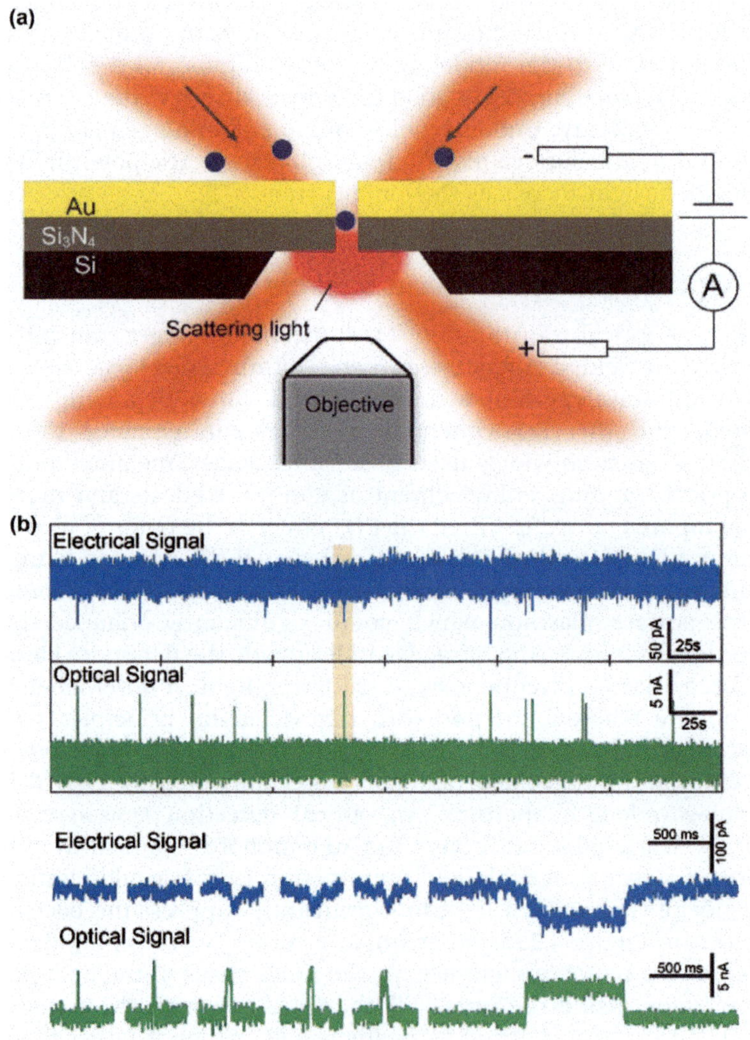

Figure 4.25 (a) Schematic of the scattering nanopore detection system. (b) Simultaneous electrical and optical recording of PS beads translocations through solid-state nanopores.
Reproduced from ref. 69 with permission from American Chemical Society, Copyright 2016.

higher refractive index. When a PS bead occupies the nanopore, the number of surface charges of a PS bead is more than the total charge of the excluded electrolyte. Therefore, a conductive ionic pulse is observed rather than a resistive ionic pulse. Similarly, considering the scattered light, the photocurrent also increases because of the higher refractive index (~1.58) of PS beads. In additional, the gold layer may significantly increase the capacitance of the membrane and reduce the bandwidth for ionic current detection. This will result in a long charging time which induces the gradual increase of ionic current baseline observed in the collected data. Therefore, the low bandwidth makes gathering the electrical signal with short dwell time difficult, even when the optical signals exhibit good performance.

Plasmonic technique facilitates the far-field communication of single molecules by the variation in the scattering intensity, making it possible to monitor a single molecule in a label-free manner. As shown in Figure 4.26a, recently, a plasmonic nanoantenna was readily combined with a solid state nanopore to form a magic label-free optical sensor. In this research, target

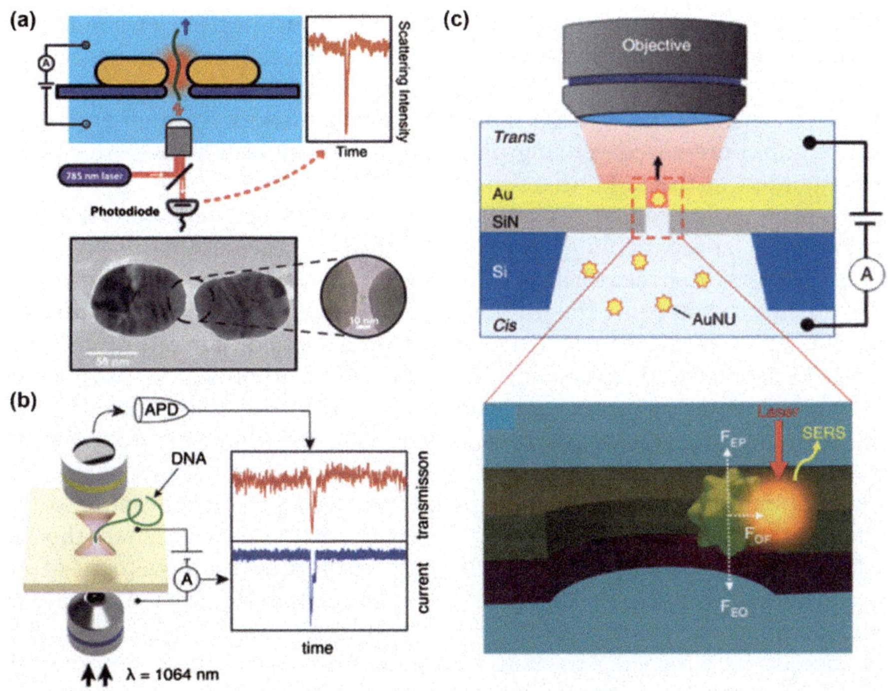

Figure 4.26 (a) Schematic of the back-scattered nanopore detection system. Reproduced from ref. 70 with permission from American Chemical Society, Copyright 2018. (b) Schematic of the inverted-bowtie plasmonic nanopore detection system. Reproduced from ref. 71 with permission from American Chemical Society, Copyright 2019. (c) Single molecule sensing based on a nanopore-SERS platform. Reproduced from ref. 72 with permission from Springer Nature, Copyright 2019.

DNA molecules can be exactly anchored into the confined hotspot at the nanoscale under the electric field. The laser excitation at 785 nm to a plasmonic nanoantenna resulted in the appearance of a nanoscale electromagnetic hotspot right above the nanopore confined space. Thus, the movement of DNA molecules through the nanopore not only caused the modulations of ionic current but also led to the changes in the scattering intensity. The sensing method based on backscattered light allows an ultrafast single molecule measurement with a better than 200 μm temporal resolution.[70] Based on the intrinsic properties of plasmon resonance biosensors, the plasmonic nanopore can also achieve purely optical, nanopore measurement, which is decoupling of the driving voltage. The optical read-out only relies on the changes in the plasmon resonance, indicating the behaviors of single molecules can be studied at any driving voltage. As shown in Figure 4.26b, an inverted-bowtie plasmonic structure was subsequently integrated into the solid-state nanopore to form a plasmonic nanopore system, which depended on the changes in the intensity of light transmission. The plasmonic resonance was extremely sensitive to the local refractive index with the hotspot, and the presence of biomolecules would alter the resonance of nanoaperture. Thus, the translocation of single biomolecules can be exactly spotted *via* an optical detection scheme. Moreover, the optical read-out was buffer concentration independent and displayed the enhanced performance of high measurement bandwidth.[71] In an emerging field of nanopore-based sensing platform, the plasmonic nanopore can be also extended to the SERS discrimination of single DNA molecules *via* the electro-plasmonic trapping method. It is widely accepted that the fast translocation speed of single molecules through the nanopore confined space hinders the acquisition of a Raman spectrum. As shown in Figure 4.26c, to overcome the limitation, a well-defined gold nanourchin (Au NU) and Au layer-coated nanopore were first designed, permitting the balance between the electroosmotic, electrophoretic, and optical forces. The Au NU was delivered to the confined space under an electric field. Subsequently, the Au NU was exposed to both electroosmotic and electrophoretic forces, because the surface charge of Au NU and nanopore were both negative. The inter-coupling between the Au NU tip and the Au wall of the nanopore would enhance the electromagnetic field when the Au NU approached the Au surface, also effectively enhancing the gradient force. The movement of Au NU can be skillfully controlled by the electro-opto-kinetic effect. The narrow space between the sharp tip of the Au NU and Au layer confines the electromagnetic field for SERS measurements. Subsequently, the discrimination of individual DNA molecules can be achieved by monitoring the Raman spectrum that occurs upon the passage of molecules through the nanopore-based confined space owing to the generation of a hotspot.[72] The scattering/plasmonic nanopore sensing system permits a great possibility for native single molecule measurements at physiological conditions. In addition, since the diameter of an array disk for a single pore is below 1 μm, the multipore sensing with higher throughput is possible for this system. And with the development of the array detector, the time resolution could be improved, such as 1 ms for the high-speed EMCCD camera, which is adequate for multipore sensing. Besides the EOT, surface plasmon also has great applications, such as surface-enhanced Raman and fluorescence methods. And plasmonic nanopore holds great potential for single molecule detection like DNA sequencing.

In addition to the plasmonic technique, the nanopore platform can be also synchronized with the fluorescence imaging technique. Recently, substantial progress has been made towards single molecule sensing *via* a nanopore-fluorescence synchronous system. Fluorescence imaging is a powerful technique, which is complementary to electrical sensing and relies on photons. The presence of a fluorescent readout can outbreak the challenges posed by the pure nanopore sensing system such as insufficient throughput, unexpected noise components, limited spatial resolution. However, the suppression of the background light is an essential requirement for most fluorescence techniques. The nanopore coincidently provides a well-defined confined space to suppress background fluorescence. Confocal imaging with solid-state nanopores was intergrated to achieve single molecule DNA methylation discrimination. In this study, the excitation laser beam was successfully confined to a diffraction-limited spot to be delivered to the sensing region. The research involved a one-step enzymatic reaction, in which dsDNA molecules were specifically reacted with DNA methyltransferases and *S*-adenosyl-Lmethionine analogs labeled with fluorescent molecules. The confined laser beam was used to quantify the unmethylated DNA molecules, and electrical readouts were used as gate signals that report the resident time of DNA molecules with the electrochemically confined space.[76] This method opened up a new way for direct sequence-specific monitoring of clinically relevant hypomethylated sites in the genome.

With the advancements in micro/nano processing technology, the use of micro/nano processing technology to prepare nanopores with a size of several nanometers for nucleotide detection has begun to be valued by the scientific community. The materials used are mainly concentrated in silicon-based materials, The processing technology of silicon-based materials has been developed for nearly 50 years, and the lithography precision of microelectronics technology has moved toward 32 nm. Therefore, fabricating nanopores using silicon-based materials has become the first choice. In addition to SiN_x, polymer nanopores are typical synthetic nanopores, which are generally confined holes drilled in polymer films such as poly(ethylene terephthalate) (PET), polycarbonate (PC) and polyimide (PI), as well as in mica. Researchers always fabricate the polymer nanopores by the track-etching method. In the process, energetic heavy ions are first focused and irradiate these dielectric films to produce local damaged zones. These focused energy heavy ions can penetrate $\sim$100 μm polymer films. Chemical etching process is then carried out to produce nanopores with the desired size and geometry. To tailor the surface chemistry of prepared polymer nanopores, researchers proposed the derivatization of the COOH groups *via* amide bonds using the coupling reagent 1-ethyl-3-[3-dimethylaminopropyl] carbodiimide hydrochloride (EDC). Moreover, the electroless plating of Au is also used to modify the sensing interfaces of these pores.[77] Besides, graphene has been widely studied in recent years and is considered to be one of the most potential alternatives to SiN_x. Researchers applied it to single molecule sequencing, using a scanning electron microscope electron beam to sculp well-defined nanopores of a few nanometres on a graphene film, because the thickness of the monoatomic layer graphene is between the two adjacent bases on the DNA strand. The distance (0.4 nm) is even smaller, so the detection accuracy of this nanopore is higher than other nanopores. At the same time, graphene is also considered to be an ideal material for the preparation of high-speed transistors, and it is hoped that the integration of single molecule sequencing with the inspection system will be achieved. In addition to graphene, a batch of 2D materials is being reported as excellent candidate substrates for the fabrication of nanopores such as molybdenum disulfide (MoS_2) and hexagonal boron nitride (h-BN). The advent of these nanopores based on 2D materials contributes to the enhanced sensing resolution of single molecule sensing.[78] We suppose that the solid-state nanopore will more easily mimic biological nanopores or be incorporated into an optical sensing platform, contributing to dynamic single molecule sensing with high accuracy, speed, and reliability in a well-defined confined space, in the near future.

References

1. C. Dekker, *Nat. Nanotechnol.*, 2007, **2**, 209–215.
2. S. Howorka and Z. Siwy, *Chem. Soc. Rev.*, 2009, **38**, 2360–2384.
3. L. Oukhaled, M. Bacri, J.-M. Pastoriza-Gallego, Betton and J. Pelta, *ACS Chem. Biol.*, 2012, **7**, 1935–1949.
4. W. Shi, A. K. Friedman and L. A. Baker, *Anal. Chem.*, 2017, **89**, 157–188.
5. B. N. Miles, A. P. Ivanov, K. A. Wilson, F. Doğan, D. Japrung and J. B. Edel, *Chem. Soc. Rev.*, 2013, **42**, 15–28.
6. N. Li, S. Yu, C. C. Harrell and C. R. Martin, *Anal. Chem.*, 2004, **76**, 2025–2030.
7. J. H. Yuan, F. Y. He, D. C. Sun and X. H. Xia, *Chem. Mater.*, 2004, **16**, 1841–1844.
8. J. Li, D. Stein, C. McMullan and D. Branton, *Nature*, 2001, **412**, 166–169.
9. J. Storm, J. H. Chen, X. S. Ling, H. W. Zandbergen and C. Dekker, *Nat. Mater.*, 2003, **2**, 537–540.
10. H. Kwok, K. Briggs and V. Tabard-Cossa, *PLoS One*, 2014, **9**, e92880.
11. T. Gibb and M. Ayub, in *Micro and Nano Technologies*, ed. J. B. Edel and T. Albrecht, William Andrew Publishing, Oxford, 2013, pp. 121–140.
12. Q. Cai, B. Ledden, E. Krueger, J. A. Golovchenko and J. Li, *J, Appl. Phys.*, 2006, **100**, 24914–249146.

13. J. Gierak, A. Madouri, A. L. Biance, E. Bourhis, G. Patriarche, C. Ulysse, D. Lucot, X. Lafosse, L. Auvray, L. Bruchhaus and R. Jede, *Microelectron. Eng.*, 2007, **84**, 779–783.
14. J. Yang, C. F. David, L. A. Stern, C. A. Sanford, J. Huang, Z. Ren, L.-C. Qin and A. R. Hall, *Nanotechnology*, 2011, **22**, 285310.
15. M. J. Kim, M. Wanunu, D. C. Bell and A. Meller, *Adv. Mater.*, 2006, **18**, 3149–3153.
16. P. Chen, T. Mitsui, D. B. Farmer, J. Golovchenko, R. G. Gordon and D. Branton, *Nano Lett.*, 2004, **4**, 1333–1337.
17. C. A. Merchant, K. Healy, M. Wanunu, V. Ray, N. Peterman, J. Bartel, M. D. Fischbein, K. Venta, Z. Luo, A. T. C. Johnson and M. Drndić, *Nano Lett.*, 2010, **10**, 2915–2921.
18. S. Habermehl, R. T. Apodaca and R. J. Kaplar, *Appl. Phys. Lett.*, 2009, **94**, 12905.
19. K. Briggs, H. Kwok and V. Tabard-Cossa, *Small*, 2014, **10**, 2077–2086.
20. R. Yanagi, T. Akahori, Hatano and K. Takeda, *Sci. Rep*, 2014, **4**, 5000.
21. R. Tahvildari, E. Beamish, V. Tabard-Cossa and M. Godin, *Lab Chip*, 2015, **15**, 1407–1411.
22. H. Kwok, M. Waugh, J. Bustamante, K. Briggs and V. Tabard-Cossa, *Adv. Funct. Mater.*, 2014, **24**, 7745–7753.
23. Y.-L. Ying and Y.-T. Long, *Sci. China: Chem.*, 2017, **60**, 1187–1190.
24. A. T. Carlsen, K. Brigss, A. R. Hall and V. Tabard-Cossa, *Nanotechnology*, 2017, **28**, 85304.
25. S. Pud, D. Verschueren, N. Vukovic, C. Plesa, M. P. Jonsson and C. Dekker, *Nano Lett.*, 2015, **15**, 7112–7117.
26. F. Sauer-Budge, J. A. Nyamwanda, D. K. Lubensky and D. Branton, *Phys. Rev. Lett.*, 2003, **90**, 238101.
27. H. Mathé, V. Visram, Y. Viasnoff, Rabin and A. Meller, *Biophys. J.*, 2004, **87**, 3205–3212.
28. Q. Jin, A. M. Fleming, C. J. Burrows and H. S. White, *J. Am. Chem. Soc.*, 2012, **134**, 11006–11011.
29. X. Zhang, N. E. Price, X. Fang, Z. Yang, L. Q. Gu and K. S. Gates, *ACS Nano*, 2015, **9**, 11812–11819.
30. B. Meller, McNally and M. Wanunu, *Nano Lett.*, 2008, **8**, 3418–3422.
31. Y. Lin, X. Shi, S.-C. Liu, Y.-L. Ying, Q. Li, R. Gao, F. Fathi, Y.-T. Long and H. Tian, *Chem. Commun.*, 2017, **53**, 3539–3542.
32. C. Plesa, S. W. Kowalczyk, R. Zinsmeester, A. Y. Grosberg, Y. Rabin and C. Dekker, *Nano Lett.*, 2013, **13**, 658–663.
33. R. Y. Larkin, M. Henley, J. K. Muthukumar, Rosenstein and M. Wanunu, *Biophys. J.*, 2014, **106**, 696–704.
34. T. Carlsen, O. K. Zahid, J. A. Ruzicka, E. W. Taylor and A. R. Hall, *Nano Lett.*, 2014, **14**, 5488–5492.
35. S. W. Kowalczyk, A. R. Hall and C. Dekker, *Nano Lett.*, 2010, **10**, 324–328.
36. C. Plesa, J. W. Ruitenberg, M. J. Witteveen and C. Dekker, *Nano Lett.*, 2015, **15**, 3153–3158.

37. J. S. Yu, M. C. Lim, D. T. N. Huynh, H. J. Kim, H. M. Kim, Y. R. Kim and K. B. Kim, *ACS Nano*, 2015, **9**, 5289–5298.
38. N. A. W. Bell and U. F. Keyser, *J. Am. Chem. Soc.*, 2015, **137**, 2035–2041.
39. J. Kong, N. A. W. Bell and U. F. Keyser, *Nano Lett.*, 2016, **16**, 3557–3562.
40. N. A. W. Bell and U. F. Keyser, *Nat. Nanotechnol.*, 2016, **11**, 1–28.
41. D. J. Niedzwiecki, R. Iyer, P. N. Borer and L. Movileanu, *ACS Nano*, 2013, **7**, 3341–3350.
42. Y. Lin, Y.-L. Ying, X. Shi, S.-C. Liu and Y.-T. Long, *Chem. Commun.*, 2017, **53**, 11564–11567.
43. S. W. Kowalczyk, D. B. Wells, A. Aksimentiev and C. Dekker, *Nano Lett.*, 2012, **2**, 1–5.
44. B. N. Anderson, M. Muthukumar and A. Meller, *ACS Nano*, 2013, **7**, 1408–1414.
45. Y. Tsutsui, M. He, S. Furuhashi, M. Rahong, Taniguchi and T. Kawai, *Sci. Rep.*, 2012, **2**, 394.
46. Y.-Q. Wang, C. Cao, Y.-L. Ying, S. Li, M.-B. Wang, J. Huang and Y.-T. Long, *ACS Sens.*, 2018, **3**, 779–783.
47. Y.-Q. Wang, M.-Y. Li, H. Qiu, C. Cao, M.-B. Wang, X.-Y. Wu, J. Huang, Y.-L. Ying and Y.-T. Long, *Anal. Chem.*, 2018, **90**, 7790–7794.
48. E. C. Yusko, J. M. Johnson, S. Majd, P. Prangkio, R. C. Rollings, J. Li, J. Yang and M. Mayer, *Nat. Nanotechnol.*, 2011, **6**, 253–260.
49. R. Wei, V. Gatterdam, R. Wieneke, R. Tampé and U. Rant, *Nat. Nanotechnol.*, 2012, **7**, 257.
50. U. F. Keyser, B. N. Koeleman, S. van Dorp, D. Krapf, R. M. M. Smeets, S. G. Lemay, N. H. Dekker and C. Dekker, *Nat. Phys.*, 2006, **2**, 473–477.
51. H. Peng and X. S. Ling, *Nanotechnology*, 2009, **20**, 185101.
52. C. Hyun, R. Rollings and J. Li, *Small*, 2012, **8**, 385–392.
53. B. Hornblower, A. Coombs, R. D. Whitaker, A. Kolomeisky, S. J. Picone, A. Meller and M. Akeson, *Nat. Methods*, 2007, **4**, 315.
54. D. Stoddart, A. J. Heron, E. Mikhailova, G. Maglia and H. Bayley, *Proc. Natl. Acad. Sci. U. S. A.*, 2009, **106**, 7702 LP–7707 LP.
55. V. Tabard-Cossa, M. Wiggin, D. Trivedi, N. N. Jetha, J. R. Dwyer and A. Marziali, *ACS Nano*, 2009, **3**, 3009–3014.
56. X. Shi, Q. Li, R. Gao, W. Si, S.-C. Liu, A. Aksimentiev and Y.-T. Long, *J. Phys. Chem. Lett*, 2018, **9**, 4689–4694.
57. G. A. T. Chansin, R. Mulero, J. Hong, M. J. Kim, A. J. DeMello and J. B. Edel, *Nano Lett.*, 2007, **7**, 2901–2906.
58. R. Y. Ivankin, J. Henley, S. Larkin, M. L. Carson, Toscano and M. Wanunu, *ACS Nano*, 2014, **8**, 10774–10781.
59. H. A. Bethe, *Phys. Rev.*, 1944, **66**, 163.
60. J. Zheng, J. Bai, Q. Zhou, J. Li, Y. Li, J. Yang and R. Yang, *Chem. Commun.*, 2015, **51**, 6552–6555.
61. Y. Deng, C. Wang, D. Liu, J. Juan, P. Hu, Ping and L. Su, *Opt. Commun.*, 2016, **372**, 113–117.
62. Q. Min, Y. Pang, D. J. Collins, N. A. Kuklev, K. Gottselig, D. W. Steuerman and R. Gordon, *Opt. Express*, 2011, **19**, 1648.

63. B. Malekian, K. Xiong, G. Emilsson, J. Andersson, C. Fager, E. Olsson, E. M. Larsson-langhammer and A. B. Dahlin, *Sensors*, 2017, **17**, 1444.
64. C. Ma, Q. Gao, W. Hong, J. Fan and J. Fang, *Adv. Funct. Mater.*, 2017, **27**, 1603233.
65. X. Shi, R. Gao, Y.-L. Ying, W. Si, Y. Chen and Y.-T. Long, *Faraday Discuss.*, 2015, **184**, 85–99.
66. J. Prikulis, P. Hanarp, L. Olofsson, D. Sutherland and M. Käll, *Nano Lett.*, 2004, **4**, 1003–1007.
67. R. M. M. Smeets, U. F. Keyser, N. H. Dekker and C. Dekker, *Proc. Natl. Acad. Sci. U. S. A.*, 2008, **105**, 417–421.
68. S. W. Kowalczyk, A. Y. Grosberg, Y. Rabin and C. Dekker, *Nanotechnology*, 2011, **22**, 315101.
69. X. Shi, R. Gao, Y. L. Ying, W. Si, Y. F. Chen and Y. T. Long, *ACS Sensors*, 2016, **1**, 1086–1090.
70. X. Shi, D. V. Verschueren and C. Dekker, *Nano Lett.*, 2018, **18**, 8003–8010.
71. D. V. Verschueren, S. Pud, X. Shi, L. D. Angelis, L. Kuipers and C. Dekker, *ACS Nano*, 2019, **13**, 61–70.
72. J.-A. Huang, M. Z. Mousavi, Y. Zhao, A. Hubarevich, G. Giovannini, M. Schutte, D. Garoli and F. D. Angelis, *Nat. Commun.*, 2019, **10**, 5321.
73. Q. Zhao, J. Comer, V. Dimitrov, S. Yemenicioglu, A. Aksimentiev and G. Timp, *Nucleic Acids Res.*, 2008, **36**, 1532–1541.
74. J. Comer, V. Dimitrov, Q. Zhao, G. Timp and A. Aksimentiev, *Biophys. J.*, 2009, **96**, 593–608.
75. J. B. Heng, A. Aksimentiev, C. Ho, P. Marks, Y. V. Grinkova, S. Sligar, K. Schulten and G. Timp, *Nano Lett.*, 2005, **5**, 1883–1888.
76. T. Gilboa, C. Torfstein, M. Juhasz, A. Grunwald, Y. Ebenstein, E. Weinhold and A. Meller, *ACS Nano*, 2016, **10**, 8861–8870.
77. S. Howorka and Z. Siwy, *Chem. Soc. Rev*, 2009, **38**, 2360–2384.
78. K. Lee, K.-B. Park, H.-Y. Kim, J.-S. Yu, H. Chae, H.-M. Kim and K.-B. Kim, *Adv. Mater*, 2018, **30**, 1704680.

CHAPTER 5

A Confined Nanopipette: From Fundamental to Application

RU-JIA YU,*[a,b] YONG-XU HU,[b] SI-MIN LU,[a] SU-WEN XU,[b] YAO LIN,[b] YI-LUN YING[a] AND YI-TAO LONG[a]

[a] Nanjing University, Nanjing, China; [b] East China University of Science and Technology, Shanghai, China
*Email: yurujia@nju.edu.cn

5.1 General Preparation and Characterization

Typically, a nanopipette is fabricated from a glass (quartz or borosilicate) capillary with an inner diameter on the micrometre scale, whose tip diameter is from several nanometres to hundreds of nanometres. Comparing these two kinds of capillary, borosilicate owns a lower melting temperature than quartz. Thus, it is difficult to prepare a borosilicate nanopipette with a very small tip diameter and short taper. A nanopipette with small tip diameter is urgently needed as it would provide good signal-to-noise ratio for single molecule/nanoparticle detection and cause negligible damage in single cell analysis. Therefore, a quartz nanopipette is superior for single entity analysis, not only because it can be prepared with very small diameter but also its rigid features make it more compatible for cell or tissue penetration.[1]

After decades of research and exploration in nanopipette fabrication, two main strategies for fabricating the conical nanotip have emerged, which are heat pulling (Figure 5.1a) and chemical etching (Figure 5.1b). Heat pulling is a direct preparation method that is advantageous for the large quantity production of nanopipettes with a uniform geometry and dimension. By using a laser-based micropipette puller to heat the middle of the capillary,

Detection Science Series No. 17
Confining Electrochemistry to Nanopores
Edited by Yi-Tao Long

Published by the Royal Society of Chemistry, www.rsc.org

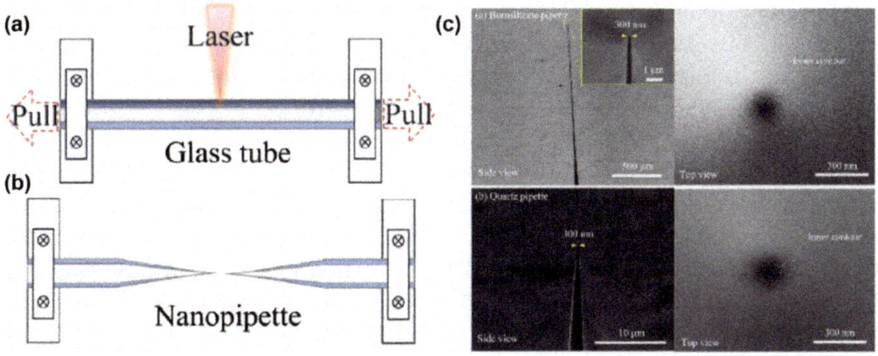

Figure 5.1 Preparation of a nanopipette by the heat pulling method. (a) The glass capillary was first mounted on a laser puller; the middle of the capillary was heated by laser and then both ends of the capillary sustained an equivalent pulling force. (b) The prepared nanopipette through this heat pulling method. (c) The SEM characterization of the prepared borosilicate and quartz nanopipette.
Reproduced from ref. 4 with permission from the IOP Publishing, Copyright 2015.

two symmetrical pipettes of the same dimensions can be simply fabricated with feasible tip opening and taper control. Parameters with their denotation are heat (HEAT), filament (FIL), velocity (VEL), delay (DEL) and pull force (PULL). Specifically, the parameters of heat and filament control the laser heated temperature and beam spot diameter, respectively. These are determined by the melting temperature of the used capillary materials. For example, 350 °C could be used for pulling a borosilicate capillary while 700 °C is more suitable for the quartz one. The parameters of delay, velocity and pull represent the delay time after laser heating, pulling velocity and pulling force, respectively. The pulling parameter plays a key role in determining the aperture size of the sharp tip of the nanopipette. In general, a higher heating temperature, pulling velocity and pulling force and a smaller pulling delay and filament are likely to yield a smaller diameter of nanopipette tip. Therefore, the pulling parameter is vital for the heat pulling method for nanopipette fabrication, and the specific parameter is dependent on the capillary materials, environmental temperature humidity and even different pullers.

As we previous demonstrated, a quartz nanopipette with controllable size can be prepared with a P-2000 CO_2-laser puller. Before pulling, the quartz capillaries were successively washed using piranha acid (3 : 1 98% H_2SO_4/30% H_2O_2), absolute ethanol and deionized water solutions and subsequently dried at 100 °C or N_2 atmosphere. The nanopipettes with diameters of 50 nm, 100 nm and 200 nm were fabricated with the following pulling programs: HEAT = 625, FIL = 4, VEL = 60, DEL = 150, PULL = 192; HEAT = 760, FIL = 3, VEL = 16, DEL = 128, PULL = 140 AND HEAT = 690, FIL = 4, VEL = 45, DEL = 190, PULL = 30, respectively.[3] It should be noted that the parameters for the nanopipette pulling could be varied by different pullers and even the same instrument in different conditions. Therefore, the pulling parameter should be adjusted accordingly in each fabrication since it is key for fabrication of the desired nanopipette.[2,3]

Alternatively, the nanopipette can be also prepared by the chemical etching method, which contains two major steps. As shown in Figure 5.2a, a micrometre sized pipette was firstly prepared and then the sharp tip fused until it was fully sealed, forming a spherical terminal with diameter of 100 ± 10 μm. The micrometre sized pipette was then back filled with electrolyte and monitored under a microscope to exclude any air bubbles. A pair of Ag/AgCl electrodes was used to simultaneously record the current response by placing them in the inside and outside solution of the micropipette cavity, respectively. Hydrofluoric acid solution was subsequently employed as the outside solution, immersing the sharp fused tip of the pipette; a sudden current increase would indicate the formation of a tiny aperture in the tip (Figure 5.2b–d).[5] The prepared nanopipette by this etching method should be thoroughly washed using an alkali solution and distilled water in order to get rid of any residue left behind. As previously demonstrated, a pipette with orifice diameter of 63 nm was prepared using this chemical etching method.[6] Compared with the previous demonstrated heat pulling method, chemical etching could be performed to fabricate a nanopipette in a common chemical laboratory without using any expensive instrumentation. However, the dimensions of the nanopipette was difficult to precisely controlled during this etching process since both the pulling and etching steps could affect the nanopipette morphology. Nevertheless, a microelectromechanical system like focused ion beam (FIB) could be used for

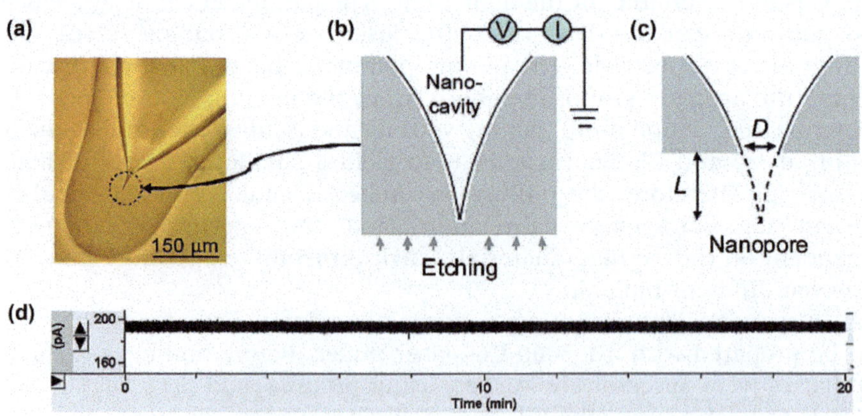

Figure 5.2 The chemical etching method for nanopipette preparation. (a) The chemical etching method from the bottom of the nanopipette with the current recording. (b)–(d) The formation of a nanopore on the bottom of the fused nanopipette.
Reproduced from ref. 5 with permission from the American Chemical Society, Copyright 2008.

further milling and fine tuning the geometry of the nanopipette to the desired scale.

The detailed characterization of a nanopipette, including the nanotip size, half cone angle, thickness and even surface charge of the glass nanopipette can be facilitated by microscopic imaging, electrochemical measurement and the simulation method (Figure 5.3).[7] Microscopic methods such as scanning electron microscopy (SEM) and transmission electron microscopy (TEM) offer the most intuitive way for nanopipette characterization, which are generally used to precisely and accurately characterize the geometry of nanopipettes, including the tip diameter, length and cone angle. Recently, focus ion beam (FIB) has also been employed to analyse the inner morphology of the metal-functionalized nanopipette by splitting the nanopipette with the direction from the tip to the other side.[7]

The electrochemical method is a direct and simple approach to estimate the inner dimensions of a nanopipette since the experiments can be easily performed with a current–voltage (i–V) measurement.[8] When the reservoirs are filled with a high-concentration electrolyte such as 0.5 M KCl, there will be a linear dependence in the i–V curve. The relationship between the inner diameter (d_i) and the conductance of the nanopipette in the solution (G) can be revealed by eqn (5.1), where l is the conical part length of the nanopipette, d_b is the original diameter of the capillary and g represents the specific conductance for the solution. Reprinted from ref. 8:

$$d_i = \frac{4Gl}{\pi g d_b}. \tag{5.1}$$

Finite-element methods (FEM) can be utilized to simulate the mass transfer and the electrochemical response.[9–11] The FEM simulation includes

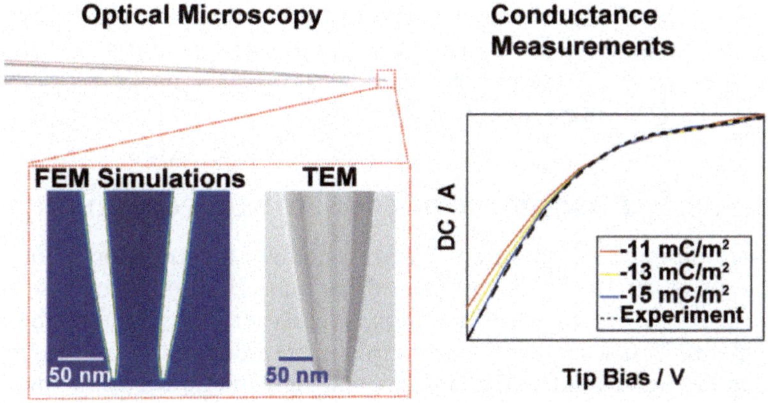

Figure 5.3 The characterization of a nanopipette by using optical microscopy and the electrochemical method.
Reproduced from ref. 7 with permission from the American Chemical Society, Copyright 2016.

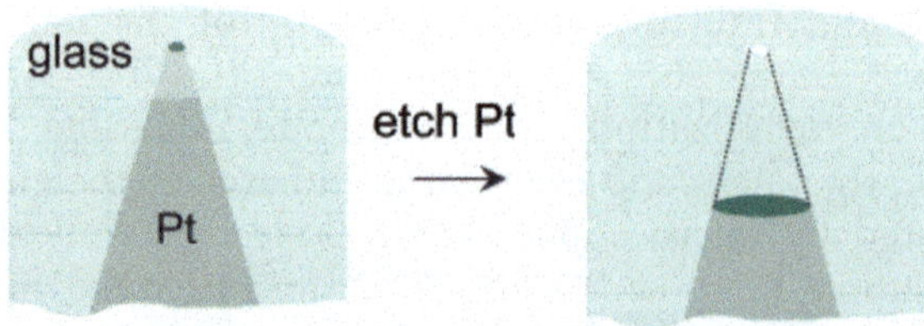

Figure 5.4 Fabrication of a nanopore electrode. Copyright 2004, American Chemical Society.
Reproduced from ref. 11 with permission from the American Chemical Society, Copyright 2004.

the following steps: (1) a sketch of the computation domain for the conical part and the solution should be demonstrated according to the measured parameter of the nanopipette; (2) the coupled equations to depict the relationship between electric field and the solution as well as the specific boundary conditions like the charge density of the quartz surface should be considered carefully; (3) an adaptive mesh refinement is of great importance for accurately revealing the features of a nanopipette like the electric double layer. After calculating using a commercial software package, the ion fluxes, current and potential profile can be directly and precisely found. Furthermore, based on the description of the nanopipette, the translocation events and the electrochemical response in nanopipette sensing could be simulated with this approach.[12,13]

Besides the nanopipette, a nanopore electrode called the truncated cone-shaped nanopore electrode was also demonstrated in a previous publication[11] with a three-step method as (1) sealing an atomical sharp Pt wire in a glass capillary; (2) polishing the capillary until a Pt disk with nanometre dimensions was exposed; and (3) electrochemically etching the exposed Pt to produce a truncated cone-shaped nanopore in glass and an embedded electrode surface (Figure 5.4). The fabrication and application of nanopore-supported Pt electrodes attract great interest from the electrochemical community since the nanocavity provides promising nanoconfinement for the localized electrochemical measurements.

5.2 Further Modification for Advanced Sensing

Bare nanopipettes are mainly utilized to analyse the translocations of analytes *via* nanopore-based methods and for mass delivery in diverse types of scanning probe microscopy techniques. Further functionalization of the nanopipette would expand the chemical and biological applications. As previous demonstrated, nanopipettes can be directly modified by specific molecules (*e.g.* polymers, peptides, proteins, DNA, *etc.*) to mimic the two main functional characteristics of natural channels: selectivity for the type of passing ions and the ability to respond to changes due to an external electric field.[14] Due to the chemical inertness of the silicon-based material, the sensing area of nanopipettes should be silanized before chemical modification with the functional groups. And the functionalized nanopipettes could be used for the detection of temperature/pH variations, ions, redox species and interactions between biomolecules by current variations and transient spikes in the signals.[6,15–21] Through this method, the functionalized nanopipettes were prepared by a two-step process, which was salinization and target modification. For example, a lysozyme binding aptamer could be modified inside the nanopipette wall through this method (Figure 5.5). Briefly, 3-amino-propyl-triethoxylsilane (APTES) in ethanol solution was used to react with the nanopipette surface after successively cleaning with piranha acid, pure water and ethanol solutions. Afterwards, glutaraldehyde aqueous solution was used to react with the quaternary ammonium group, leaving the bare aldehyde group in the other terminal. The target aptamer molecule was thus bound onto the inner wall of pipette due to the reaction between aldehyde group and amine group on the aptamer.[22] Otherwise, nanopipette could be functionalized by electrostatic immobilization. As previous demonstrated, the surface of quartz nanopipette is negatively charged in neutral solution and could thus adsorb positive charged molecules. Poly-l-lysine (PLL) is commonly used as a coating reagent, which is strong hydrophilic and provides an amine-reactive surface on the nanopipette tip. The obvious current rectification in *i–V* curves could be used to confirm the adsorption of PLL onto the nanopipette wall. In detail, the nanopipette was first filled and wetted with PLL aqueous solution for several minutes forming a PLL film on the nanopipette surface, followed by a baking process at ~120 °C for 1 h. Other molecules (DNA or protein) could bind to the nanopipette surface by means of amine facilitated covalent bond or electrostatic adsorption. Biotin-BSA molecules were directly self-assembled on the nanopipette surface by immersion in protein solutions for 1 h at room temperature to enable full physisorption.[15] Alternatively, as

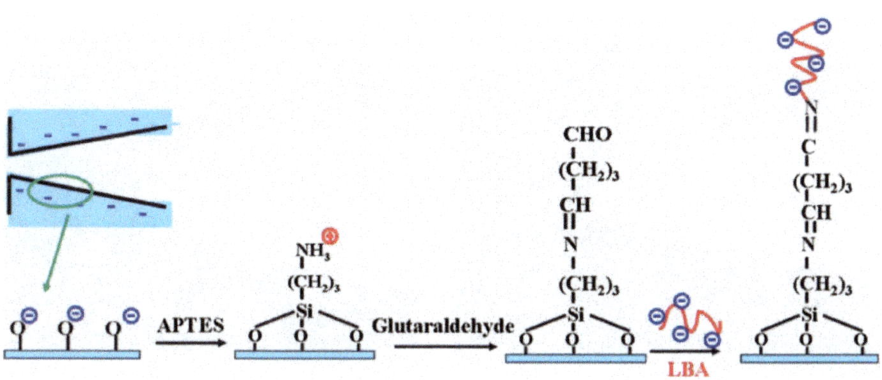

Figure 5.5 Schematic illustration of the modification of a glass nanopipette surface with lysozyme binding aptamer.
Reproduced from ref. 23 with permission from the Elsevier, Copyright 2011.

previously demonstrated, aptamer molecules were immobilize on the PLL coated nanopipette surface by successively employing polyacrylic acid, EDC/NHS and aptamer solutions to form the covalent bond.[23]

Although direct chemical modifications provide a sensitive zone for pore–analyte interactions, the poor continuity and conducting of the modified layer by functional groups may limit the ability of nanopipettes to monitor electrochemical processes. Therefore, the coating of conductive materials, such as carbon and metal, on the inner wall of nanopipette is needed to form an electroactive interface for studying an intrinsic reaction–electron transfer during the electrochemical reactions.

Carbon-coated nanopipettes are generally prepared by chemical vapour deposition (CVD), which could deposit a carbon film on the inner wall (Figure 5.6). Both the length and thickness of the carbon film can be precisely controlled by adjusting the deposition conditions.[24] For example, a CVD time of 2 h would develop a carbon film of ~30 nm and 4 h would result in a film of ~80 nm. The outer surface of quartz was dissolved by immersing

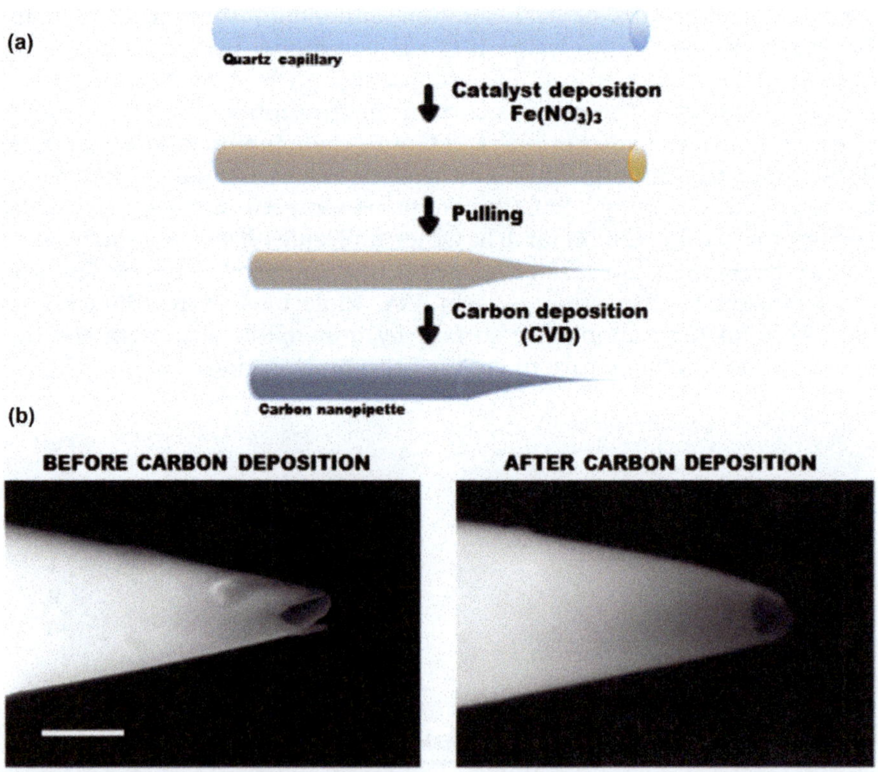

Figure 5.6 (a) The preparation of a carbon coated nanopipette through the chemical vapour deposition method. (b) SEM images of the nanopipette before and after carbon coating.
Reproduced from ref. 26 with permission from the Public Library of Science, Copyright 2013.

it into a hydrofluoric acid buffer solution and thus the inner carbon nanopipette was exposed at the tip. The length of the exposed carbon tip at the nanopipette was mostly determined by the etching time and temperature in the HF solutions. Further plasma oxidation was employed to decrease the diameter of the carbon nanotip and the tip diameter was reduced ranging from several tens to hundreds of nanometres. The prepared carbon nanopipettes could act not only as a resistive-pulse sensor, rectification sensor, and electrochemical probe for monitoring catalytic reactions, energy storage processes, and gated electron transfer processes, but also as a single cell probe and a injector for examining cellular activities.[25]

Compared with carbon-coated nanopipettes, metal-functionalized nanopipettes can be fabricated by many strategies. Through a specific chemical reaction such as an enzyme-catalysed reaction, an ultrathin gold film was formed on the inner wall.[27] Taking advantage of the photochemical property of metals, a metal layer was also deposited on the inner surface.[28] To optimize the homogeneity and compactness of the metal-based electroactive interface, in our group, we have mainly fabricated metal coated nanopipettes by magnetron sputtering and ion/electron beam evaporation, which are well controlled by adjusting the rate and duration of evaporation.[29,30] A detailed characterization of a metal-functionalized nanopipette is shown in Figure 5.7. The conductive layer provides many new detection mechanisms for electrochemical processes and new applications. For example, a silver-coated nanopipette, as a nanopore electrode, was employed to monitor the

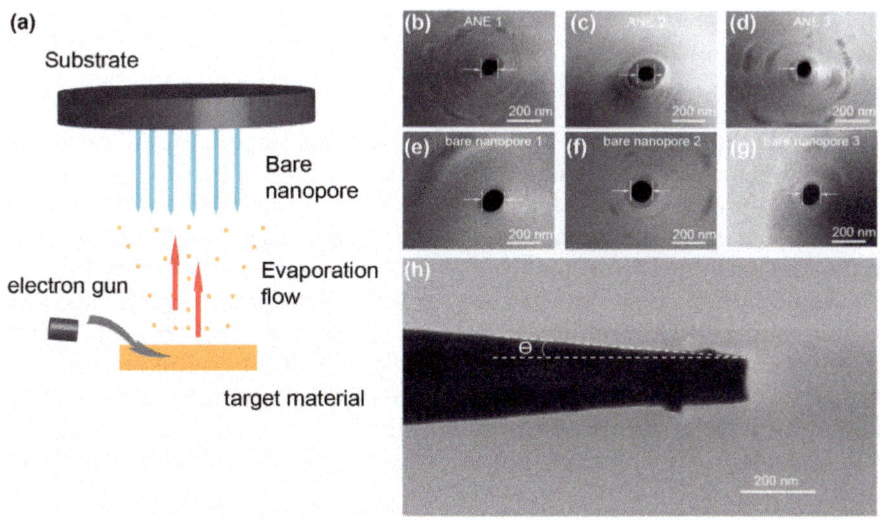

Figure 5.7 (a) The schematic representation of metal coating on nanopipette though electron beam evaporation. (b)–(g) Top view of SEM images of the metal coated nanopipette (b)–(d) compared with the bare nanopipette (e)–(g). (h) The side view of the metal coated nanopipette. Reproduced from ref. 30 with permission from the American Chemical Society, Copyright 2018.

generation of small molecules and ions.[13] Further, this metal coated nanopipette achieved nanoparticle self-assembly in solution *via* the template effect of nanopores and facilitated chemical modifications that were used for ion detection.[13] The detailed mechanism of this nanopore electrode will be explained in the next section.

Except for the conductive features of the metal coated film, silver or gold coating nanopipette facilitates further chemical modification, which enables various kinds of specific sensing platforms. For example, a silver coated nanopipette was employed for the fabrication of a nanopore electrode and further used for the detection of mercury ions (Figure 5.8).[13] Moreover, the silver coated nanopipette was functionalized with macrocyclic dioxotetraamines and used for selective sensing of target metal ions.[29] The modification process was performed as follows: first dissolve the molecules into degassed ethanol solution and fill them into a nanopipette, incubate for 24 h at N_2 atmosphere and then wash the nanopipette with pure solvent. The nanopipette was used for electrical sensing of mercury ions after it was fully dried with N_2.

A functionalized gold coated nanopipette was prepared using a similar procedure. A 4-thiol-catechol-modified gold nanopipette was prepared by filling the nanopipette cavity with catechol solutions and washing repeatedly using ethanol and water solutions. Successive modification was facilitated by the Au–S bond. Note that in this research, catechol modification only occurred at the inner side surface of nanopipette, which was implemented by removing the outer gold on the nanopipette. Specifically, a gold coated nanopipette was firstly deposited with 0.1 M HCl solution before catechol modification and thus only the modification on the inner side surface remained.[30] Moreover, the metal coated nanopipette could be functionalized with biological molecules, such as protein, DNA, *etc.* by a covalent bond. For example, target antibody was modified on the nanopipette tip followed by a step-by-step procedure (Figure 5.9).[18] First the nanopipette was modified with mercaptoacetic acid to introduce an active carboxyl group on the other terminal of the molecules. A mixed solution of 1-ethyl-3-(3-dimethylaminopropyl) carbodiimide hydrochloride (EDC) and *N*-hydroxysuccinimide (NHS) was employed to activate the carboxyl group and promote the reaction with amine group on the antibody. A thoroughly washing was performed before adding each new solution into the nanopipette. Through this method, the antibody was immobilized on the gold coated nanopipette uniformly and anchored target antigen molecules specifically. The modification process was confirmed by the current–voltage response and other mass methods.

5.3 The Nanopipette-based Wireless Nanopore Electrode (WNE)

The study of bipolar electrochemistry has grown extensively in the last two decades.[30] In comparison with the conventional electrochemical cell with

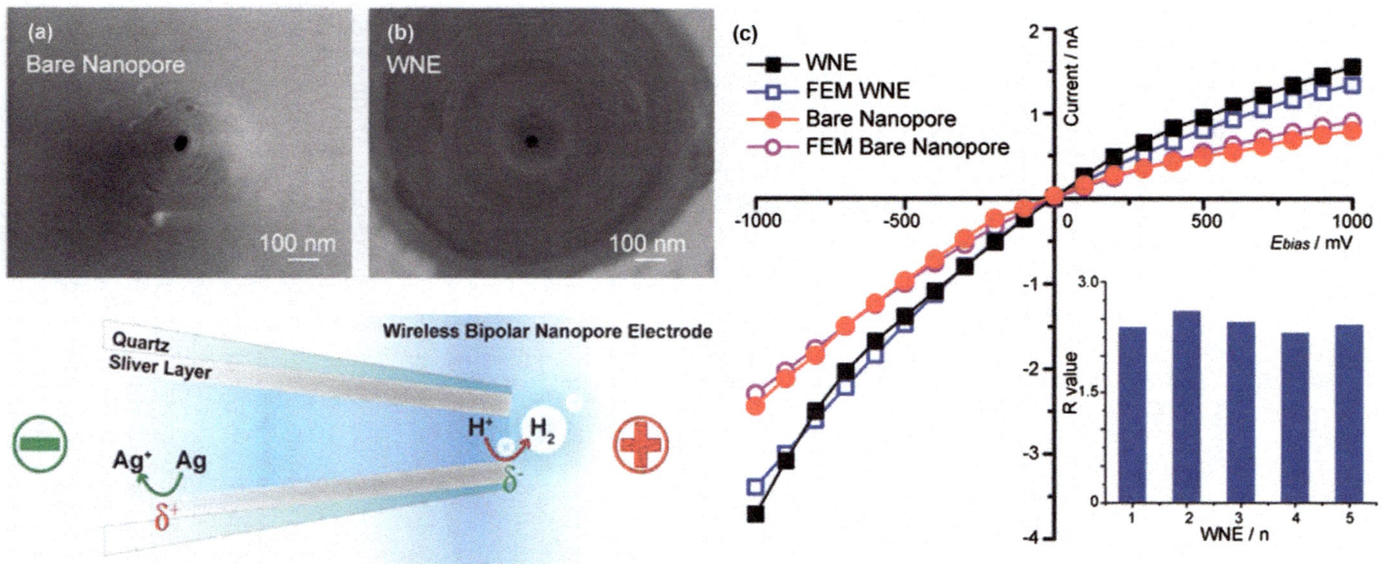

Figure 5.8 The characterization of wireless nanopore electrode. It was performed by coating the nanopipette tip with silver film using the electron beam evaporation technique. The SEM images of the prepared (a) bare glass nanopore and (b) wireless nanopore electrode. (c) The I-V curves of these two kinds of nanopores and the corresponding FEM simulations. Inset: R values of five independent wireless nanopore electrodes.
Reproduced from ref. 13 with permission from the American Chemical Society, Copyright 2017.

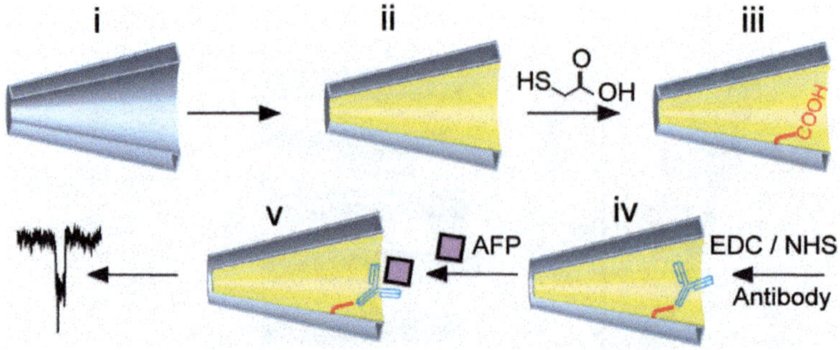

Figure 5.9 The preparation of an antibody modified gold nanopipette followed a step by step modification procedure.
Reproduced from ref. 18 with permission from the Royal Society of Chemistry, Copyright 2017.

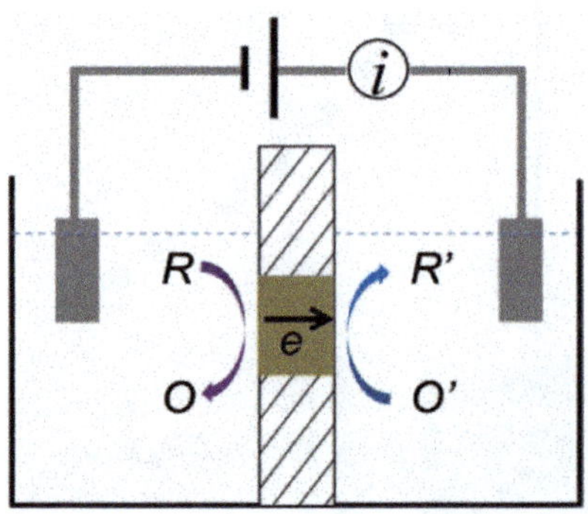

Figure 5.10 Comparison of a closed bipolar cell.
Reproduced from ref. 32 with permission from the American Chemical Society, Copyright 2012.

two or three electrodes connected to the power supply, the bipolar electrode is placed in the electric field and its two sides are polarized to cathodic pole and anodic pole respectively.[31] Thus, a pair of redox reactions occur at the two polarized sides under the electric field between the driving electrodes.[32] At present, bipolar cells are specifically classified into two main categories: open bipolar and closed bipolar.[33] In the open bipolar cell, the electrode is placed inside the channel or in bulk electrolyte solution. Both ionic current and electronic current are driven by an electric field due to coupled faradaic reactions, which are in the electrolyte solution and inside the bipolar electrode, respectively. As such, the rate of chemical reaction that occurs on the bipolar electrode cannot be measured in the external circuit. In another bipolar configuration, the two sides of the closed bipolar electrode are separated in two different compartments by eliminating the microfluidic path, and only the electronic current can be measured between the driving electrodes. A typical closed bipolar cell is illustrated in Figure 5.10. The cell contains a pair of driving electrodes that could apply a potential across the two separated solutions. When two ideal none polarizable electrodes were used, *i.e.* Ag/AgCl electrodes, most of the voltage was dropped at the two ends of the bipolar electrode. A relatively small voltage (less than 1 V) was applied to drive the electrochemical reactions on the bipolar electrode. In this case, the redox reaction rate on the bipolar electrode could be monitored by the entire electrical current that passed through the bipolar electrochemistry (BPE).

With the advantage of bipolar electrochemistry, our group proposed a wireless nanopore electrode (WNE) that did not require the sealing of a traditional metal wire inside a nanoelectrode. The prepared WNE further broadened the applications of nanopipettes to single entity sensing including single molecules, single nanoparticles and single cells (Figure 5.11).[34] With the metal deposition onto the nanopipette tip, two kinds of WNEs were prepared including both open-type WNE and closed-type WNE. In this kind of bipolar electrochemistry, when an external electric field was applied in the solution across the metalized nanopore, the two terminals of the WNE were polarized and thus generated the polarization

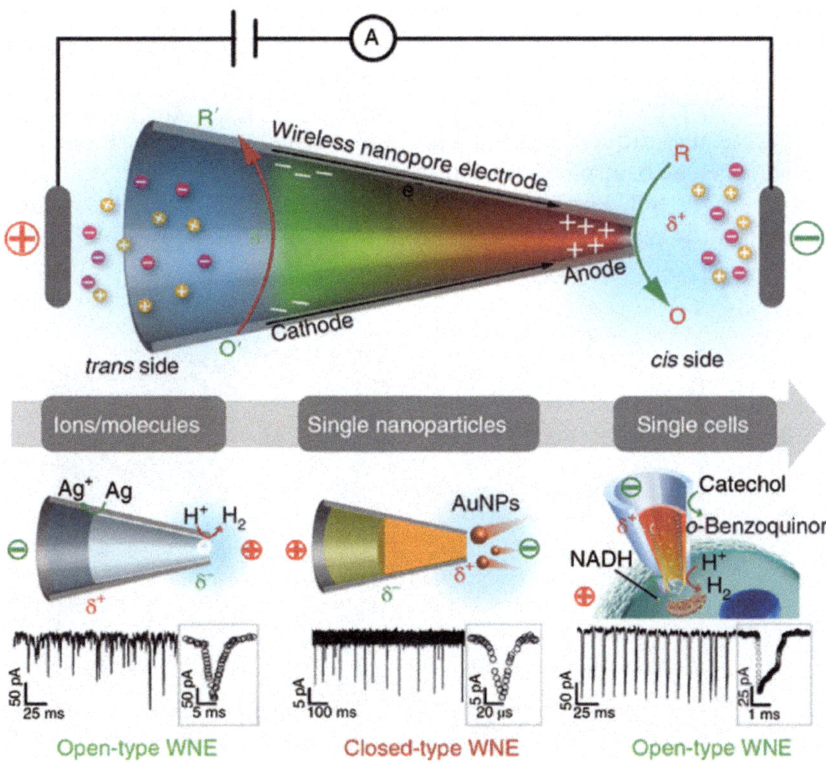

Figure 5.11 Schematic illustration of the presented wireless nanopore electrode, including the principles and the broad applications.
Reproduced from ref. 34 with permission from the Springer Nature, Copyright 2019.

potential difference. Subsequently, the two sides of WNE behaved as anode and cathode respectively, which enabled the bipolar redox reactions at the two terminals. This presented WNE provides a highly confined electric field at the metal pipette tip, which would further broaden the electrochemical applications of nanopipette.

With a driving voltage applied across a wireless nanopore electrode, an almost voltage drop occurred within the narrowest region at the tip of the nanopore (Figure 5.12). The potential decrease caused by the conductance from the electrolyte solution was very limited. Under an external electric field, two terminals of the WNE were polarized by taking advantages of bipolar electrochemistry. There was a polarization voltage difference between the two ends of the WNE along the orientation of the external electric field. Electrochemical reactions occurred at the anode and cathode of the WNE when the difference in solution potential across the WNE was sufficient for a pair of faradaic reactions. In detail, at the end that acted as the cathode, the molecules were reduced at the electrodes, and the molecules at the other end that acted as the anode were oxidized. Importantly, the reduction reaction was correlated to the oxidation reaction with the same rate. Thus, the redox reaction could occur under a relative low bias potential which benefitted from the geometry advantage of nanopores. Moreover, different bipolar reactions could be achieved by directly regulating the bias potential. Therefore, electrochemical measurement was enabled by the inexpensive, safe and simple WNE techniques. Apart from all the advantages mentioned above, the WNE was also applied to acquire complicated dynamic redox information a single cell. For example, the asymmetric nanopore electrode was used to monitor the respiration chain (*i.e.* NADH) in living cells.[30]

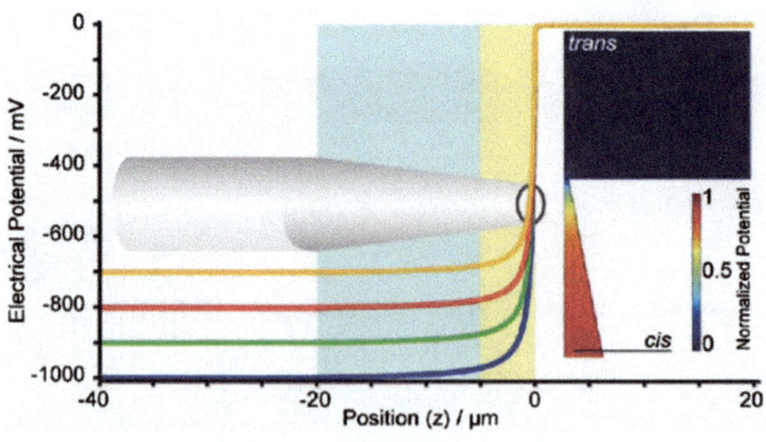

Figure 5.12 The FEM simulation of the distributed biased potential along the position of the nanopore.
Reproduced from ref. 13 with permission from the American Chemical Society, Copyright 2017.

In the past two decades, solid-state nanopores have been widely utilized for the detection of diverse single entities,[36] generally based on the resistive-pulse sensing mechanism. In detail, when a single molecule is driven across the confined nanopore by a bias potential,[37] the flux of ions in the nanopore will be blocked, with a temporary pulse in the current baseline. The pulse signals in the current are correlated with the translocation of single molecules, and thus facilitate the single molecule detection.[13,38] The comparable dimensions of nanopores and the target analytes would improve the signal-to-noise ratio in nanopore sensing. However, it is still challenging to prepare a solid-state nanopore with a diameter of less than 2 nm, which limits the sensing of small molecules. In recent years, a wireless nanopore electrode was proposed and extended as a versatile tool for the detection of single molecules and ions (Figure 5.13).[13] H_2 nanobubbles were firstly detected by a silver-coated WNE. A pair of Ag/AgCl electrodes was used as working electrode and reference electrode respectively to supply a millivolt bias potential. As above-mentioned, a pair of redox reactions would occur at the nanopore tip due to the high electric field at the very tip of nanopore. Under this electric field, the following two reactions [eqn (5.2) and (5.3)] occurred at the two terminals of the silver coated WNE:

$$2H^+ + 2e \rightarrow H_2(g) \qquad E = 0 \text{ mV } \textit{versus} \text{ SHE} \tag{5.2}$$

$$Ag(s) \rightarrow Ag^+ + 2e \qquad E = 799.6 \text{ mV } \textit{versus} \text{ SHE.} \tag{5.3}$$

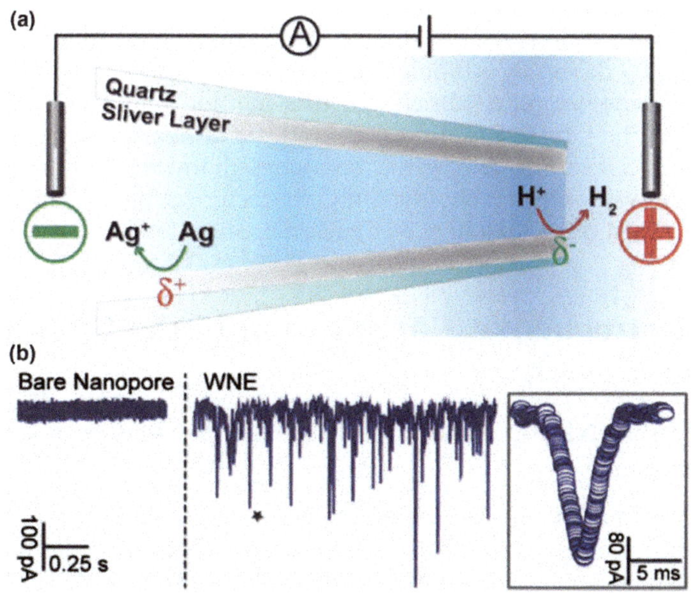

Figure 5.13 (a) Illustration of the redox reactions at the WNE under a bias potential. (b) The raw current traces of the bare nanopore and the WNE at −800 mV. Reproduced from ref. 13 with permission from the American Chemical Society, Copyright 2017.

The translocation of H_2 nanobubbles through the tip of conical nanopore resulted in the increase of ionic conductivity due to the high density of charges at the bubble surface. Thus, enhanced spike signals were observed with the generation of H_2 in solution. This showed that there were distinct signals in the ionic current with the bias potential higher than −799.6 mV, which revealed the events of redox reaction at the wireless electrode. It was demonstrated that 14 H_2 molecules and 28 Ag^+ were distinguished from each spike signals at the potential of −800 mV. While with positive potentials applied, H_2 nanobubbles were generated at the other terminal of the WNE since it polarized as the cathodic electrode. The generation of H_2 at the terminal away from nanopore hardly contributed to the ionic conductivity through the nanopore, and thus few spike signals were observed on the ionic current. Therefore, the wireless nanopore electrode proposed a new sensing mechanism with simple, label-free and ultra-sensitive features for various small molecules detection.

Nanopores with smaller dimensions enable more possibilities for precise analysing of small-size objects. However, it still remains a challenge for conventional approaches for coating or modifying the nanopipettes with diameters smaller than 50 nm. Recently, a bipolar electrochemical based method for dynamically fabricating a metal nanowire inside a nanopipette has attracted increasing attention in both the nanopore and electrochemistry fields.[39–41] As shown in Figure 5.14, a metal precursor and reducing agent were placed on each side of a nanopipette, and the reduction of metal precursor ion effectively promoted the growth of metal, blocking the nanopore orifice. Thus, the confined nanoelectrode was prepared with high reproducibility driven by a bipolar electrochemical process. Compared with traditional nanoelectrode fabrication, this method had advantaged in the simplicity of the fabrication procedure and the dimensions of the electrode could be strictly controlled by the nanopore diameter. This prepared nanoelectrode was used for the dynamic characterization of single gold nanoparticles by a new capacitive feedback response detection mechanism.[42]

5.4 Electrochemical Application of the Confined Nanopipette

5.4.1 Single Nanoparticle Sensing in the Nanopore Confined Space

Nanoparticles have wide applications in various nanoscale research because of their high surface-to-volume ratio, excellent electroactivity and optical features, which are highly related to their shape, size, surface charge.[11,43] Despite numerous traditional ensemble measurements being developed, it is still challenging to monitor the intrinsic features of single nanoparticles. New methods for high-efficiency and high-sensitivity measurement need to be developed to understand the fundamental features of individual

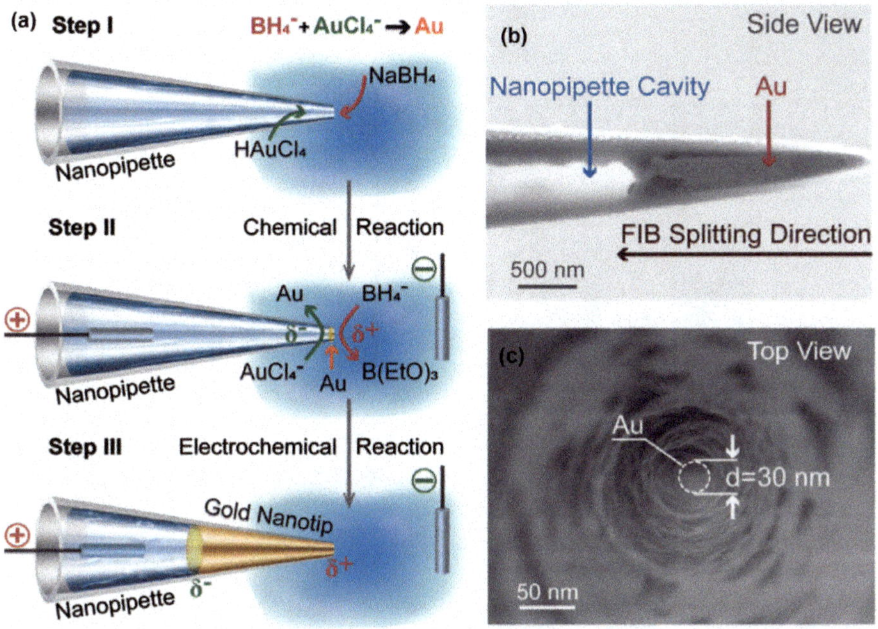

Figure 5.14 The preparation of confined nanopore electrode by an electrochemical reduction process.
Reproduced from ref. 42 with permission from the John Wiley and Sons, Copyright 2018.

nanoparticle in catalysis, cellular analysis, environmental monitoring, *etc.*[44] In this section, we review the new nanopipette-based methods for the analysis of single nanoparticles.

Researchers have paid great attention on the sensing of single nanoparticles by nanopipette due to the benefits of ease and low-cost fabrication process, controllable diameters and the electrochemistry confinement at the tip. The challenge lies in how to drive particles throughout the nanopore with an unchanged surface. Based on the characteristics of the tapered channel, the nanoparticles could be driven by the applied pressure (Figure 5.15a). By changing the direction of the pressure, individual nanoparticles with diameters of 120–160 nm were repeatedly translocated through the nanopore, which could be further used to study the relationship between the diffusion behaviour of nanoparticles under various forces and applied pressures (Figure 5.15b).[45] For surface-charged nanoparticles, the electrostatic interaction with the pore surface was investigated by analysing the current signal of the pressure-induced particle translocation. It was proved that the blocking current was greatly affected by the surface charges of both the channel material and the nanoparticles. Nanopipettes have been previously used to distinguish polystyrene spheres with radii of 60 nm and 160 nm.[46] A symmetric blockade current was observed when the particles

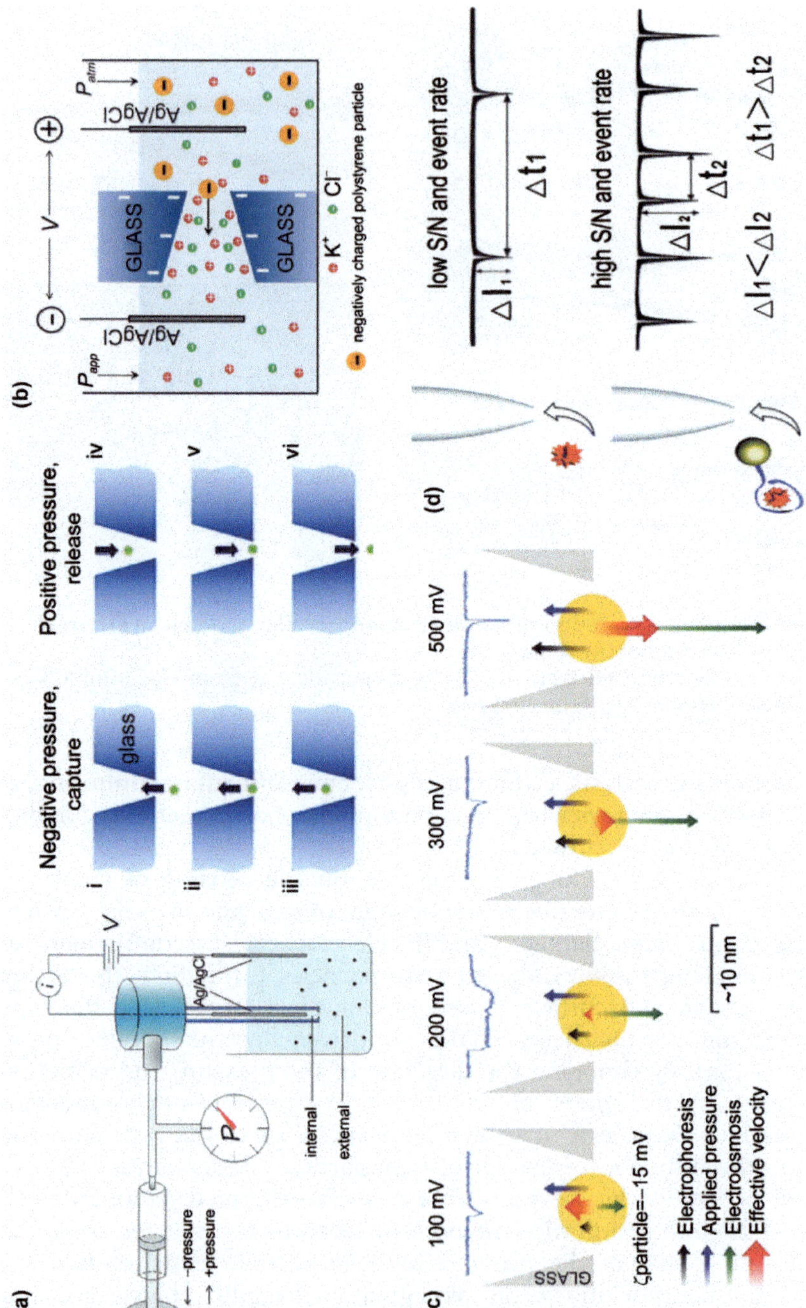
(a)
← −pressure
→ +pressure
P
i
V
Ag/AgCl
internal
external
Negative pressure, capture
Positive pressure, release
glass
i
ii
iii
iv
v
vi
(b)
P_app
P_atm
Ag/AgCl
Ag/AgCl
V
GLASS
GLASS
K+
Cl−
negatively charged polystyrene particle
(c)
100 mV
200 mV
300 mV
500 mV
GLASS
ζparticle=−15 mV
~10 nm
Electrophoresis
Applied pressure
Electroosmosis
Effective velocity
(d)
low S/N and event rate
ΔI1
Δt1
high S/N and event rate
ΔI2
Δt2
ΔI1<ΔI2
Δt1>Δt2

were traversing through the nanopore, which indicated that the passing polystyrene beads were symmetrical spherical. Further studies found that a negative voltage as a driving force in this system resulted in double pulse shaped signals that was first enhancement in the ionic current followed by a current blockage. This phenomenon was mainly caused by the surface charge of polystyrene beads. When the negatively charged polystyrene beads passed through the nanopore, the current enhancement caused by the surface charge of particles surpassed the volume-induced ion exclusion, resulting in a current-enhanced signal. As the charged particles slowly flowed toward the inside of the pore, the current blockage caused by ion exclusion gradually became dominate in sensing; the role of current enhancement decreased, resulting in a double pulse signal with a current increase firstly and then decrease.

The intensity of the bidirectional signal can be increased by increasing the ionic strength and the applied pressure.[12] Although the translocation of nanoparticles was difficult to drive under low voltages, the employment of pressure increased the signal frequency greatly.[47] In addition, the concentration of nanoparticles also played an important role in the frequency of translocation. It showed that for nanoparticles with sizes between 80 and 160 nm, the logarithm of the concentration was linear with the logarithm of the event frequency with the particle concentration in the range of 10^7 to 10^{11} mL^{-1}.[46] The force of the pressure applied in this system was mainly determined by the size of the nanoparticles. The electroosmotic flow of the particles in the orifice was enhanced by the particle size increasing, which resulted in larger current amplitude of the particles. Under this discipline, the translocation rate of the nanoparticles could be slowed by applying an external force (pressure) opposite to the electric field force, and the translocation time could be extended, which improved the resolution of nanopore sensors (Figure 5.15c).[48] On the other hand, modification of the translocated nanoparticles with a specific nucleic acid aptamer could be used to improve the selectivity of the nanopore. The targeted lysozyme molecules that interacted with the surface aptamer could be easily trapped by the pore,

Figure 5.15 (a) Schematic illustration of the nanopipette membrane (GNM), and particle capture and release method using a three-part pressure waveform and the schematic of the particle translocation and resulting *i–t* recording. Reproduced from ref. 12 with permission from the American Chemical Society, Copyright 2014. (b) Schematic illustration of negatively charged polystyrene particles translocating through a nanopipette (not drawn to scale) in a KCl solution. Reproduced from ref. 45 with permission from the American Chemical Society, Copyright 2012. (c) Schematic depicting control of nanoparticle velocity in conical nanopores. Reproduced from ref. 48 with permission from the American Chemical Society, Copyright 2013. (d) Schematic demonstrating the enhanced detection rate for the AuNP–LBA/lysozyme complex in comparison to that of free lysozyme. Reproduced from ref. 49 with permission from the Royal Society of Chemistry, Copyright 2017.

whereas the non-specific molecules were difficult to capture (Figure 5.15d). More importantly, the target molecules adsorbed on the nanoparticle surface after interaction and traversed the nanopore together with particles. Thus, the modification of nanoparticles could enhance the amplitude of the signal, and further benefit the signal amplification in nanopore sensing.[49]

Besides the classical translocation events generated by the particles passing through the nanopore, particles with larger diameters than the nanopores can also be identified by the corresponding characteristic signals in the ionic current. It was studied that the particles with slightly larger diameter than the nanopores were driven near the tip orifice by the applied voltage. And a stepped current blocking signal was observed as the particles were trapped and adhered by the nanopores. While for the nanoparticle with diameter much larger than the pore size, it left from the pore orifice after collision, thus producing a transient current-blocking pulse signal (Figure 5.16a)[50]

Due to the controllable diameter of the nanopipette, many kinds of single entities could be studied using this method such as vesicles,[51] nanobubbles[52] and nanoprecipitations.[53] In 2008, a solid state nanopore was first employed to study the formation and dissolution of nanoprecipitates in the

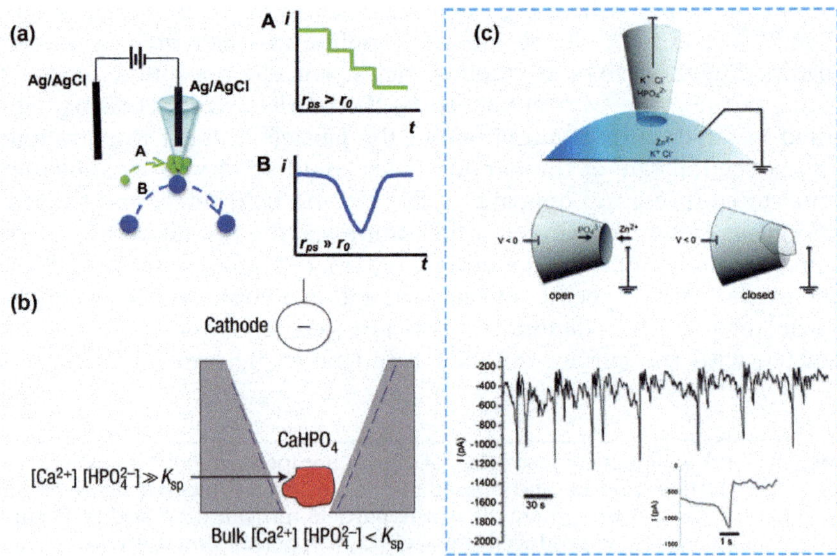

Figure 5.16 (a) The application of a glass nanopipette for the sensing of single nanoparticles with diameter larger than the pore. Reproduced from ref. 50 with permission from the Royal Society of Chemistry, Copyright 2016. (b) The formation and dissolution of nanoprecipitates inside the confined nanopore with the addition of phosphoric acid. Reproduced from ref. 53 with permission from the Springer Nature, Copyright 2008. (c) Measurement of ion current oscillations in a nanopipette to reveal the nanoscale reaction information. Reproduced from ref. 54 with permission from the American Chemical Society, Copyright 2011.

presence of phosphoric acid on both sides of a 2–6 nm tapered Teflon nanopore (Figure 5.16b).[53] With the addition of buffer solutions into the chamber and 0.1 mM calcium ions, both calcium and phosphate ions were driven to flow toward the nanopore under an applied negative voltage. These two kinds of ions were concentrated and precipitated at the confined nanopore orifice even though these two ions in the solution were insufficient to reach the precipitate solubility. A distinct current blockage was consequently observed during transient ionic current recording. The formation of pore-induced precipitate decreased the further flow of these two ions toward the orifice, thus lowering the ion concentration near the precipitate. The precipitate dissolved gradually in order to achieve the equilibrium of dissolution near the nanopore until the following precipitation occurs. Subsequently, zinc phosphate precipitation was prepared with a 40–60 nm nanopipette, demonstrating the same current oscillation in the current time traces.[54] It further indicated that nanopore technology could be used to detect nanoscale reaction information, which could hardly be obtained by the conventional methods (Figure 5.16c).

Nanopipettes are also a powerful approach to monitor the dynamic process of nanobubble formation and dissolution with high spatial and temporal resolution. A confined nanopipette was employed to track the single hydrogen nanobubble induced by the chemical reaction between $NaBH_4$ and H_2O.[55] The diagram of this experiment is shown in Figure 5.17. A cylindrical nanopipette with a diameter of ~90 nm was employed in this method, which was capable of tracking single entities with high sensitivity. The $NaBH_4$ dissolved in ethanol was injected into the nanopore lumen in order to confine the generation of nanobubbles at the nanopipette orifice. Tetrabutylammonium hexafluorophosphate ($TBAPF_6$) was used as electrolyte to conduct the ionic current in nanopore. The tip of nanopipette was immersed into a PBS solution to induce the reaction of the hydrogen nanobubble generation. The spontaneous generation of nanobubbles at the nanopore tip induced the periodic ionic current oscillation. The sharp increase in ionic current and the current blockage revealed the growth dynamic of the nanobubbles. In each period, the increased pulse was defined as conductive current (I_C), and the blockage pulse was defined as resistive current (I_R). There were no pulse-like signals in the absence of $NaBH_4$, which verified that the current oscillation originated from the hydrogen nanobubbles. The growth time and formation time of a single hydrogen nanobubble were estimated by analysing the statistic results of the ionic current oscillation. It showed that both the growth time and formation time were independent of the applied potential. In addition, the growth process of the nanobubble can be modulated by varying the nanopore size. In summary, the nanopipette has the capability of tracking the dynamic behaviour of a single bubble at the nanoscale with high spatial and temporal resolution.

In recent years, the electrochemical method has been employed to further acquire the individual particle properties, and the two common approaches

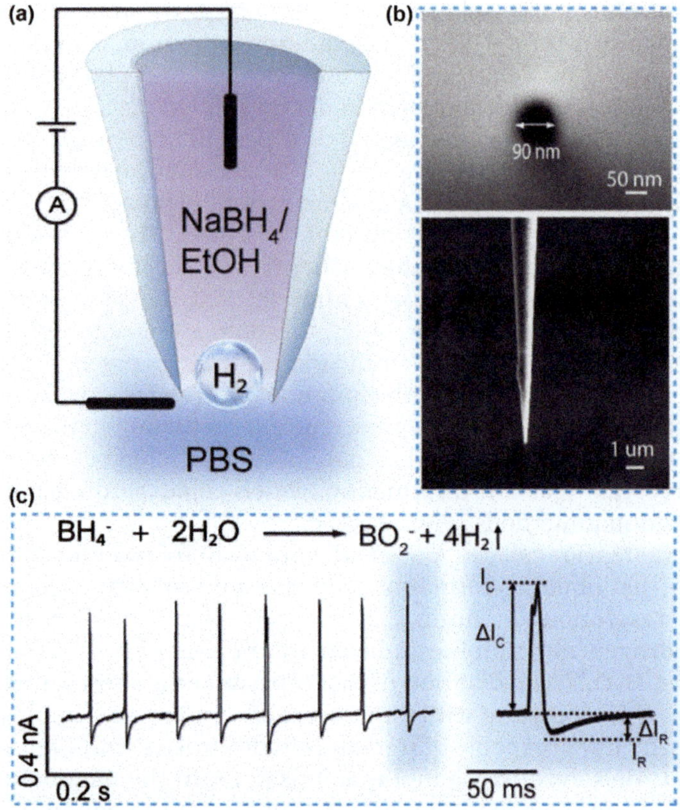

Figure 5.17 (a) Setup for nanopore characterization of the dynamic growth of nanobubbles. (b) Top and side views of SEM characterization of the nanopipette. (c) The biphasic pulse with both conductive current (I_C) and resistive current (I_R) at an applied potential of 0 mV.
Reproduced from ref. 55 with permission from the American Chemical Society, Copyright 2018.

are single nanoparticle collision and single nanoparticle immobilization respectively. The former showed the redox properties of single nanoparticles, and the latter studied the electrocatalytic properties by immobilizing a single nanoparticle on the ultramicroelectrodes surface.[105] Both of these methods can facilitate the study of the electron transfer process of a single particle. However, there are still several challenges in this field. First, high current noise submerges the current fluctuation information which reveals the dynamic interactions of nanoparticles. Second, each electrode has a difference in morphology and dimensions, which induce inconsistent ability of the current response. The uncontrollable interface also causes the deviation for each measurement due to the inconsistent interactions between the single nanoparticle and the electrodes. Third, a number of transient events are undetectable due to limited measurement bandwidth, which results in the relatively low frequency of single particle collision. High throughput

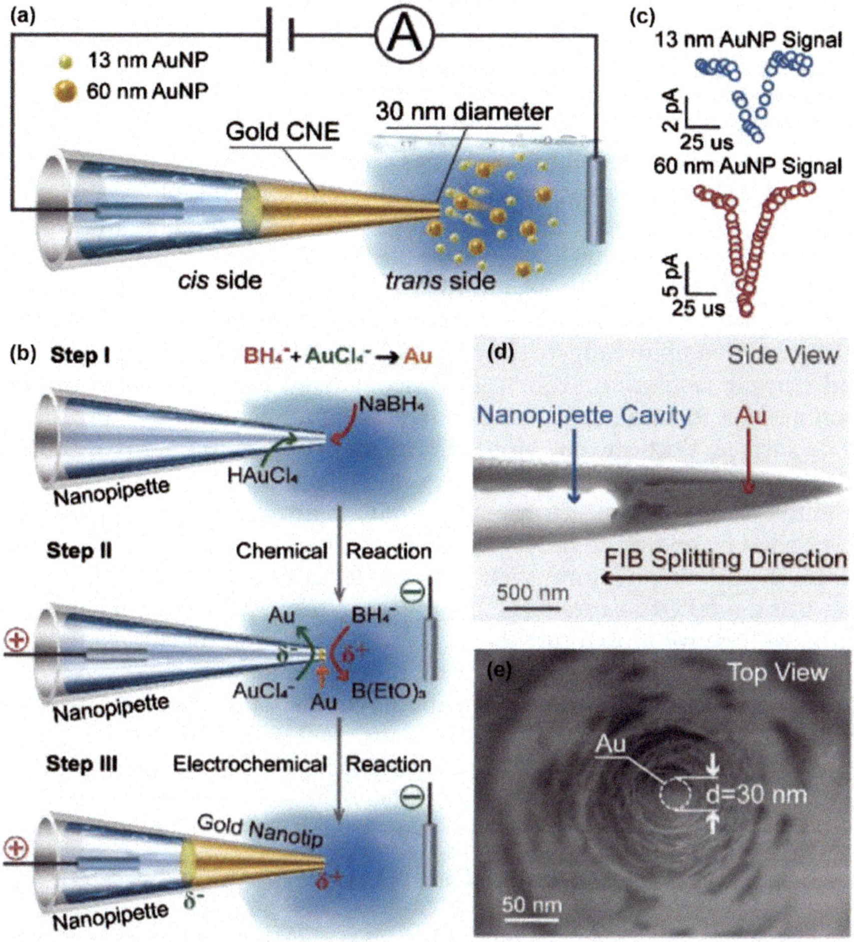

Figure 5.18 (a) The scheme of the CNE for revealing the intrinsic and dynamic collisions of AuNPs in the mixture. A pair of Ag/AgCl electrodes were used to apply the bias potential. (b) The controllable fabrication process of the CNE. (c) The typical single nanoparticle collision events for 13 nm (blue) and 60 nm AuNPs (red), respectively. (d) The side view SEM image of the CNE after FIB splitting along the nanotip. The scale bar is 500 nm. (e) The top view SEM image of the CNE with a diameter of 30 nm. The scale bar is 50 nm.
Reproduced from ref. 42 with permission from the John Wiley and Sons, Copyright 2018.

detection of individual nanoparticles at large scale is hardly achieved under the obstacle of this restriction.

The wireless nanopore electrode is employed to overcome these difficulties, which reveals the interactions at nanoscale in a high temporal and current resolution. The WNE, a 30 nm gold nanoelectrode was fabricated by a chemical-electrochemical method (Figure 5.18a). This rapid fabrication process can be measured in real time by simultaneously observing the dark-field images/scattering spectra and the ionic current variation. The gold nanopore tip was *in situ* generated *via* the chemical reaction between chloroauric acid and sodium borohydride by means of the successive electrochemical reduction (Figure 5.18b). After fabrication, the current returned to 0 pA since the tip of nanopipette was completely blocked by the gold. The WNE achieved high sensitivity detection for single entity collision with a low noise level with a root mean square value of 0.6 ± 0.1 pA. For single particle sensing, the WNE was utilized to monitor the intrinsic and dynamic features of collision between particles and the nano-interface of the WNE. Due to the excellent morphology and the capacitive feedback response (CFR) effect, this WNE was used to investigate the complicated interactions in a high temporal and current resolution. The WNE was polarized under a positive bias potential, thus forming a negative-charged double electric layer on the electrode surface. With the collision of a charged particle to the tip of the WNE, the charging–discharging process would be changed and a spike signal was monitored on the ionic current. Different diameters of particles were distinguished by analysing the signal current amplitudes and the current peak width. By utilizing a 30 nm WNE, the single nanoparticles collision process was distinguished to reveal the complicated process in the collision. Compared with conventional methods, the WNE detection mechanism achieved high sensitivity and temporal resolution detection which overcame the limitation of bandwidth and the inhomogeneous interface. In conclusion, the WNE with the small dimensions and well-defined interface could be used to for the detection of cells and particles with less mechanical damage, high spatial resolution and high sensitivity.

5.4.2 Small Biomolecules Detection with a Nanopipette

As small-sized nanopipettes were successfully prepared, biomolecules such as DNA could be studied thoroughly at single molecule level. The folded and unfolded morphology of the λ-DNA molecules could be identified by analysing the characteristic ionic current signals.[8] It was found that the ionic strength of the electrolyte had an important effect on the DNA translocation. In a high concentration of KCl solution (400 mM to 2 M), the translocation of the DNA molecule resulted in a typical current blockage. By contrast, a distinct increase in the ionic current was observed instead as the electrolyte concentration was lowered to about 10–300 mM. This indicated that the effect of surface charge in individual DNA molecules and inside the channel greatly surpassed the effect of volume exclusion, becoming the dominate factor for current increase under such low salt concentration.[56] Moreover, a nanopipette array was prepared to achieve high-throughput detection. Nanopipettes were enclosed in a ring shape and shared a liquid pool by designing a special PDMS detection cell (Figure 5.19a).[57]

The combination of nanopipette with the microfluidic technology could increase the capture rate in sensing, with the microfluidic channel containing the DNA molecule being perpendicular to the nanopore. Thus, high-speed flowing DNA molecules were captured directly by the electric field force, which shortened the detection time significantly (Figure 5.19b).[58] By introducing an alternating electric field into a conventional DC electric field, the gold-plated conical nanopores were used to "electrically capture" the DNA molecules, enabling real-time measurement of ultra-low concentration samples (Figure 5.19c). The sensing of DNA molecules at a concentration of 5 fM was reached by this method, which was at least three orders of magnitude lower than the detection limit of the existing methods.[59] By utilizing the confinement effect of the conical nanopipette, the transport of the target substance could be precisely controlled by applying an asymmetric pulse voltage. The dynamic process during transport was further observed with the integration of fluorescence imaging methods (Figure 5.19d).[60] In recent years, the θ-type quartz capillary has been increasingly used for electrochemical sensing, which offers two identical channels at the nanopore tip. As shown in Figure 5.19e, by applying opposite voltages in the two channels, the detailed migration of DNA molecules was well regulated from one channel into the other through the outside solution. With the electric field force attracting back, a "nanobridge" was constructed on the nanopore by using DNA molecules.[61]

Given the difficulty of detecting small molecules such as short-chain DNA in the glass nanopipette, the advantages of highly controllable DNA sequences are utilized for small molecular sensing. By designing a special structure of DNA carriers, target molecules were bound by interactions such as covalent interactions. Based on this DNA side chain, small molecule DNA was accurately measured by analysing the characteristic current signals with translocation through the nanopipette (Figure 5.20). For example, 7200 bp (base pair) single-stranded DNA was employed as carrier, and a central 152 bp was selected to design a pair of complementary stranded DNA. With specific design and the principle of complementary, biotin-labelled prominent double-stranded DNA was formed on the side chain. With the addition of a non-specific wild-type DNA molecule, the long-chain DNA carrier with a prominent double-stranded molecule translocated through the nanopore orifice. The existence of biotin molecules on the DNA carrier showed a distinct exclusion effect and resulted in a secondary current tip at the middle of the stepped blocking current (probe present). While in the presence of a mutated DNA molecule in solution, the attached biotin molecule on the prominent double-stand molecule was complemented and displaced, inducing a simple stepped blocking current during translocation (probe displaced). The number of mutations in the DNA molecule were thus measured by the ratio of these two current signals.[62] Moreover, special structures such as a short-chain DNA-motif could be directly measured by using this DNA carrier induced signal amplification.[63]

Nanopipettes have also been used for protein analysis and detection at single molecule level. It was reported that the label-free protein molecules were directly detected by using this method. A glass nanopipette with a tip

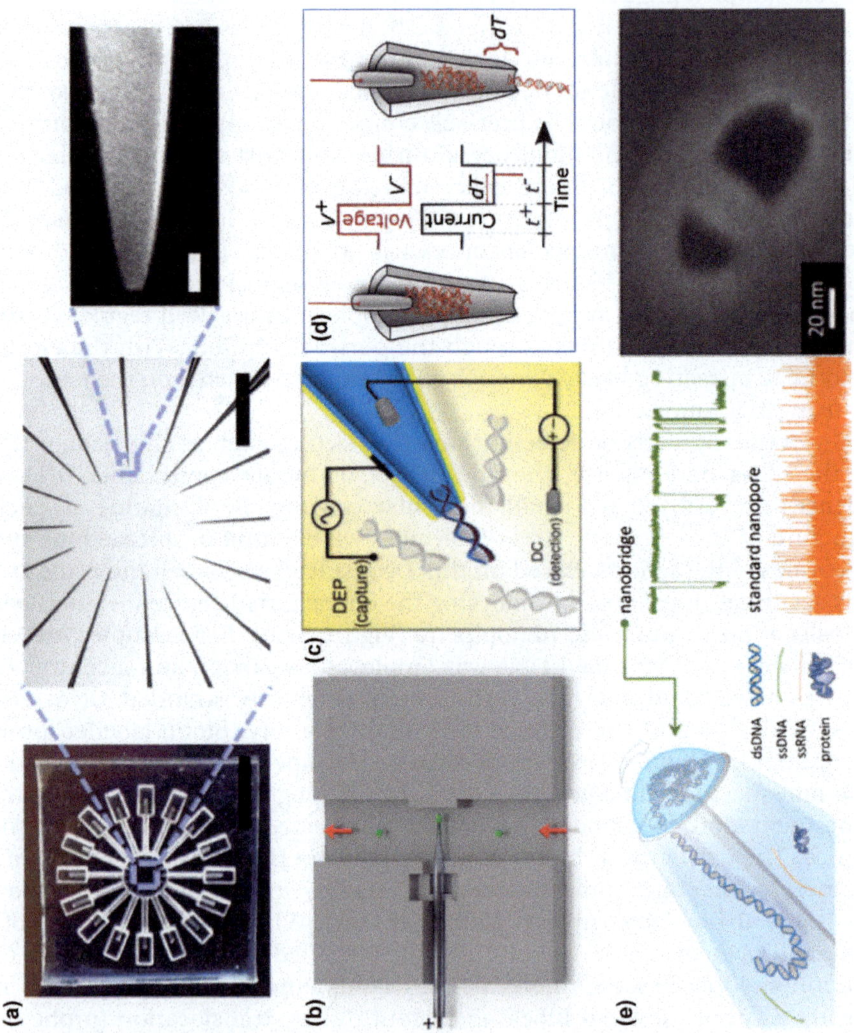

(a)
(b)
(c)
DEP
(capture)
DC
(detection)
(d)
Voltage
Current
Time
dT
(e)
nanobridge
standard nanopore
dsDNA
ssDNA
ssRNA
protein
20 nm

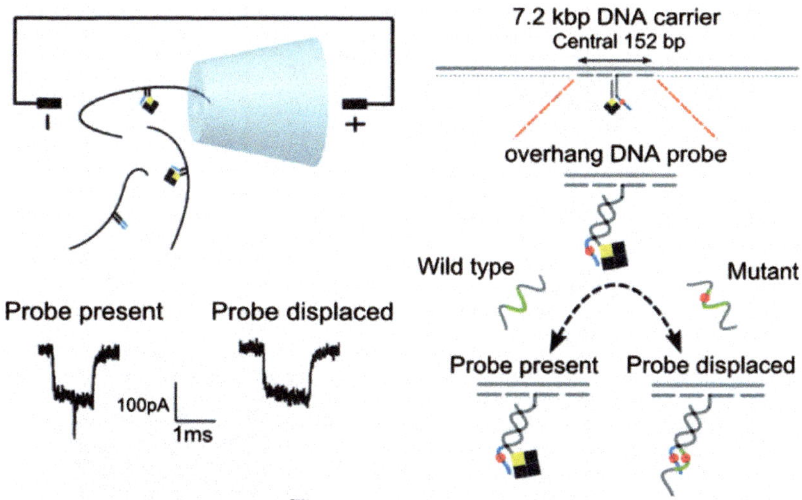

Figure 5.20 DNA detection using DNA carriers with nanopore sensing. Reproduced from ref. 62 with permission from the Royal Society of Chemistry, Copyright 2017.

diameter of 65 nm was used to directly distinguish the β-galactosidase, bovine serum albumin, β-lactoglobulin, immunoglobulin, avidin, lysin and ovalbumin in solution (Figure 5.21a).[64] Different from the previously mentioned detection of DNA molecules by nanopipette, the surface charge of protein differs greatly regarding the protein types, which indicates a significant difference in hydrophilicity and hydrophobicity. Therefore, the translocation rates of proteins are significantly different. Further considering the electrical and dielectric constants of the protein, the interaction between the translocated protein and nanopore is affected extremely, inducing the inconsistence of translocated signals of the same protein. Thus, accurate protein sensing using a nanopipette still remains challenging. Pore modification is firstly employed to optimize the detection of proteins in a nanopipette. The ability to sense immunoglobulins,

Figure 5.19 (a) Multichannel detection of double-stranded DNA. Reproduced from ref. 57 with permission from the Royal Society of Chemistry, Copyright 2013. (b) Schematic diagram illustrating the nanopipette integrated microfluidic device. Reproduced from ref. 58 with permission from the American Chemical Society, Copyright 2013. (c) Schematic of DNA being threaded through the tip of a gold-coated nanopipette. Reproduced from ref. 59 with permission from Springer Nature, Copyright 2016. (d) Schematic illustration showing the nanopipette delivery system. Reproduced from ref. 60 with permission from the American Chemical Society, Copyright 2015. (e) Schematic representation of the nanobridge formed at the tip of a nanopipette. Reproduced from ref. 61 with permission from the American Chemical Society, Copyright 2017.

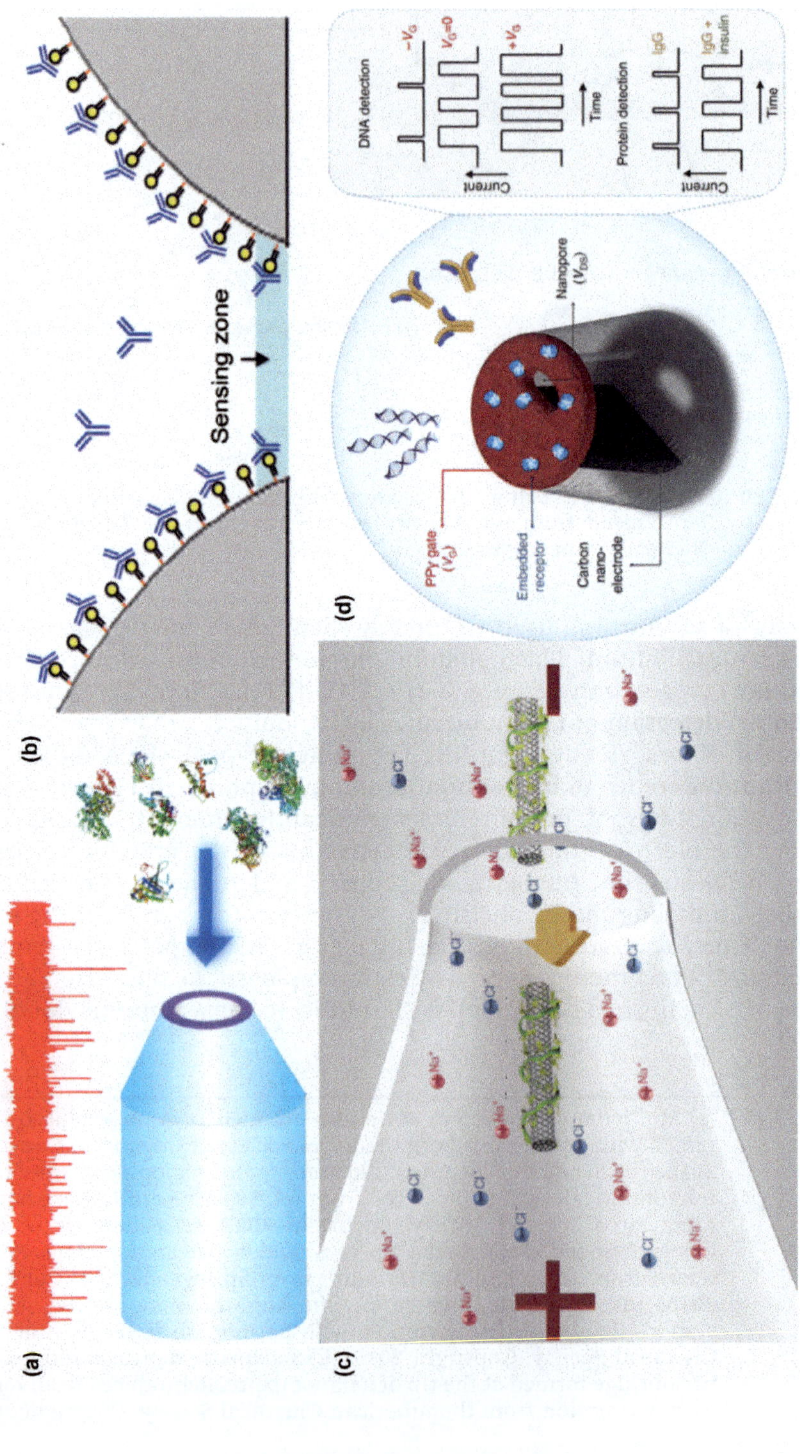
(a)
(b)
Sensing zone
(c)
(d)
PPy gate (V_G)
Embedded receptor
Carbon nano-electrode
Nanopore (V_{DS})
DNA detection
Protein detection
Current
Time
IgG
IgG + insulin

ricin, *etc.* was enhanced by modifying the aptamer at the tip of nanopipette (Figure 5.21b).[6] Likewise, biocompatible carbon nanotubes were employed for protein sensing, which could encapsulate target proteins on the tube surface *via* interaction. And the signal amplification was overserved with this interaction (Figure 5.21c).[65] A nanopore extended field effect transistor at the nanopore was developed and functionalized with a high molecular polymer, which enabled the selective sensing of target molecules such as an insulin antibody. The controllable gate voltage featured in this transistor was used to slow down the transportation of single molecules. This strategy shows great potential for selective sensing in the biological system (Figure 5.21d).[66]

On the other hand, the selectivity for protein sensing with a nanopipette could be easily improved by employing a DNA carrier. Similar to the previous DNA sensing, target proteins could be covalently bound to the long chain DNA carrier by delicate DNA design. DNA carriers with biotin groups were employed for specific sensing of streptavidin and the translocation of this complex through the nanopore induced an additional current spike at the middle of the stepped translocation event. The current amplitude and site of the spiked signals depended on the number of protein molecules and the binding location onto the long chain DNA carrier, respectively (Figure 5.22a).[67] This DNA carrier induced technique offered a versatile method for specific protein sensing at the single molecule level. It further enabled the simultaneous and quantitative detection of multiple proteins *via* DNA conjugation chemistry and a deep understanding of the characteristic current signals (Figure 5.22b, c).[68] This made a great forward step for specific nanopore sensing from mixtures for potential diagnostic applications.

Understanding the interactions between an antibody and an antigen has great significance in the fields of immunology and biochemistry. The present popular methods for qualitative and quantitative protein analysis are Western blotting and enzyme-linked immunosorbent assay, which are based on the specific interaction between an antibody and an antigen. And the understanding of binding affinities between target antigens and antibodies plays a vital role in improving the sensitivity and selectivity in measurement. The single molecule techniques provide a statistical view of molecules,

Figure 5.21 (a) Detection of a series of proteins by nanopipette. Reproduced from ref. 64 with permission from the American Chemical Society, Copyright 2013. (b) Sensing zone at the narrow opening of the functionalized nanopipette. Reproduced from ref. 6 with permission from the American Chemical Society, Copyright 2009. (c) The nanotube-based detection of proteins by a nanopipette. Reproduced from ref. 65 with permission from the American Chemical Society, Copyright 2013. (d) Schematic of the nanopipette FET biosensor for DNA and protein detection. Reproduced from ref. 66 with permission from the Springer Nature, Copyright 2017.

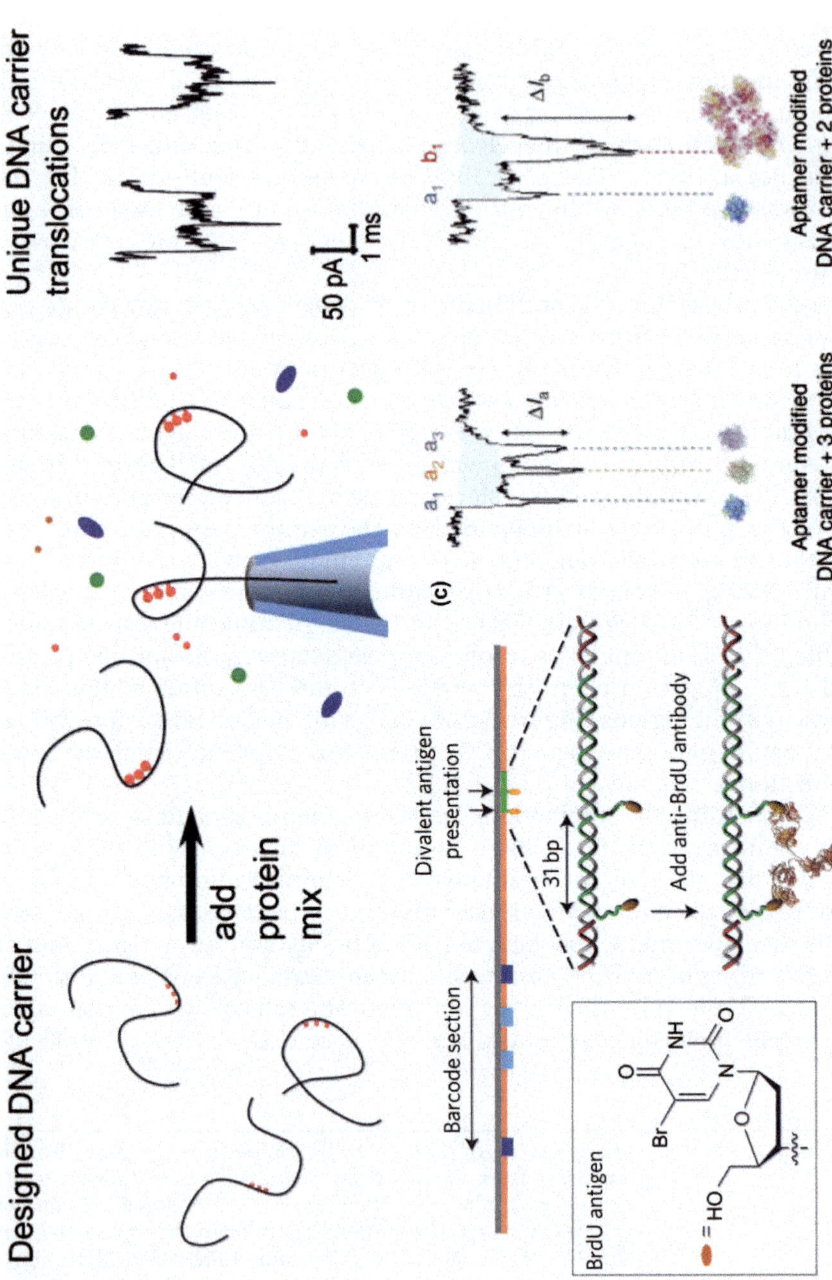
(a)
Designed DNA carrier
add protein mix
Unique DNA carrier translocations
50 pA
1 ms
(b)
BrdU antigen
Barcode section
Divalent antigen presentation
31 bp
Add anti-BrdU antibody
(c)
a_1 a_2 a_3
ΔI_a
Aptamer modified DNA carrier + 3 proteins
a_1 b_1
ΔI_b
Aptamer modified DNA carrier + 2 proteins

revealing unsynchronized and rare events. However, these techniques are mostly limited by various labeling of target biomolecules or dry sample conditions. Therefore, there is an urgent demand to develop a label-free single molecular method to characterize the single molecular interaction between an antibody and an antigen at aqueous solution. To achieve this, a nanopipette with the confined effect was employed for sensing. The nanopipette exhibited the superior features of low dissipation factor and dielectric constant, facilitating wide applications in single cell injection[25,26] and ion detection.[51] It was employed for the specific sensing of vascular endothelial growth factor C by employing ~10 nm gold nanoparticles functionalized with the antibody of vascular endothelial growth factor C on the surface.[69] The diameter of functionalized gold nanoparticles increased to ~14.5 nm after incubation with target antigen, showing a relatively small difference with the antibody-only nanoparticles. By careful modulating the diameter and geometry of the nanopipette, the small difference was distinguished by the corresponding resistive pulse signals in ionic current. Thus, the vascular endothelial growth factor C was finally measured with the nanopipette.

Recently, antibody functionalized nanopipette with a diameter of 50 nm was used for monitoring the antigen–antibody interaction *in situ*. A gold coated nanopipette[70] was obtained easily by the electron beam evaporation method,[71] and employed for further modification. An antibody of alpha-fetoprotein was utilized to prepare a functionalized interface at the nanopore, which was performed by a step-by-step modification procedure, as shown in Figure 5.9. The successful immobilization of the antibody on a gold surface was confirmed by the current–voltage responses and the images of time-of-flight secondary ion mass spectrometry. Alpha-fetoprotein, negatively charged, was driven to transverse through the nanopipette under the applied positive bias potential, generating a spike-like blockage signal on the ionic current trace. The specific interactions between antibody and the translocated antigen was obtained by comparing the characteristic signals both on the bare gold coated nanopipette and the antibody functionalized nanopipette respectively. This method provided a label-free method for better investigation and understanding of protein–protein interactions in the heterogenetic state, held significant potential for the development of immunoassays for diagnostics.[18]

Figure 5.22 (a) Schematic of DNA carrier and target protein translocation through a conical nanopipette. Reproduced from ref. 67 with permission from the American Chemical Society, Copyright 2015. (b) Binding-site presentation and analysis for the bound antibody in translocations. (c) Design concept for using DNA aptamers and a carrier for protein sensing. Reproduced from ref. 68 with permission from the Springer Nature, Copyright 2016.

5.5 Application of a Nanopipette by Combining with Optical Methods

5.5.1 Visualization of Single Molecule Translocation Through the Nanopipette

The dynamic behaviour and unique information of a single molecule can be investigated by characteristic ionic current signals with translocation through a nanopore channel. By contrast, it is still challenging to acquire detailed information of single molecules due to the noise and bandwidth limitations in current recording. Optical-spectroscopic methods are widely used for molecular identifying, providing multiple information on molecular structure and functional groups. Therefore, the combination of electrochemical nanopore measurement with optical-spectroscopic methods could provide multidimensional information at the single molecule level.

A nanopipette shows superiority in combination with optical methods due to its high transparency features. Fluorescence spectroscopy, one of the most common optical methods was employed to be integrated with the conical nanopipette, which provided single molecule information from multiple angles. A λ-DNA molecule in solution was lit up by embedding fluorescent dyes onto the DNA double helix structure and the translocation behaviour was directly monitored in real time. This observation provided the most direct evidence for single molecule translocation events (Figure 5.23a, b).[60,72] The translocation of DNA through a nanopipette was analysed and controlled by employing the optical tweezer setup. λ-DNA coated microspheres were placed in front of the nanopipette tip with an accurate distance by this novel setup. With the DNA molecules on the surface of the microspheres driven into the pores under the biased potential, corresponding electrical information was obtained through the ion current response. This method enabled full control over individual DNA molecules, profiling the movement of the microsphere by the combination of force and ionic current measurement (Figure 5.23c).[73] These findings paved the way for better understanding of long macromolecules and entities.

5.5.2 Electrochemically Controlling Small-volume Delivery with a Nanopipette

The advantages of controllable size features and high spatial resolution of the nanopipette has promoted wide application for sensing of single entities ranging from single molecules to single particles. In addition, nanopipettes have been further integrated with a three-dimensional micromanipulation system for cell analysis by means of its needle-like geometry with a sharp nanoscale tip and length of centimetre scale. It has been reported that very small volume (attolitre–femtolitre) of solutions could be accurately manipulated and injected into a single living cell without causing fatal cellular

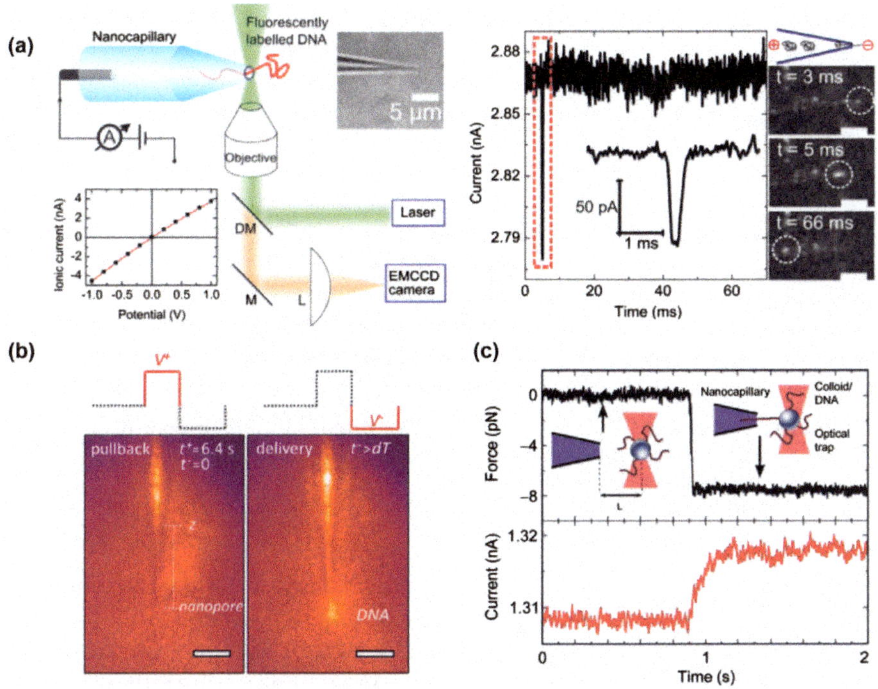

Figure 5.23 (a) DNA manipulation by optical tweezers. (b) Fluorescence imaging of DNA molecules at the nanopipette tip. Reproduced from ref. 72 with permission from the American Institute of Physics, Copyright 2012. (c) DNA manipulation by ionic current detection and fluorescent imaging. Reproduced from ref. 73 with permission from the American Institute of Physics, Copyright 2011.

damage.[74] Nanopipette shows great advantages in analysis at single cell level with high spatial resolution and accurate manipulation in a heterogeneous cell population.

Single cell analysis has great prospect in the fields of biology and biochemistry, and helps to further understand disease processes such as cancer progression.[75] It plays an important role in exploring the fundamental life process and physiological mechanism, and investigating cell signalling pathways.[2] Recently, various methods have been developed for single cell detection, like fluorescence methods and enzyme-linked methods.[76] Among these, a single-cell fluorescence imaging method has attracted increasing attention due to its advantages of a superior imaging effect, simple operation and easy measurement process.[77] However, the precise delivery of fluorescent probe to a certain location in living cells still remains a great challenge, in other words, probe delivery with a subcellular resolution is difficult in traditional fluorescence imaging. Featured with high spatial resolution, nanopipettes were employed as a promising method for the transportation of fluorescent molecules directly into a

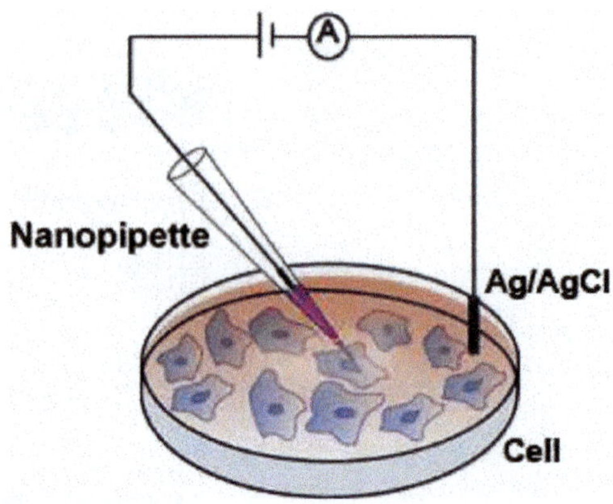

Figure 5.24 Schematic illustration of the RhB-filled nanopipette connected with a DC voltage source for the electroosmotic flow-triggered injection of RhB into a single MCF-7 cell.
Reproduced from ref. 77 with permission from the Royal Society of Chemistry, Copyright 2016.

designated location in single cells by electroosmotic flow. As shown in Figure 5.24, the insertion of a nanopipette enabled accessibility to the inside cell, which was then connected with a pair of Ag/AgCl electrodes. The counter electrode was immersed in the extracellular solutions connected with the electrode inside the nanopipette. Under the applied electric field, fluorescent labelled molecules inside the nanopipette were injected into the given place in a target living cell,[78] and the overall injecting process was monitored under fluorescence microscopy in real time. In this work, a nanopore based method was developed to inject a Rhodamine B(RhB) fluorescent probe into a single living cell *via* a home-made nanopipette.[79] In detail, as shown in Figure 5.24, a nanopipette backfilled with RhB aggregated inside the pipette tip was inserted into a living cell (MCF-7). The diameter of the nanopipette tip was about 100 nm, which was specifically favourable for cell injection. Then, a DC voltage was applied on the Ag/AgCl electrodes, placed in the inside of nanopipette solution and the culture dish respectively. With the tip of the nanopipette contacting the cell membrane, there was a distinct current signal in the time–current trace. Different voltages were subsequently employed on the electrodes to drive the RhB solution into the cell *via* electroosmotic flow triggered translocation, with the whole process monitored under the microscope. Therefore, the nanopipette enabled the precise positioning of fluorescent molecules in a single living cell, which could be utilized as a promising and powerful approach for single-cell analysis. Compared with conventional microinjectors, the use of a nanopipette was much more flexible, and the size of the nanopipette could be easily controlled and selected according to different experimental requirements.[7,103] This method provides a convenient and effective method for fluorescence labelling and is promising for applications in biological and biomedical fields.

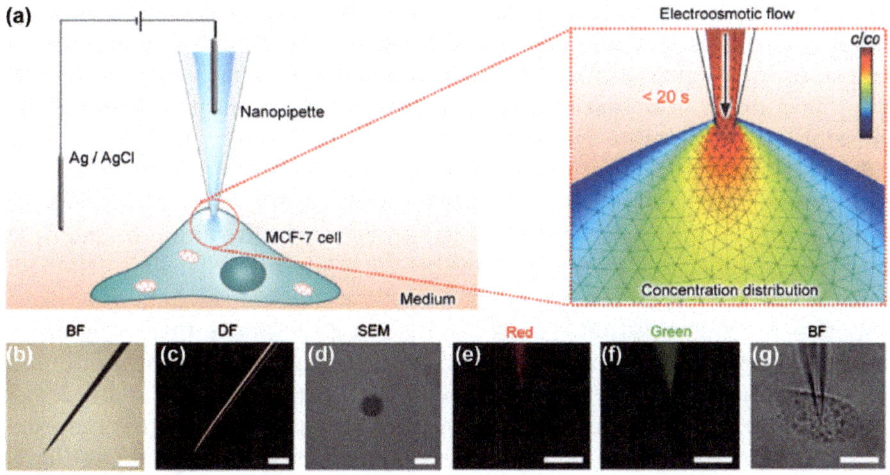

Figure 5.25 Schematic illustration of the electoosmotic delivery of multi-fluorescent molecules into a single living cell.[104] (a) Electroosmotically modulated imaging in a single MCF-7 cell *via* the nanopipette-based controllable injection (right: the solute distribution profile around the tip with finite element mesh; colour bar from up to down: c/c0 from 1 to 0). (b) and (c) Microscopic images of the nanopipette (BF: bright field; DF: dark field). (d) Image showing the pore on the tip of the nanopipette. (e) Fluorescent image of nanopipette filled with Cy5 at the tip. (f) Fluorescent image of nanopipette filled with FITC at the tip. (g) Bright field of the nanopipette inserted into a cell. Scale bars in (b) and (c): 1 μm; scale bar in (d): 100 nm; scale bars in (e), (f) and (g): 10 μm.
Reproduced from ref. 104 with permission from the American Chemical Society, Copyright 2018.

Further use of nanopipettes with microoperation control devices enables ultrafast imaging of subcellular structures. As previous reported, an ultrasmall nanopipette with a tip diameter of about 100 nm was used to achieve the delivery of multiple kinds of fluorescent molecules into the cellular area, with a DC voltage applied in the nanopipette and the culture dish through a pair of Ag/AgCl electrodes.[105] Different potentials were applied for adjusting subcellular physiological processes. An ultrafast and targeted imaging method was thus proposed using a nanopipette, which realized the modulated, non-destructive and one-step injection of multi-fluorescence molecules into a single cell and successfully delivered different fluorescent probes to the specific subcellular regions (Figure 5.25). In addition, this method was further employed for monitoring the actions of a mitochondrial membrane potential with the incubation metformin in a single MCF-7 cell. The driven force in the nanopipette was also simulated by theory calculation and finite element simulations. It indicated a confined distribution profile in the surrounding of the tip, which was especially promising and convenient for multi-channel imaging, showing better understanding and monitoring of the biological processes from cellular to subcellular area.

A glass nanopipette provides an elegant confined space at the very tip, which enables electrochemical controlling of the "turn-off" of fluorescence molecule. In recent years, aggregation-induced emission luminogens (AIE-gens) have drawn much attention of researchers for their greatly enhanced emission efficiency after aggregation, with additional merits of high photo stability and large Stokes' shift.[80,81] The well-known mechanism for the aggregation-induced emission characteristics is that the intramolecular motions can be rationally restricted in the aggregation state. For example, the fluorescent emission of 1,1-dimethyl-2,3,4,5-tetraphenylsilole (DMTPS) could be effectively turned on by changing the organic solvents of acetonitrile into aqueous solution (Figure 5.25). Using the above mechanism, the exciton energy could be consumed by the specific intramolecular motions of phenyl peripheries in acetonitrile, making DMTPS non-emissive or weakly emissive. While in aqueous solutions, with the restriction of such motions, the fluorescence emission of molecules would be lit up through the radiative decay of excited-state energy. The controllable fluorescent emission of AIE-gens facilitates a variety of applications in bio/chemosensing,[82] bioimaging,[83] and optoelectronic devices.[84,85] The strategies of the turn-on principle include (1) self-assembly with the target to form aggregates,[86] (2) restrict intermolecular motion by the interaction with specific chemical species,[87] (3) cleavage of dissolution-promoting ligands to induce the aggregates formation of the AIE residue in the poor solvent.[88,89] The presenting principles are mainly based on the fluorescence measurement of the final dispersed and aggregated states of AIEgens, without knowing the unique changing process. It is also difficult to reverse the aggregated state into a dispersed state or cycle these two states of AIEgens, which limits the further applications to reversible or reproducible sensing. Additionally, there is still a great challenge for directly monitoring the dynamic emission of AIEgen in real time since it is difficult to manipulate the molecules movement in an interface between two kinds of solution.

A nanopipette provides a promising confined space at the nanoscale, which could separate from the surrounding bulk space and thus contribute to the high sensitivity and high spatial resolution.[3,90,91] This confined space at the inner cavity of the nanopore tip enables the confinement of the chemical reactions, intermolecular interactions as well as the fluorescence emission.[92,93] More importantly, the solutions and molecules in the confined space of a nanopore could be manipulated and characterized by an electrochemical method.[35] Therefore, nanopipettes hold great potential for controlling and monitoring the aggregation of AIEgens, which could hardly be realized through traditional methods in bulk systems.

In this work,[94] the confined nanopipette was employed to manipulate the movement of 1,1-dimethyl-2,3,4,5-tetraphenylsilole dicyanovinyl (DMTPS-DCV) solution inward and outward of the nanopore by applying a bias voltage. The outside acetonitrile solution with DMTPS-DCV was driven to flow inward the inside aqueous solution under the optimal bias voltage of 1 V. Under irradiation with UV light, the fluorescent emission of

DMTPS-DCV was intensified with the encounter of aqueous solution at the nanopore confined space (Figure 5.26). This was monitored by the EMCCD camera in real time. The volume of the inward acetonitrile solution could be calculated to be 26 fL in the very tip of nanopore (position I in Figure 5.26). Consequently, the number of DMTPS-DCV molecules in the tip of the nanopore was estimated to be 161 000. With a continuous voltage applied, the neutral DMTPS-DCV aggregates were driven to flow in the deep inside of the nanopore cavity, which was further confirmed by the corresponding shifts of a brightest pixel in the individual frame of the fluorescence image to the deep inside of the nanopore (position II in Figure 5.26). Subsequently, the DMTPS-DCV solution was driven to flow outwards in the nanopore with the applied biased voltage switched to −1 V (Figure 5.26c). And the traversing of DMTPS-DCV through the nanopore confined space (position II in Figure 5.26) induced a higher fluorescent emission. It demonstrated dynamic "off-to-on" emissions of the DMTPS-DCV, along with the gradual lighting from the deep inside toward the very tip of nanopore. These results further demonstrated that the confinement of nanopore promote the emission of AIEgens while the wide cavity suppressed its fluorescence behaviours.

The electrochemical modulation of AIEgens by a confined nanopore was further utilized for imaging at the single cell level. The nanopipette cavity was filled with DMTPS-DCV in a concentration of 10 mM and further integrated with cell injection system. It demonstrated that the applied voltage of −1 V drove the DMTPS-DCV aggregate flow outwards and exhibited high emission in the tip region with the existence of confinement. While DMTPS-DCV showed weak emission with the applied voltage of 1 V, it flowed back the nanopore cavity. Thus, the target cell was specifically lit up in the optical micrograph without the interference of other adjacent cells. This method showed negligible damage to the living cell due to the relatively small nanopore diameter of ∼100 nm.

5.5.3 Real-time Detection in a Single Living Cell with a Nanopipette

The real-time monitoring of electron transfer, molecular dynamics, enzyme activity, and biochemical messengers in a single living cell is essential for investigating cellular communications and cellular heterogeneity.[95,96] At present, there is still an urgent demand to establish a generalizable method for the detection of redox-active species in a single living cell with high spatial resolution. Due to the remarkable advantages of sensitivity and selectivity, the electrochemical method is applied to the dynamics study of redox metabolism in cells.[97] Compared to micron-sized probes, the nanoscale of the nanopipette exhibits excellent performance for maintaining cell viability and achieving more precise cell injection. Further, nanopore technology is used for the detection of specific ions or small molecules within cells. In 2015, a nanopipette with diameter around 100 nm was used

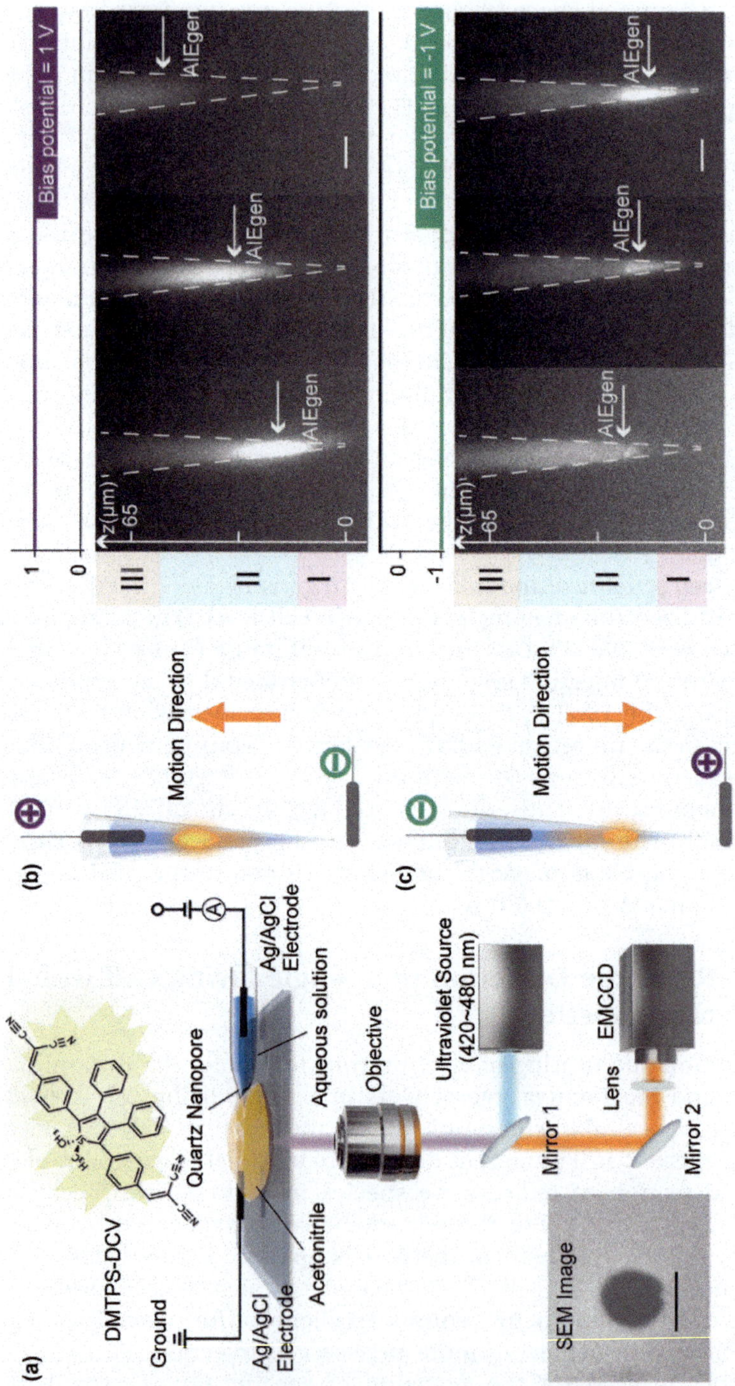

(a)
DMTPS-DCV
Ground
Ag/AgCl Electrode
Quartz Nanopore
Acetonitrile
Aqueous solution
Ag/AgCl Electrode
Objective
Ultraviolet Source (420~480 nm)
Mirror 1
Lens
EMCCD
Mirror 2
SEM Image
(b)
Motion Direction
Bias potential = 1 V
z(μm)
65
0
AIEgen
III
II
I
(c)
Motion Direction
Bias potential = -1 V

to prepare nano pH probes for sensing of different intracellular pH levels (Figure 5.27a, b).[98] By modifying the chitosan layer, which was sensitive to pH changes on the inner wall of the quartz channel, the prepared pH probe achieved a sensitive response in the range of pH 6–8 *in vitro*. At the same time, this pH probe also showed a good response in the detection of pH levels of cancer cells and normal cells. By detecting 7 normal human fibroblasts and 14 cervical cancer cells, the average pH distribution was 7.37 ± 0.29 and 6.75 ± 0.27, respectively. Further, in order to verify the capability of a chitosan-modified nanopipette for recording the change in intracellular pH in real time, NPPB (chloride channel blocker) was used to increase the pH of the cell. It was found that this nanopore probe captured the local acidity increase effectively and continuously during NPPB stimulation to cells. In addition, using the principle of current response caused by the change of inner wall surface, it was also possible to detect other important small molecules in the cell. For example, by modifying glucose oxidase on the inner wall of the nanopipette, glucose differences were determined directly in three cells of fibroblasts, breast cancer cells MCF-7, and MDA-MB-231 cells.[99]

In order to further improve the spatial resolution of the nanopipette in single cell detection, a nanopipette with a small diameter of 30 nm was prepared and used for hydrogen peroxide detection at different sites in the cell (Figure 5.27c, d).[100] First, G-quadruplex-DNAzyme was modified on the inner wall of the nanopipette, catalysing the reaction of 2,2′-azinobis(3-ethylbenzothiazoline-6-sulfonic acid) diammonium salt (ABTS) and hydrogen peroxide. The free radical intermediate $ABTS^{\bullet -}$ produced by the ABTS oxidation,

Figure 5.26 Visualizing the dynamic emission of DMTPS-DCV regulated by the electrochemically confinement of a quartz nanopipette. (a) The instrumentation for the visualization of DMTPS-DCV emission inside a quartz nanopipette. The quartz nanopipette is placed on the designed cover glass. The inside and outside of the nanopore are filled with aqueous solution and 10 μM DMTPS-DCV acetonitrile solution, respectively. The 10 mM TBAPF6 and 10 mM KCl were used as an organic and aqueous electrolyte to conduct the ionic flow. Two Ag/AgCl electrodes were used to apply the potential while the acetonitrile side is defined as virtual ground. Light of $\lambda = 450$–490 nm was used to excite the DMTPS-DCV. (b) The electrochemical manipulation of DMTPS-DCV at applied bias potential of 1 V. Left: the DMTPS-DCV solution moves from outside nanopore to the inside nanopore at 1 V, leading to an intensive emission. Right: the fluorescent images of the "on-to-off" motion of DMTPS-DCV inside a confined nanopore, which are recorded within 5 s. Scale bars, 10 μm. (c) The electrochemical manipulation of DMTPS-DCV at applied bias potential of −1 V. Left: the applied potential of −1 V drove the movement of DMTPS-DCV aggregates from the wider lumen to the confined tip. Right: Fluorescent images recorded at −1 V, illustrating the "off-to-on" motion of DMTPS-DCV. The bright pixel moves from upper to tip within 35 s. Scale bars, 10 μm.

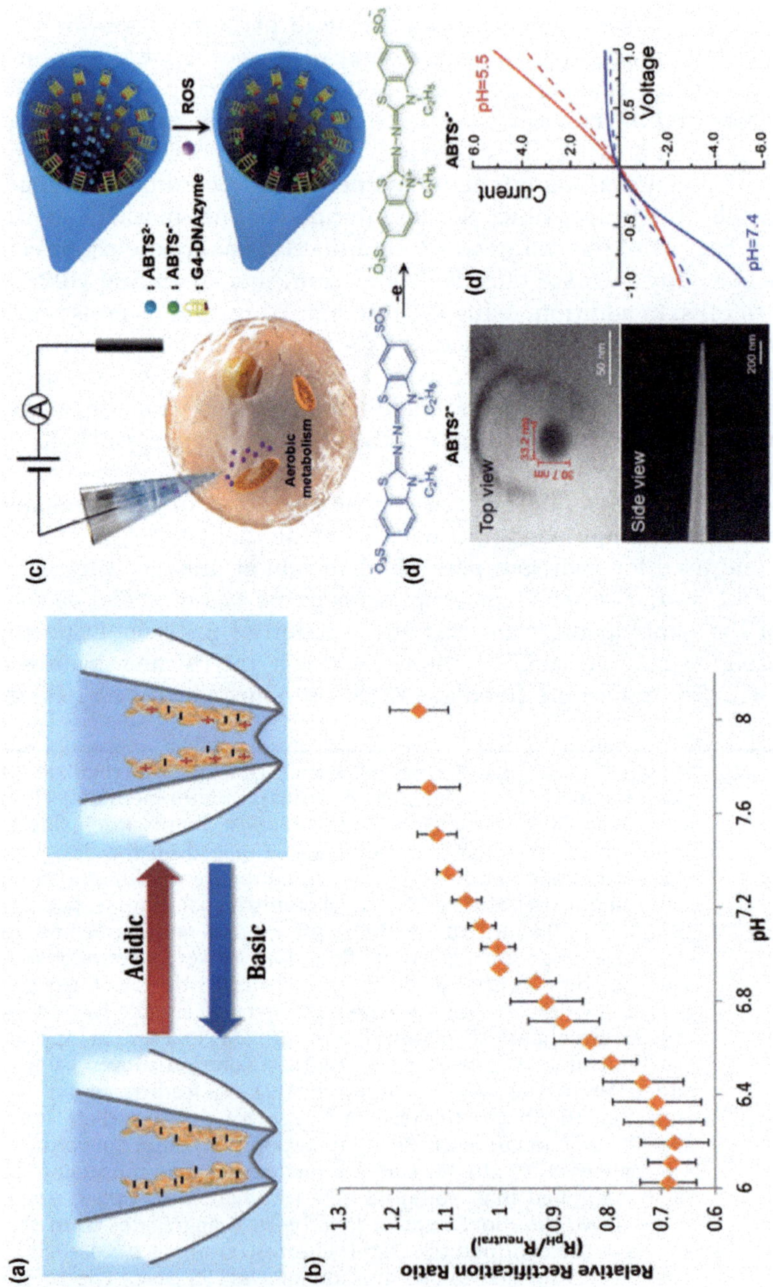
(a)
Acidic
Basic
(b)
Relative Rectification Ratio
(R_pH/R_neutral)
1.3
1.2
1.1
1
0.9
0.8
0.7
0.6
6
6.4
6.8
7.2
7.6
8
pH
(c)
A
Aerobic metabolism
: ABTS2-
: ABTS•-
: G4-DNAzyme
ROS
(d)
ABTS2-
-e
ABTS•-
Top view
33.2 nm
30.7 nm
50 nm
Side view
200 nm
(d)
Current
Voltage
pH=5.5
pH=7.4

could adsorb on DNA, increasing the negative charge of the nanopore and enhancing the ionic current rectification. Based on this principle, linear detection of hydrogen peroxide in the range of 0.2–5.0 mM was achieved *in vitro*. Furthermore, since hydrogen peroxide was closely related to cellular respiratory metabolism, mainly concentrated near the mitochondria, fluorochrome (Mitotracker Green FM) was used to label hamster ovary cells and PMA to stimulate cell metabolism to increase ROS levels. After about 25 min, the change in the rectification ratio relative to that of the control group was measured at different positions using the nanopipette probe. This demonstrated that in the region where the mitochondria were more concentrated in the fluorescent label, the measured rectification ratio value was significantly larger than that in the control region, indicating that a higher ROS content. Thus, the ultrasmall nanopipette enabled the intracellular sensing at subcellular level, demonstrating the target distribution in a single cell.

Recent nanopore detection is usually based on the resistive pulse mechanism, which could hardly realize the intracellular detection of redox metabolism with high spatial and temporal resolution and less damage to the living cell, and nanoelectrodes less than 100 nm in diameter are applied in the analysis.[101,102] The implementation of the electrochemical method is greatly dependent on the fabrication of the nanoelectrode, which is crucial for the signal-to-noise ratio and electrochemical performance. In addition, due to the poor signal responses, this proposed electrochemical method could hardly be applied for the sensing of target species with relative low concentration (<1 μM). To overcome these problems, a wireless asymmetric nanopore electrode (ANE) was constructed by coating the interior with gold film. This wireless ANE with a small diameter of 90 nm was used for the detection of redox metabolism in living cells by increasing the current resolution from the nanoampere to the picoampere level. Attributed to the high signal-to-noise ratio, a trace amount of nicotinamide adenine nucleotide (NADH) could be detected with high sensitivity and selectivity. Furthermore, ANEs were used for probing of NADH in a single living cell and the investigation of the efficacy of anticancer drugs (Figure 5.28).[30]

Figure 5.27 (a) Schematic of the reversible changes in surface charge as a result of pH. (b) Calibration of chitosan-modified nanopipettes within the physiologically relevant pH range from 6.02 to 8.04. All data points are represented as relative rectification ratios at +/-0.5 V *vs.* Ag/AgCl reference electrode. Reproduced from ref. 98 with permission from the Royal Society of Chemistry, Copyright 2016. (c) Schematic illustration of the G4-DNAzyme modified nanopipette for single-cell analysis using ABTS as the substrate. (d) Chemical formulae of ABTS before and after oxidation including the SEM images of the nanopipette and typical *I–V* curves obtained at the G4-DNAzyme modified nanopipettes treated with different concentrations of H_2O_2 in 20 mm pH 7.4 Tris–HCl (added 40 mm KCl). Reproduced from ref. 100 with permission from the John Wiley and Sons, Copyright 2018.

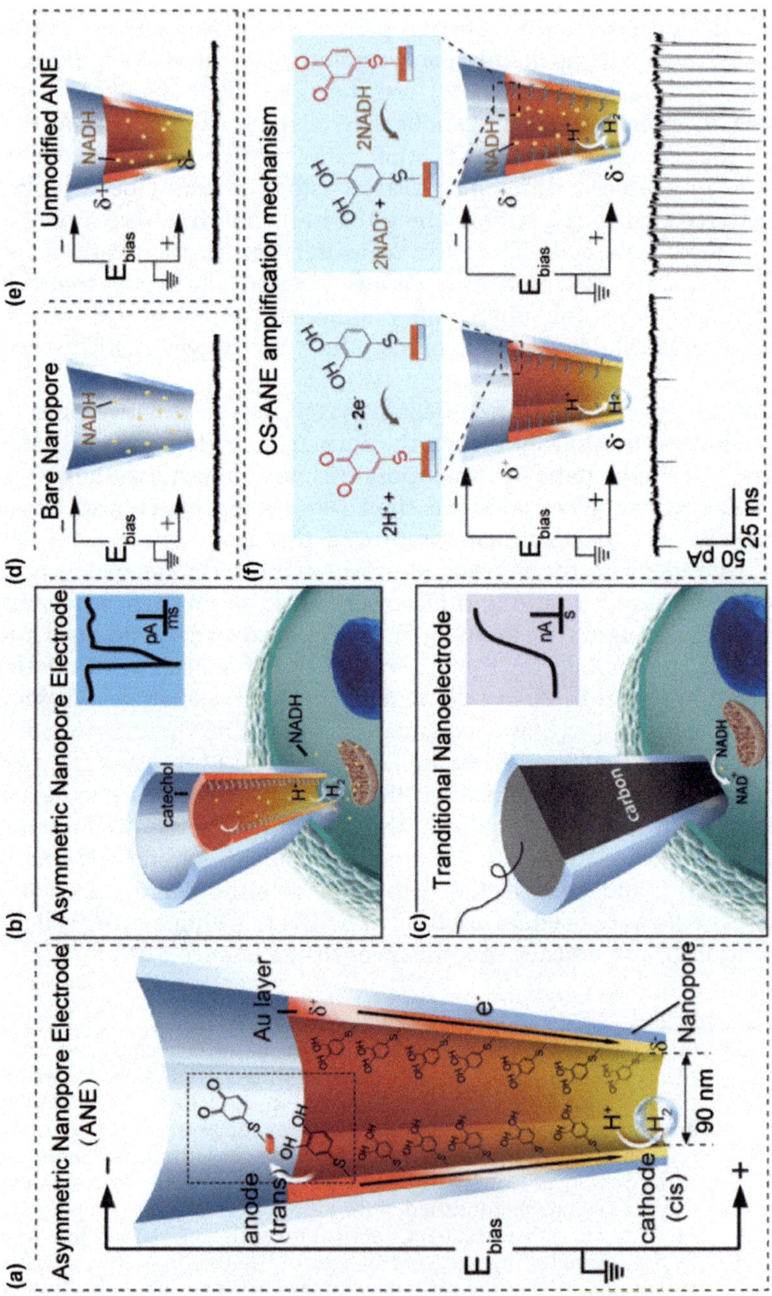
(a)
Asymmetric Nanopore Electrode (ANE)
anode (trans)
cathode (cis)
Au layer
Nanopore
90 nm
E
bias
(b)
Asymmetric Nanopore Electrode
catechol
NADH
pA
ms
(c)
Tranditional Nanoelectrode
carbon
NADH
NAD+
nA
s
(d)
Bare Nanopore
NADH
(e)
Unmodified ANE
NADH
(f)
CS-ANE amplification mechanism
2H+ +
-2e-
2NAD+ +
2NADH
50 pA
25 ms

In detail, an Au layer was coated on the ~110 nm nanopipette through electron-beam evaporation. To ensure that the Au layer only existed inside the nanopore, the coated Au on the outside wall of the nanopore was removed by HCl solution under N_2 flow. The presented ANE was characterized by SEM images after sculpting the sidewall using FIB. The coated gold layer inside the nanopore showed high robustness, which was confirmed by the $I-V$ curve and ion current rectification property. The gold coated nanopore was functionalized with a derivative of catechol, which was characterized by both XPS and TOF-SIMS analysis. Notably, this prepared catechol functionalized ANE could be used to acquire a picoampere level ionic response produced by redox reactions and the content noise performance could be demonstrated by the spectrum of the PSD (power spectral density). Most of the potential drop (nearly 90%) occurred at both ends of the ANEs owing to the asymmetric geometry of the electrode. That is to say, the applied potential was mostly at the very end of the nanopore electrode. Thus, the redox reaction at the WNE could be effectively regulated by varying the applied potential. Under the external electric field, the oxidation of catechol occurred at the anodic pole, meanwhile the reduction of H^+ occurred at the cathodic pole with the following equilibrium equations:

$$\text{Catechol} \rightarrow 2\text{H}^+ + 2\text{e}^- + \text{o} - \text{Benzoquinone} \quad E = 400 \text{ mV } vs. \text{ SHE} \tag{5.4}$$

$$2\text{H}^+ + 2\text{e}^- \rightarrow \text{H}_2 \quad E = 0 \text{ mV } vs. \text{ SHE.} \tag{5.5}$$

This indicates that the above-mentioned reactions occurred when the applied potential was above 400 mV. Hydrogen nanobubbles were generated at the cathode, which was located at the very tip of the nanopore. Due to the sensitivity of the cathode geometry of ANEs, the faradic current generated by H_2 generation is enhanced into distinct pulse-like signals at the ionic current.[77] And the WNEs achieved the amplification of the current signal by more than three orders of magnitude, which enabled NADH detection with concentrations down to 1 pM. Furthermore, because the diameter of the WNEs could be controlled to 30 nm, they could be inserted into single living cells to record intracellular electrochemical processes. By modifying with different redox probes, a broad range of redox species (*e.g.*, peroxiredoxin,

Figure 5.28 Asymmetric nanopore electrode for monitoring electron transfer dynamics in live cells. (a) Illustration of an asymmetric nanopore electrode (ANE) for probing redox-active species. (b) The ANE for single-cell probing. (c) Traditional nanoelectrode (*e.g.*, carbon nanoelectrode) for probing intracellular redox species. (d) The bare nanopore does not produce any transient responses to NADH. (e) The unmodified ANE generates a stable baseline without any signals. (f) The 4-thiol-catechol-modified ANE generates an enhanced current signal due to the generation of H_2 nanobubbles at the cathodic pole.
Reproduced from ref. 30 with permission from the American Chemical Society, Copyright 2018.

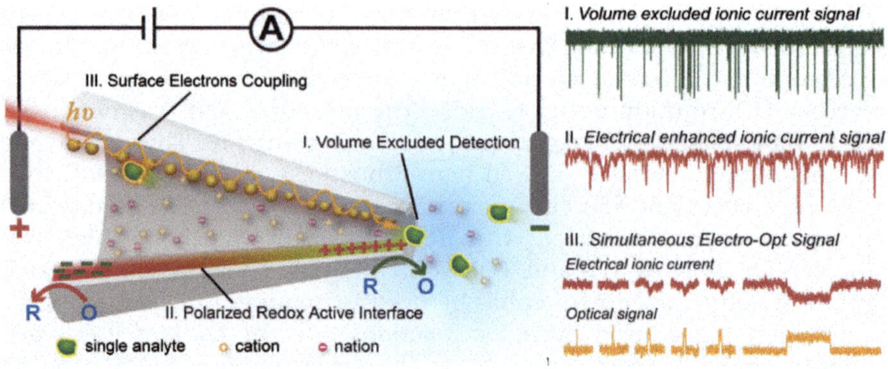

Figure 5.29 Electrochemical confinement effects for the nanopore sensing mechanisms.
Reproduced from ref. 35 with permission from the John Wiley and Sons, Copyright 2018.

glutaredoxin, neurotransmitters and catalase) could be monitored by ANEs. Consequently, the ANEs play a key role in understanding the critical bodily processes in single cellular and multicellular organisms.

5.6 Conclusion and Outlook

Nanopipettes shows broader application due to their superior geometry compared with traditional protein and silicon nitride (SiN_x) nanopores. This chapter summarises the fundamental preparation and the various applications of nanopipettes and nanopore electrodes. Three dominating mechanisms for nanopipette sensing are discussed, which are the volume excluded ionic current signal, the electrical enhanced ionic current signal and the simultaneous electro-opt signal, respectively (Figure 5.29).[35] Still there are both unprecedented opportunities and challenges existing in the nanopipette techniques. With improved sensitivity of the nanopipette, more hidden properties of a single molecule could be "seen" from the current signatures.[105] By uniting the characteristics of both high spatial resolution and new detection mechanism, the nanopipette brings together compelling capabilities for probing every component in the cell with superior performance. Not limited by these, the confined nanopipette may include the gaseous, magnetic and quantum mechanism, which may bring forward new sensing mechanism. These features of nanopipette will promote it to display more impressive futures in biological and chemical applications.

Acknowledgements

We would like to thank Rui Gao and Yuan-Jie Li for their help in the section on nanopore electrodes and nanopipette fabrication.

References

1. M. G. Schrlau, N. J. Dun and H. H. Bau, *ACS Nano*, 2009, **3**, 563–568.
2. J. A. Bafna and G. V. Soni, *PLoS One*, 2016, **11**, e0157399.
3. R. Gao, Y. Lin, Y. Ying, X. Liu, X. Shi, Y. Hu, Y. Long and H. Tian, *Small*, 2017, **13**, 1700234.
4. Y. Shen, Z. Zhang and T. Fukuda, *Nanotechnology*, 2015, **26**, 155702.
5. C. Gao, S. Ding, Q. Tan and L.-Q. Gu, *Anal. Chem.*, 2009, **81**, 80–86.
6. S. Ding, C. Gao and L. Q. Gu, *Anal. Chem.*, 2009, **81**, 6649–6655.
7. D. Perry, D. Momotenko, R. A. Lazenby, M. Kang and P. R. Unwin, *Anal. Chem.*, 2016, **88**, 5523–5530.
8. L. J. Steinbock, O. Otto, C. Chimerel, J. Gornall and U. F. Keyser, *Nano Lett.*, 2010, **10**, 2493–2497.
9. J. Kwak and A. J. Bard, *Anal. Chem.*, 1989, **61**, 1221–1227.
10. H. S. White and A. Bund, *Langmuir*, 2008, **24**, 2212–2218.
11. B. Zhang, Y. Zhang and H. S. White, *Anal. Chem.*, 2004, **76**, 6229–6238.
12. W. J. Lan, C. Kubeil, J. W. Xiong, A. Bund and H. S. White, *J. Phys. Chem. C*, 2014, **118**, 2726–2734.
13. R. Gao, Y. Ying, Y. Hu, Y. Li and Y. Long, *Anal. Chem.*, 2017, **89**, 7382–7387.
14. S. Howorka and Z. Siwy, *Chem. Soc. Rev.*, 2009, **38**(8), 2360–2384.
15. S. Umehara, M. Karhanek, R. W. Davis and N. Pourmand, *Proc. Natl. Acad. Sci. U. S. A.*, 2009, **106**, 4611–4616.
16. P. Actis, A. McDonald, D. Beeler, B. Vilozny, G. Millhauser and N. Pourmand, *RSC Adv.*, 2012, **2**, 11638.
17. R. J. Yu, Y. L. Ying, Y. X. Hu, R. Gao and Y. T. Long, *Anal. Chem.*, 2017, **89**, 8203–8206.
18. Y.-L. Ying, R.-J. Yu, Y.-X. Hu, R. Gao and Y.-T. Long, *Chem. Commun.*, 2017, **53**, 8620–8623.
19. L. X. Zhang, S. L. Cai, Y. Bin Zheng, X. H. Cao and Y. Q. Li, *Adv. Funct. Mater.*, 2011, **21**, 2103–2107.
20. Y.-B. Zheng, S. Zhao, S.-H. Cao, S.-L. Cai, X.-H. Cai and Y.-Q. Li, *Nanoscale*, 2017, **9**, 433–439.
21. T. Plett, W. Shi, Y. Zeng, W. Mann, I. Vlassiouk, L. A. Baker and Z. S. Siwy, *Nanoscale*, 2015, **7**, 19080–19091.
22. S.-L. Cai, S.-H. Cao, Y.-B. Zheng, S. Zhao, J.-L. Yang and Y.-Q. Li, *Biosens. Bioelectron.*, 2015, **71**, 37–43.
23. P. Actis, A. Rogers, J. Nivala, B. Vilozny, R. A. Seger, O. Jejelowo and N. Pourmand, *Biosens. Bioelectron.*, 2011, **26**, 4503–4507.
24. E. A. Vitol, M. G. Schrlau, S. Bhattacharyya, P. Ducheyne, H. H. Bau, G. Friedman and Y. Gogotsi, *Chem. Vap. Deposition*, 2009, **15**, 204–208.
25. M. G. Schrlau and H. H. Bau, *JALA*, 2010, **15**, 145–151.
26. L. D. Brennan, T. Roland, D. G. Morton, S. M. Fellman, S. Y. Chung, M. Soltani, J. W. Kevek, P. M. McEuen, K. J. Kemphues and M. D. Wang, *PLoS One*, 2013, **8**, 1–13.
27. H. He, X. Xu and Y. Jin, *Anal. Chem.*, 2014, **86**, 4815–4821.

28. X. Xu, H. He and Y. Jin, *Anal. Chem.*, 2015, **87**, 3216–3221.
29. R. Gao, Y. L. Ying, B. Y. Yan, P. Iqbal, J. A. Preece and X. Wu, *Microchim. Acta*, 2016, **183**, 491–495.
30. Y. Ying, Y. Hu, R. Gao, R. Yu, Z. Gu, L. P. Lee and Y. Long, *J. Am. Chem. Soc.*, 2018, **140**, 5385–5392.
31. C. A. C. Sequeira, D. S. P. Cardoso and M. L. F. Gameiro, *Chem. Eng. Commun.*, 2016, **203**, 1001–1008.
32. J. P. Guerrette, S. M. Oja and B. Zhang, *Anal. Chem.*, 2012, **84**, 1609–1616.
33. S. E. Fosdick, K. N. Knust, K. Scida and R. M. Crooks, *Angew. Chem., Int. Ed.*, 2013, **52**, 10438–10456.
34. R. Gao, Y. Lin, Y. L. Ying, Y. X. Hu, S. W. Xu, L. Q. Ruan, R. J. Yu, Y. J. Li, H. W. Li, L. F. Cui and Y. T. Long, *Nat. Protoc.*, 2019, **14**, 2015–2035.
35. Y. Ying, R. Gao, Y. Hu and Y. Long, *Small Methods*, 2018, **2**, 1700390.
36. J. J. Kasianowicz, E. Brandin, D. Branton and D. W. Deamer, *Proc. Natl. Acad. Sci. U. S. A.*, 1996, **93**, 13770–13773.
37. Y. L. Ying and Y. T. Long, *Sci. China: Chem.*, 2017, **60**(9), 1187–1190.
38. S. Vajda, M. J. Pellin, J. P. Greeley, C. L. Marshall, L. A. Curtiss, G. A. Ballentine, J. W. Elam, S. Catillonmucherie, P. C. Redfern and F. Mehmood, *Nat. Mater.*, 2009, **8**, 213–216.
39. Y. L. Ying, C. Chao, Y. X. Hu and Y. T. Long, *Natl. Sci. Rev.*, 2018, **5**(4), 450–452.
40. M. Wood and B. Zhang, *ACS Nano*, 2015, **9**, 2454–2464.
41. K. Fu and P. W. Bohn, *ACS Cent. Sci.*, 2018, **4**, 20–29.
42. R. Gao, Y.-L. Ying, Y.-J. Li, Y.-X. Hu, R.-J. Yu, Y. Lin and Y.-T. Long, *Angew. Chem., Int. Ed.*, 2018, **57**, 1011–1015.
43. N. A. Kotov and P. S. Weiss, *ACS Nano*, 2014, **8**, 3101–3103.
44. J. Ustarroz, M. Kang, E. Bullions and P. R. Unwin, *Chem. Sci.*, 2017, **8**, 1841–1853.
45. W.-J. Lan and H. S. White, *ACS Nano*, 2012, **6**(2), 1757–1765.
46. W.-J. Lan, D. A. Holden, B. Zhang and H. S. White, *Anal. Chem.*, 2011, **83**(10), 3840–3847.
47. W. Lan, D. A. Holden, J. Liu and H. S. White, *J. Phys. Chem. C*, 2011, **115**, 18445–18452.
48. S. R. German, L. Luo, H. S. White and T. L. Mega, *J. Phys. Chem. C*, 2013, **117**, 703–711.
49. X. Lin, A. P. Ivanov and J. B. Edel, *Chem. Sci.*, 2017, **8**, 3905–3912.
50. T. Li, X. He, K. Zhang, K. Wang, P. Yu and L. Mao, *Chem. Sci.*, 2016, **7**, 6365–6368.
51. L. Chen, H. He and Y. Jin, *Anal. Chem.*, 2014, **87**(1), 522–529.
52. R. Smeets, U. Keyser, M. Wu, N. Dekker and C. Dekker, *Phy. Rev. Lett.*, 2006, **97**(8), 088101.
53. M. R. Powell, M. Sullivan, I. Vlassiouk, D. Constantin, O. Sudre, C. C. Martens, R. S. Eisenberg and Z. S. Siwy, *Nat. Nanotechnol.*, 2008, **3**(1), 51–57.

54. B. Vilozny, P. Actis, R. A. Seger and N. Pourmand, *ACS Nano*, 2011, **5**, 3191–3197.
55. Y.-X. Hu, Y.-L. Ying, R. Gao, R.-J. Yu and Y.-T. Long, *Anal. Chem.*, 2018, **90**, 12352–12355.
56. L. J. Steinbock, A. Lucas, O. Otto and U. F. Keyser, *Electrophoresis*, 2012, **33**(23), 3480–3487.
57. N. A. Bell, V. V. Thacker, S. Hernández-Ainsa, M. E. Fuentes-Perez, F. Moreno-Herrero, T. Liedl and U. F. Keyser, *Lab Chip*, 2013, **13**(10), 1859–1862.
58. X. Gong, A. V. Patil, A. P. Ivanov, Q. Kong, T. Gibb, F. Dogan, A. J. deMello and J. B. Edel, *Anal. Chem.*, 2013, **86**(1), 835–841.
59. K. J. Freedman, L. M. Otto, A. P. Ivanov, A. Barik, S.-H. Oh and J. B. Edel, *Nat. Commun.*, 2016, **7**, 10217.
60. A. P. Ivanov, P. Actis, P. Jönsson, D. Klenerman, Y. Korchev and J. B. Edel, *ACS Nano*, 2015, **9**(4), 3587–3595.
61. P. Cadinu, B. Paulose Nadappuram, D. J. Lee, J. Y. Sze, G. Campolo, Y. Zhang, A. Shevchuk, S. Ladame, T. Albrecht and Y. Korchev, *Nano Lett.*, 2017, **17**(10), 6376–6384.
62. J. Kong, J. Zhu and U. F. Keyser, *Chem. Commun.*, 2017, **53**(2), 436–439.
63. K. Chen, M. Juhasz, F. Gularek, E. Weinhold, Y. Tian, U. F. Keyser and N. A. Bell, *Nano Lett.*, 2017, **17**(9), 5199–5205.
64. W. Li, N. A. Bell, S. Hernández-Ainsa, V. V. Thacker, A. M. Thackray, R. Bujdoso and U. F. Keyser, *ACS Nano*, 2013, **7**(5), 4129–4134.
65. J. Sha, T. Hasan, S. Milana, C. Bertulli, N. A. Bell, G. Privitera, Z. Ni, Y. Chen, F. Bonaccorso and A. C. Ferrari, *ACS Nano*, 2013, **7**(10), 8857–8869.
66. R. Ren, Y. Zhang, B. P. Nadappuram, B. Akpinar, D. Klenerman, A. P. Ivanov, J. B. Edel and Y. Korchev, *Nat. Commun.*, 2017, **8**(1), 586.
67. N. A. Bell and U. F. Keyser, *J. Am. Chem. Soc.*, 2015, **137**(5), 2035–2041.
68. N. A. Bell and U. F. Keyser, *Nat. Nanotechnol.*, 2016, **11**(7), 645.
69. H. Cai, Y. Wang, Y. Yu, M. V. Mirkin, S. Bhakta, G. W. Bishop, A. A. Joshi and J. F. Rusling, *Anal. Chem.*, 2015, **87**, 6403–6410.
70. J. Sha, W. Si, W. Xu, Y. Zou and Y. Chen, *Sci. China: Technol. Sci.*, 2015, **58**, 803–812.
71. R. Wei, D. Pedone, A. Zürner, M. Döblinger and U. Rant, *Small*, 2010, **6**, 1406–1414.
72. V. V. Thacker, S. Ghosal, S. Hernández-Ainsa, N. A. Bell and U. F. Keyser, *Appl. Phy. Lett.*, 2012, **101**(22), 223704.
73. O. Otto, L. Steinbock, D. Wong, J. Gornall and U. Keyser, *Rev. Sci. Instrum.*, 2011, **82**(8), 086102.
74. W. Kang, R. L. McNaughton and H. D. Espinosa, *Trends Biotechnol.*, 2016, **34**, 665–678.
75. L. Yang, Z. Wang, Y. Deng, Y. Li, W. Wei and Q. Shi, *Anal. Chem.*, 2016, **88**(22), 11077–11083.
76. J. T. Aerts, K. R. Louis, S. R. Crandall, G. Govindaiah, C. L. Cox and J. V. Sweedler, *Anal. Chem.*, 2014, **86**, 3203–3208.

77. A. Hayashi, D. Asanuma, M. Kamiya, Y. Urano and S. Okabe, *Neuropharmacology*, 2016, **100**, 66–75.
78. S. Hennig, S. Van De Linde, M. Lummer, M. Simonis, T. Huser and M. Sauer, *Nano Lett.*, 2015, **15**, 1374–1381.
79. J. Lv, R.-C. Qian, Y.-X. Hu, S.-C. Liu, Y. Cao, Y.-J. Zheng and Y.-T. Long, *Chem. Commun.*, 2016, **52**, 13909–13911.
80. J. Luo, Z. Xie, J. W. Lam, L. Cheng, H. Chen, C. Qiu, H. S. Kwok, X. Zhan, Y. Liu and D. Zhu, *Chem. Commun.*, 2001, **18**, 1740–1741.
81. J. Mei, N. L. Leung, R. T. Kwok, J. W. Lam and B. Z. Tang, *Chem. Rev.*, 2015, **115**(21), 11718–11940.
82. J. Mei, Y. Hong, J. W. Lam, A. Qin, Y. Tang and B. Z. Tang, *Adv. Mater.*, 2014, **26**(31), 5429–5479.
83. R. T. Kwok, C. W. Leung, J. W. Lam and B. Z. Tang, *Chem. Soc. Rev.*, 2015, **44**(13), 4228–4238.
84. N. Zhao, M. Li, Y. Yan, J. W. Lam, Y. L. Zhang, Y. S. Zhao, K. S. Wong and B. Z. Tang, *J. Mater. Chem. C*, 2013, **1**(31), 4640–4646.
85. J. Liu, H. Su, L. Meng, Y. Zhao, C. Deng, J. C. Ng, P. Lu, M. Faisal, J. W. Lam and X. Huang, *Chem. Sci.*, 2012, **3**(9), 2737–2747.
86. Y. Hang, J. Wang, T. Jiang, N. Lu and J. Hua, *Anal. Chem.*, 2016, **88**(3), 1696–1703.
87. M. Gao, L. Wang, J. Chen, S. Li, G. Lu, L. Wang, Y. Wang, L. Ren, A. Qin and B. Z. Tang, *Chem. – Eur. J.*, 2016, **22**(15), 5107–5112.
88. J. Liang, R. T. Kwok, H. Shi, B. Z. Tang and B. Liu, *ACS Appl. Mater. Interfaces*, 2013, **5**(17), 8784–8789.
89. X. Gu, G. Zhang, Z. Wang, W. Liu, L. Xiao and D. Zhang, *Analyst*, 2013, **138**(8), 2427–2431.
90. C. Cao, Y.-L. Ying, Z.-L. Hu, D.-F. Liao, H. Tian and Y.-T. Long, *Nat. Nanotechnol.*, 2016, **11**(8), 713–718.
91. K. Jayant, J. J. Hirtz, I. Jen-La Plante, D. M. Tsai, W. D. De Boer, A. Semonche, D. S. Peterka, J. S. Owen, O. Sahin and K. L. Shepard, *Nat. Nanotechnol.*, 2017, **12**(4), 335.
92. S. Huang, M. Romero-Ruiz, O. K. Castell, H. Bayley and M. I. Wallace, *Nat. Nanotechnol.*, 2015, **10**(11), 986.
93. P. Waduge, R. Hu, P. Bandarkar, H. Yamazaki, B. Cressiot, Q. Zhao, P. C. Whitford and M. Wanunu, *ACS Nano*, 2017, **11**(6), 5706–5716.
94. Y. L. Ying, Y. J. Li, J. Mei, R. Gao, Y. X. Hu, Y. T. Long and H. Tian, *Nat. Commun.*, 2018, **9**, 3657.
95. X. T. Zheng and C. M. Li, *Chem. Soc. Rev.*, 2012, **41**(6), 2061–2071.
96. D. Evanko, *Nat. Methods*, 2008, **5**(1), 25–26.
97. C. L. Bentley, M. Kang, F. M. Maddar, F. Li, M. Walker, J. Zhang and P. R. Unwin, *Chem. Sci.*, 2017, **8**(9), 6583–6593.
98. R. E. Özel, A. Lohith, W. H. Mak and N. Pourmand, *RSC Adv*, 2015, **5**(65), 52436–52443.
99. R. A. Nascimento, R. E. Özel, W. H. Mak, M. Mulato, B. Singaram and N. Pourmand, *Nano Lett.*, 2016, **16**(2), 1194–1200.

100. J. Song, C. H. Xu, S. Z. Huang, W. Lei, Y. F. Ruan, H. J. Lu, W. Zhao, J. J. Xu and H. Y. Chen, *Angew. Chem., Int. Ed.*, 2018, **57**(40), 13226–13230.
101. Y. Li, K. Hu, Y. Yu, S. A. Rotenberg, C. Amatore and M. V. Mirkin, *J. Am. Chem. Soc.*, 2017, **139**(37), 13055–13062.
102. X. Li, S. Majdi, J. Dunevall, H. Fathali and A. G. Ewing, *Angew. Chem., Int. Ed.*, 2015, **54**(41), 11978–11982.
103. C. A. Morris, A. K. Friedman and L. A. Baker, *Analyst*, 2010, **135**(9), 2190–2202.
104. R.-C. Qian, J. Lv and Y.-T. Long, *Anal. Chem.*, 2018, **90**(22), 13744–13750.
105. S.-M. Lu, Y.-Y. Peng, Y.-L. Ying and Y.-T. Long, *Anal. Chem.*, 2020, **92**, 5621–5644.

CHAPTER 6

Instrumentation and Data Algorithm for Single Molecule Sensing

JIAJUN WANG,[a,b] YI-LUN YING[a] AND YI-TAO LONG*[a]

[a] Nanjing University, Nanjing, China; [b] East China University of Science and Technology, Shanghai, China
*Email: yitaolong@nju.edu.cn

6.1 Challenges in Nanopore Instrumentation and Data Processing

Nanopore electrochemistry is a powerful tool for single molecule detection. Tremendous efforts in experimental design have meant that nanopore electrochemistry has wide application in nanotechnology and biotechnology. However, the detection instrumentation for nanopore analysis faces the difficulties of how to amplify the acquired ionic current, as well as how to efficiently process the current data at massive handling high throughput for extracting the correct information. The limitations of the detection system, such as high noise and low frequency response, may lead to corruption or distortion of information, which is a bottleneck preventing new discoveries in not only nanopore studies, but also confined/localized electrochemical measurements. The latest achievements in nanopore instrumentation and data algorithms take a significant step forward in real applications for biological analysis. In this chapter, we focus on the challenges of realizing low current measurement with high temporal resolution

Detection Science Series No. 17
Confining Electrochemistry to Nanopores
Edited by Yi-Tao Long

Published by the Royal Society of Chemistry, www.rsc.org

in nanopore detection and the demand for algorithms developed for recognition, signal distortion recovery, noise suppression and high-throughput processing of the nanopore events. Ionic current across the nanopore acquires as the analogue signal at the level of pico-amperes, which is hard to accurately measure.[1] One of the challenges is the relatively low signal-to-noise ratio in current measurements, because the weak signal almost reaches the bottom of physical limitations. A variety of background noise can contribute to the challenges such as thermal noise from the nanopore, dielectric noise from electronics, quantum noise and some external electromagnetic interference.[2,3] Another challenge is how to measure the current at high bandwidth to ensure the events with short duration are detected. The bandwidth determines the time resolution or frequency response for the detection. A low bandwidth means a long rise time of the recording system, which distorts the current events by increasing their duration time and attenuating the amplitude. Hence, short-lived events will be submerged when the amplitude is attenuated lower than the noise intensity. The distortion induced by the limits of recording bandwidth and noise is particularly true for inorganic based nanopores such as SiN_x nanopores, which exhibit insufficient signal-to-noise ratio and low temporal resolution.[4] The limitation for recording short current blockage causes the loss of information for analysing the proteins, short DNA/RNA fragments,[5–7] which have short-lived interactions with the nanopore. The low time resolution is also major barrier for reading the DNA sequence through the electrical signal of a nanopore sensor.[8] Therefore, there is an urgent demand to enhance the current resolution and the time resolution in nanopore detection to obtain more detailed information on single molecules.[9] Moreover, since the study of nanopores is based on the analysis of massive events recognized from continuous and long-time recording of current signals, experiments producing a large volume of data should be processed. So analysis of the raw data is really time-consuming work.[10] Therefore, an automatic and accurate data process is also an important part of the nanopore system.

Herein, this chapter mainly describes the basic knowledge of the nanopore instrumentation, and presents some optimized circuit design for high bandwidth and low noise current recording. Then, light isshed on the selected examples for data nanopore data processing including the algorithm for recovering of distorted signals and the automatic processing methods. The instrumentation and algorithm discussed in this chapter are not only be applied in biological nanopore, but also could be adopted for single entity analysis with nanopores composed from silicon derivatives, pipettes, *etc.*

6.2 Design Principles for Nanopore Electrochemical Instrumentation

The measurement system for nanopore sensing is based on a two-electrode system with a working electrode and a countering electrode.[11] Similar to a patch clamp system, the device used in nanopore studies has multiple parts,[12] including a trans-impedance amplifier (TIA), a differential amplifier (Diff), a low-pass filter (LPF), an analogue-to-digital converter (ADC), a voltage apply circuit, a digital-to-analogue converter (DAC), a digital signal processing circuit and a computer for data recording/analysis, as shown in Figure 6.1.

The TIA is a sensitive current-to-voltage (I–V) converter for converting the pico-ampere currents into micro-volt voltage signals that are large enough during the signal processing to boost the signal-to-noise ratio (SNR).[13] As shown in Figure 6.2a, the biased voltage at the inverting input of the operational amplifier (OPA) $V-$ is controlled by the voltage source at the non-inverting input $V+$ due to the virtual short of the negative feedback of the circuit. The high input impedance of the OPA ensures a low biased current into the inverting input; therefore, the ionic current is almost equal to the current through the feedback resistor. The gain for converting the ionic current into a voltage is adjusted by the feedback resistor. Then, the output voltage is an addition of the voltage drop across the feedback resistor and the voltage at the inverting input. The current flowing through the nanopore (i_{pore}) measured by the electrode surface can be described as:

$$V_{\text{out}} = V_{\text{apply}} - i_{\text{pore}}\, R_{\text{f}}. \tag{6.1}$$

To convert the pico-ampere current to a micro-volt signal, the feedback resistor should be more than 100 MΩ, even 1 GΩ. The noise mainly added to the current outcome from the nanopore system is generated by the I–V converter.[2] Eqn (6.2) gives the power spectral density (PSD) of the noise of an I–V converter $S_{\text{A}}^{2}(f)$ ($\text{A}^2\,\text{Hz}^{-1}$) and the RMS noise of an I–V converter $I_{\text{A}}(f_{\text{c}})$ (A RMS) is given by eqn (6.3):[14]

$$S_{\text{A}}^{2}(f) = 2qI_{\text{fet}} + \frac{4akT_h}{R_{\text{f}}} + e_{\text{n}}^{2}\left(\frac{1}{R_{\text{f}}^{2}} + 4\pi^2 C_{\text{t}}^{2} f^{2}\right) \tag{6.2}$$

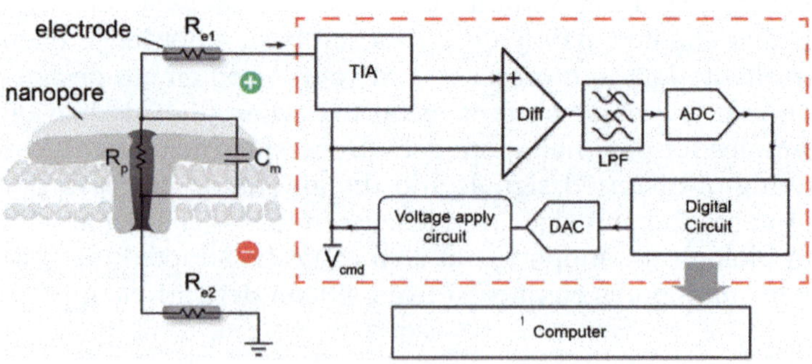

Figure 6.1 The components in the instrument system for nanopore sensing. The nanopore is placed in the membrane measured by two electrodes. The working electrode is connected to the headstage to acquire single molecule signal and the ground electrode is connected to the reference. Reproduced from ref. 15 with permission from Springer Nature, Copyright 2018.

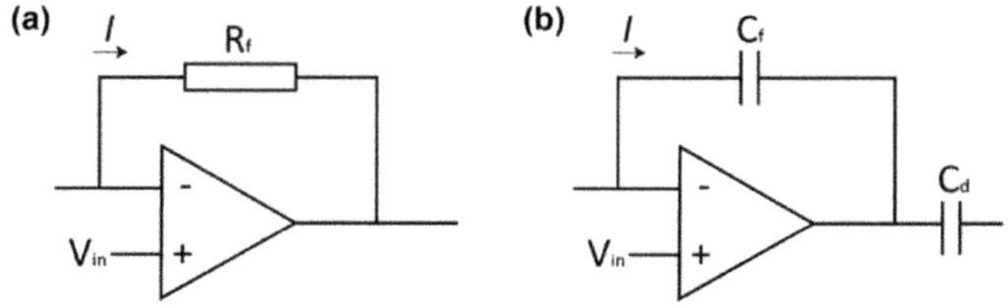

Figure 6.2 Schematics of two different circuits to measure current. (a) Illustration of an OPA that uses a resistor feedback circuit to measure current. In general usage, one can assume that there is no difference between voltage through the nanopore and the voltage applied from V_{in} (inverting input terminal). (b) Illustration of an OPA that uses capacitor feedback circuit to measure minimal current (<100 pA). This way the thermal noise is non-existent because there is no feedback resistor and it has a better dynamic range than the resistor feedback circuit. However, this circuit often needs the discharging of the capacitor with resets, and hence exhibits the low current limit.

$$I_{\text{A}}(f_{\text{c}}) = \sqrt{2qI_{\text{fet}}C_1f_{\text{c}} + \frac{4akT_hc_1f_{\text{c}}}{R_{\text{f}}} + e_{\text{n}}^2\left(\frac{c_1f_{\text{c}}}{R_{\text{f}}^2} + \frac{4}{3}\pi^2C_{\text{t}}^2c_3f_{\text{c}}^3\right)}. \tag{6.3}$$

Throughout the equation, q (C) is the elementary charge of an electron, I_{fet} (A) refers to the current leakage into the gate of the field effect transistor (FET) of the OPA, a is a constant factor that represents excess noise, T_h (K) is the temperature of the feedback resistor, k is Boltzmann's constant, c_1 and c_3 are correction coefficients, e_{n} (V Hz$^{-1/2}$) is the input voltage noise at the input stage of the I–V converter, and C_{t} (F) is the total capacitance at the input of the headstage. According to eqn (6.2) and (6.3), it is obvious that the signal-to-noise ratio increases with increasing feedback resistance (R_{f}).

However, the usage of a large resistor has two-sided effects. It may not only limit the dynamic range of current sensing due to the high gain, but also cause low frequency response of the current signal induced by the parasitic capacitor. The parasitic capacitor in parallel with the feedback resistor will become a resistor–capacitor (RC) filter to the signal. If the OPA used in the I–V converter has a high gain-bandwidth product, the cut-off frequency of the I–V converter is dominated by the time constant of the RC filter that can be given by eqn (6.4):[14]

$$f_{\text{c}} = \frac{1}{2\pi R_{\text{f}}C_{\text{p}}}. \tag{6.4}$$

As an example, a feedback resistor of 100 MΩ with 0.5 pF parasitic capacitor will have a low bandwidth of only about 3.1 kHz. It is supposed to measure the integration of charge over time to determine the current. As shown in Figure 6.2b,[13] this is fulfilled by replacing the feedback resistor with a capacitor as a conventional approach, theoretically increasing the frequency response of the system and ensuring no thermal noise. In a capacitor feedback circuit, the settling time is only affected by the speed of the OPA. Therefore, a capacitor feedback circuit is often faster than a resistor feedback circuit, since the RC time constant limits the settling time of resistor feedback circuits, resulting in milliseconds of settling time or longer. The capacitor feedback circuit has the disadvantage of the inability of handling steady-state currents that are larger than the commonly accepted 100 pA. This is the result of output saturation caused by a large, constant current. In typical commercial amplifiers, the feedback capacitor is discharged every 50 μs with switches to reset the output voltage back to zero again. The exact period of discharge can vary based on different pore current and capacitor value, but every reset will unavoidably create noise in the collected data. For the reasons above, the capacitor feedback circuit is better used in measuring the current of biological nanopores with open pore current less than 100 pA.

The nanopore system could be modelled as a resistor (R_p) parallel with a capacitor (C_m), while the noise of the resistors arises from the thermal agitation of charge carriers. The power spectral density of thermal noise $S_T{}^2(f)$ is given by eqn (6.5) and the RMS (root mean square) thermal noise $I_T(f_c)$ is given by eqn (6.6):[13,14]

$$S_T^2(f) = \frac{4kT}{R_p} \tag{6.5}$$

$$I_T(f_c) = \sqrt{\frac{4kTc_1 f_c}{R_p}} \tag{6.6}$$

where R_p (Ω) is the resistance of the pore structure, k is Boltzmann's constant, and T (K) is the temperature. c_1 is the correction coefficient of 1.05. f_c is the cut-off frequency of the filter. The resistive model has been used in many studies to estimate the thermal noise on the condition that the system has a small membrane capacitance (less than 100 pF), small pore resistance (less than 100 MΩ), and small electrode resistance (less than 500 Ω). If the membrane capacitance is large, the condition would be much more complicated. Because the membrane capacitance has a significant effect on the noise of the TIA and amplifier by amplifying the noise of the applied voltage. The relationship between the system capacitance, the applied bias voltage and the signal-to-noise ratio were focused and compared by experimental analysis, in circuit simulation and analysis models.[15]

The outcome suggested that the noise of the TIA increases with the increase in either membrane capacitance (Figure 6.3a) or the applied voltage (Figure 6.3b). Consequently, the improvement of the signal-to-noise ratio can be achieved by lowering the voltage induced noise instead of optimizing the membrane capacitance. Therefore, suppression of applied voltage noise is an efficient approach for reducing the noise in nanopore detection. Therefore, a low noise voltage apply circuit was designed to achieve

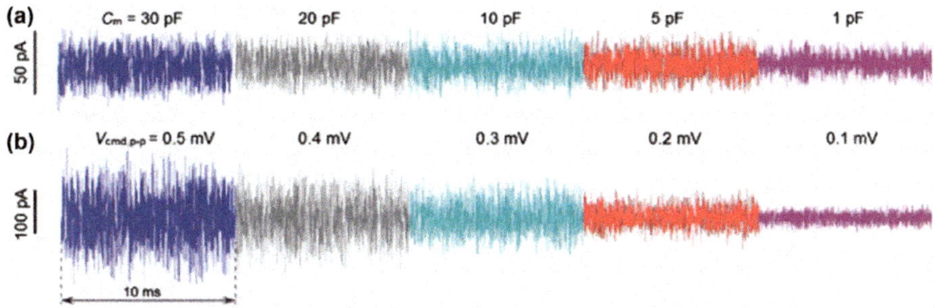

Figure 6.3 Outcomes of the noise simulation analysis. (a) Stimulated current noise with 0.1 mV V_{cmd} p-p noise, the membrane capacitance ranges from 1 to 30 pF; (b) simulated current noise under 10 pF membrane capacitance with p-p noise of V_{cmd} ranging from 0.1 to 0.5 mV.
Reproduced from ref. 12 with permission from Elsevier, Copyright 2017, Science China Press.

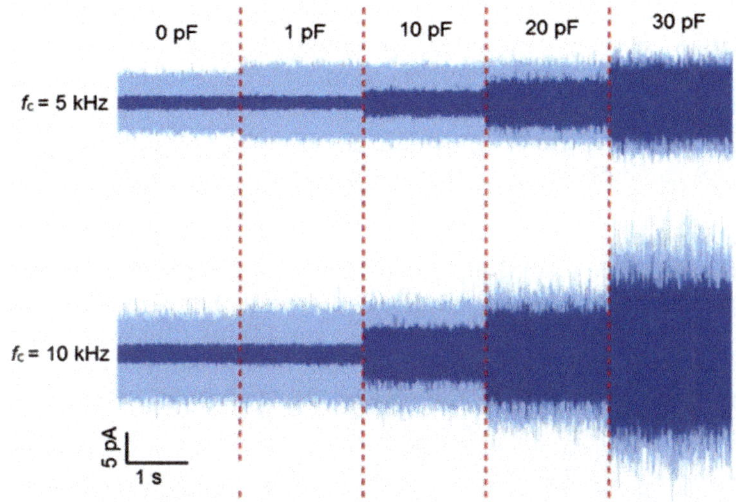

Figure 6.4 Test of current recording noise of the developed instrument with low voltage applied noise (dark colour) and Axopatch 200B (light colour).
Reproduced from ref. 12 with permission from Elsevier, Copyright 2017, Science China Press.

low-noise current recording without narrowing the measurement range or reducing the time resolution. The developed circuit was tested at bandwidths of 5 kHz and 10 kHz to investigate the noise caused by the membrane capacitance ranging from 0 to 30 pF and applied voltage ranging from 0.1 mV p-p noise to 0.5 mV p-p noise. The result showed that the signal noise could be highly reduced by lowering the noise of applied voltage (Figure 6.4).

The reference zero of the output of the TIA is the applied voltage as it is tied to the $V+$. A differential amplifier is introduced to deduct the applied voltage and set the reference zero to be the ground of the circuit. To further suppress the noise, the low pass filter is applied to attenuate the unwanted high frequency noise. The cut-off frequency of the filter should be carefully selected as the filter also limits the bandwidth of signal. As depicted in Figure 6.5, the dominant sources of noise are various at different frequency regimes. In the low-frequency domain (<100 Hz), the nanopore noise rise was dominated by the ionic friction inside the nanopore lumen.[16] At moderate frequencies (ranging from 100 to 10 000 Hz), the raised noise could be contributed by white noise[17] and the electronic components, and also background noise generated by energy dissipation in a non-ideal dielectric circuit. Finally, at high frequencies more than 10 kHz, the dominant noise is attributed to the interaction of the amplifier's voltage noise with the total capacitance at the input. Under such high frequency, a low-pass filter with cut-off frequency of 10 kHz is commonly used in nanopore studies.

A high-speed analogue-to-digital converter (ADC) is used to sample the signal for further recording and data analysis. Based on the Nyquist–Shannon theory, the sampling rate should always be twice larger than the bandwidth.[18] It is recommended to use a sampling rate of more than 100 kHz at a filter frequency of 5 kHz or 10 kHz.[19] The digital circuit contains a digital signal processor and a communication chip for interfacing with the ADC, DAC and computer.[20] To ensure that the data processing speed is fast enough for the high-speed sampling, the FPGA is always used in the digital circuit.

In the series of nanopore single molecule experiments, the acquired current data points are always along with the noise from both instrumentation and the ambient. In order to verify whether the system could accurately record a real signal under different noise conditions, the wave generator is used to produce artificial noise, as illustrated in Figure 6.6A. The power spectrum analysis as well as the typical single molecule event at different

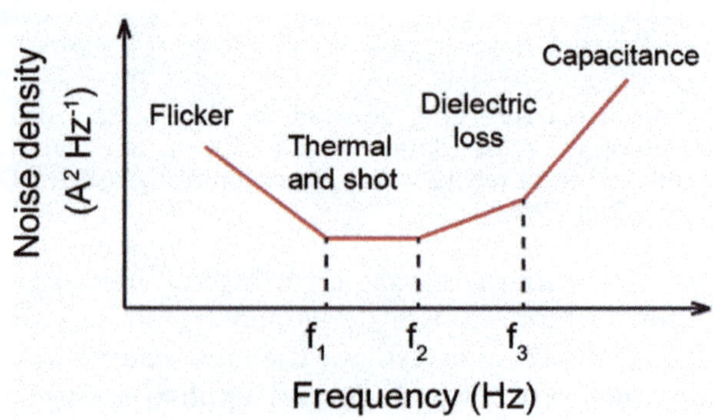

Figure 6.5 Dominant sources of noise power spectral density, illustrated as a function of frequency.
Reproduced from ref. 5 with permission from Springer Nature, Copyright 2012.

bandwidth (*i.e.* 3 kHz (black), 5 kHz (red), 10 kHz (blue), 50 kHz (purple), 100 kHz (green)) are illustrated in Figure 6.6B and C.

6.3 Setups for Nanopore Electrochemical Instrumentation

6.3.1 Setups Based on Patch-clamp System

The conventional nanopore electrochemistry instrumentation based on a patch-clamp system comprised of a polytetrafluoroethylene (PTFE) electrolyte chamber that has two fluidic-tight chambers, was designed and built. The nanopore is fabricated on the membrane between the two reservoirs of electrolytes. This system was designed so that the only connection is between the two reservoirs. Ag/AgCl electrodes are placed on both sides of the nanopore to apply voltage at millivolt scale and measure the current at picoampere scale. The outside of the system is surrounded by a metallic enclosure to shield any electromagnetic (EM) noise. The shielding factor can be further enhanced with an additional copper Faraday cage.[21] To protect the integrity of the lipid bilayer, mechanical vibrations have to be minimal, and this can be achieved by setting up the system on an isolation table. This can also limit the electrical noise caused by electrodes vibrating. The digitized data are collected from the analogue signals from the current amplifier using the data acquisition (DAQ) card; this card is controlled by the computer and is also used to control the electric potential through changing the amplifier's voltage.

The simplest way to measure currents is using a *trans*-impedance amplifier (TIA) with a large feedback resistor to convert the current into a voltage signal. Such a circuit is typically found in most electrometers, pico-ammeters

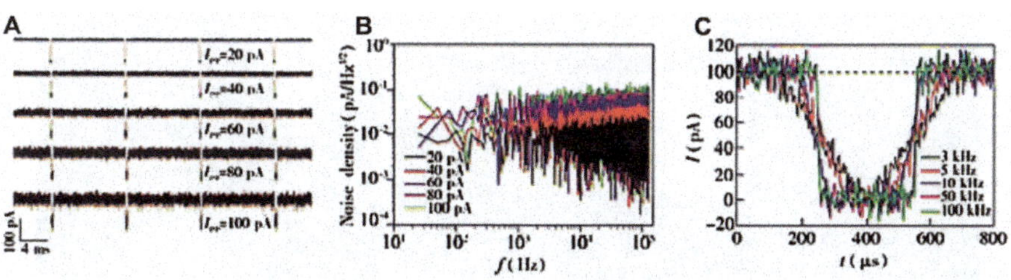

Figure 6.6 (A) Simulation signals at different noise levels: 20, 40, 60, 80 and 100 pA, blockade time of the blockade is 300 μs; (B) Power spectral density curves for the simulation signals at the noise level of 20 pA (black), 40 pA (red), 60 pA (blue), 80 pA (purple) and 100 pA (green); (C) Simulation signals in different bandwidth environments:, noise peak–peak of the blockade is 40 pA, blockade time is 300 μs.
Reproduced from ref. 23 with permission from Elsevier, Copyright 2018 Changchun Institute of Applied Chemistry, Chinese Academy of Sciences.

or electrophysiology amplifiers, including the Axopatch200B (Molecular Devices) and the EPC-10 (HEKA Instruments Inc.).

6.3.2 Setups Based on "Cube" Electrochemical Instrumentation

The home-designed electrochemical instrument named "Cube" is designed typically for the purpose of nanopore sensing. It is a low noise and portable device for rapid nanopore sensing (Figure 6.7).[22] It could also be applied in other nanoscale electrochemical measurement, such as single particle collision and single cell analysis with nanoelectrodes. This device contains a signal processor and a preamplifier module with a Faraday cage. The signal processor is powered directly from a PC USB port. A DC–DC converter is integrated to apply the voltage though the USB with a range ±12 V to ensure a higher dynamic range of the signal. Moreover, the signal processor is also integrated with a Bessel filter, an analogue-to-digital converter (ADC), a digital-to-analogue converter (DAC) and a voltage apply circuit. The preamplifier is powered by the battery to avoid the noise crosstalk from the power supply unit. A frequency compensation circuit is introduced for the TIA to boost the bandwidth as high as 100 kHz with a gain of 1 GΩ.[22]

For a detailed view of how to perform the nanopore electrochemistry experiment using the Cube instrumentation, the assembly of the detection cage of the Cube is given in Figure 6.8.[42] The whole detection cage setup is mainly comprised of the headstage, the experimental chamber, the DC power supply, all surrounded by the Faraday cage for the purpose of shielding the ambient noise. The Faraday cage is a cubic cage *ca.* 7.5 cm×12.6 cm×4.5 cm dimensions to place the headstage in, as well as leave enough room for experiment performing. To acquire nanopore electrochemical data, the strategy to separate the signal acquisition unit and

Figure 6.7 Photos of a "Cube" electrochemical instrumentation for nanopore sensing. Left: assembly of the portable "Cube" instrumentation in a general laboratory. Right: One practical "Cube" kit for nanopore experiments including the membrane formation setups and electrochemical instrumentation.

filtering processing unit is adapted. The signal acquisition unit, here the headstage, is firstly sealed into a metal cage with only the electrode pinhole and the I/O port exposed to other interfaces. The electrode pinhole is used to connect the measuring electrode, usually Ag/AgCl, to acquire the low amperometric signal. The I/O port is designed to connect the post-processing unit. The least electronic component exposure to ambient provides a high signal-to-noise ratio as well as eliminates the electric noise generated from the PCB. Apart from the signal acquisition unit, powering is also essential. Herein, the Cube setup makes use of the DC power supply to cancel the sinusoidal interference from the input. The simplest approach is to choose a DC battery. Without using a DC regulator, the cost and space occupancy is minimized, leaving the DC power supply farthest from the measuring chamber, providing the least change in noise interference as well as making it waterproof. Then the most important part for performing the nanopore single molecule electrochemistry is the measuring chamber. The size of the sit for the chamber is manufactured to perfectly fit for the purpose of avoiding mechanical vibration. The measuring chamber is comprised of two segregated cups with a small hole placed in the middle. This small hole is the only connection between the two cups; to ensure this, water sealing

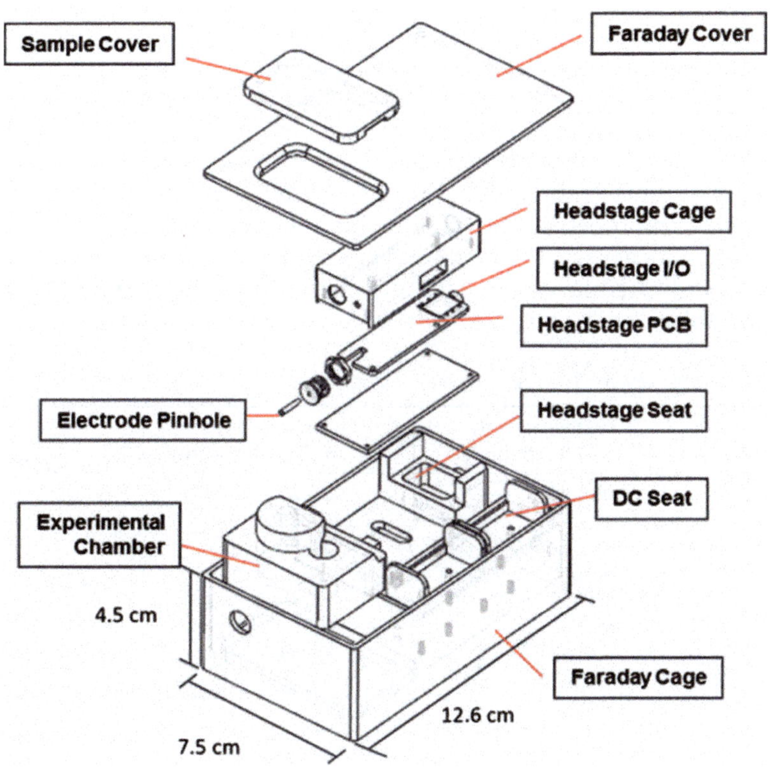

Figure 6.8 An expanded view of the Cube nanopore electrochemistry assembly.

between the two cups is achieved by an external screw tightening. The small aperture is used to form the BLM then self-assembly of the nanopore, which is described in Chapter 2. Finally, after placing the headstage, the DC power and the experimental chamber, the Faraday cover is closed. There is a small window in the Faraday cage for ease of sample preparation. Without the opening of the whole Faraday cage, the open window on the experimental chamber allows the continuous measurement while adding the analyte of interest.

After the assembly of the Cube system, a series of calibrations should be made to ensure that the real measurements are accurate. The standard current characterization of the headstage is firstly calculated. The applied voltage is first set to 0 mV, then applying Ohm's law yields the current. The resistor used herein is 1.069 G, a series amplified current is acquired and summarized in Table 6.1.[42] The error rate is calculated considering the real time current from the oscilloscope as well as the current from instrumentation. According to the calculated error rate, the measurement from the Cube system is less than 1% compared to a high-resolution oscilloscope. The calibration is an essential handling process every time before the on-site experiments.

To analyse the accuracy of the single molecule analysis under different noise conditions, the error rate is plotted in Figure 6.9a. In the noise analysis, a set of single molecule events (*ca.* 3000 blockades) was collected at 10 kHz in addition artificial noise, then the system accuracy is analysed by reading from the nanopore instrumentation compared to the oscilloscope.[23] When the $I_{\text{p-p}}$ is less than 60 pA, the error rate is less than 0.2%, although the increase in the noise level increases the error rate, the overall rate remain less than 1%. During the simulation, the baseline of the nanopore experiments is always proceeded by the FIR filter, which has been previously described. In an ideal case, the nanopore electrochemical simulation is fixed at 100 pA, as indicated in Figure 6.9b, the error rate of the current baseline at different noise level was also investigated, and the overall error rate

Table 6.1 Accuracy of the Cube current amplifier.

Voltage (mV)	Current from oscilloscope (pA)	Current from instrumentation (pA)	Error
1000	935.45	935.25	− 0.02%
500	467.73	467.53	− 0.04%
200	187.10	186.86	− 0.12%
100	93.55	93.38	− 0.18%
50	46.77	46.51	− 0.56%
10	9.35	9.29	− 0.69%
−10	−9.35	−9.40	0.49%
−50	−46.77	−46.8	0.06%
−100	−93.55	−93.63	0.09%
−200	−187.10	−187.18	0.05%
−500	−467.73	−467.91	0.04%
−1000	−935.45	−935.76	0.03%

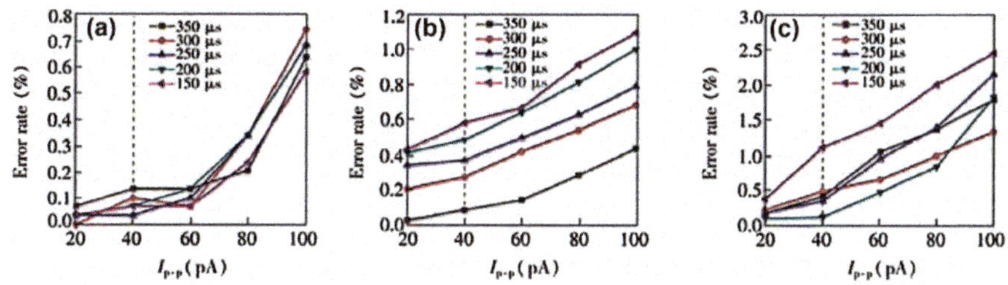

Figure 6.9 Influence of noise on system performance. (a) Error rate of blockade event detection number. (b) Error rate of current baseline detection. (c) Error rate of blockade amplitude detection; $I_{p-p} \leq 40$ pA (left side of the black dashed line) is the actual experimental noise range.
Reproduced from ref. 23 with permission from Elsevier, Copyright 2018 Changchun Institute of Applied Chemistry, Chinese Academy of Sciences.

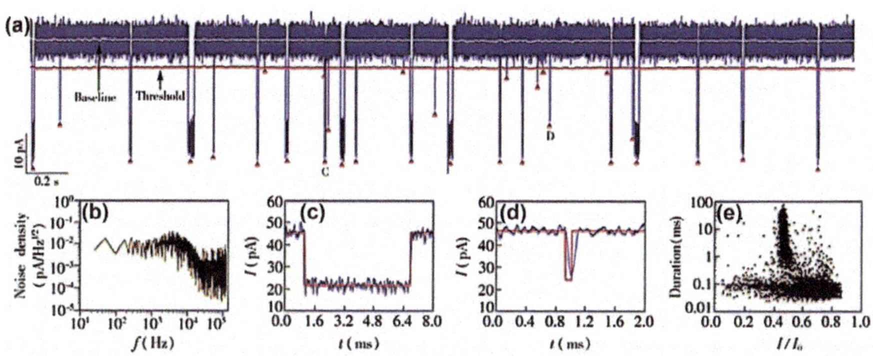

Figure 6.10 (a) Raw current data collected and recorded during the experiment (blue) and blockade events detected by the proposed algorithm (red mark). (b) Noise spectral density curve of the experimental current data. (c) Typical blockade event detected in the experiment (blue) and blockade event plotted according to analysis results (red). (d) Short blockade event detected in the experiment (blue) and blockade event plotted according to analysis results (red). (e) Blockade time-blockade current scatter plot.
Reproduced from ref. 23 with permission from Elsevier, Copyright 2018 Changchun Institute of Applied Chemistry, Chinese Academy of Sciences.

remained less than 1%. Finally, when analysing the single molecule event, there would be a blockade occur in the current recording; this would inevitably cause fluctuation in the current baseline. Figure 6.9c also indicates the error rate of single molecule event detection by comparing the ideal and estimated value of the blockade amplitude.

Then, the Cube system will be connected to the experimental chamber. After the formation of the bilayer and assembly of a single nanopore the current noise is analysed. As given in Figure 6.10, the aerolysin is used to investigate the dA4, under a low pass filter at 5 kHz, and the outcome from the Cube system is only 3.26 pA peak to peak. The single molecule current signal gives a quite good uniformity as given in Figure 6.9a and b, even for much shorter cases (Figure 6.10c, $I_{res} = 4.65$ pA), the current resolution remains stable and high quality.

6.3.3 Other Commercialized Setups

The Oxford Nanopore Technologies (ONT) has developed a highly integrated device for nanopore-based DNA sequencing with a large number of channels which is shown in their start-up device (Flongle). Then, the Minion is the mostly used in laboratory device and it could be easily integrated into Gridion for commercial standard. Finally, the Promethion is the largest DNA sequencing station for on-site analysis. The main characteristics of these instruments are techniques of high-throughput nanopore preparation and DNA deceleration, which enables the reduction of the DNA single base translocation time to milliseconds. With 2048 holes on each chip, they can achieve high-throughput data acquisition using CsgG as a specifically designed nanopore compatible with this device (www.nanoporetech.com).

Nanion Technologies GmbH (Nanion) is among the first multi-bilayer platform developers. Their Orbit 16 allows a maximum of 16 membranes to be painted automatically at the same time. Each membrane is supported over a well, which is connected to a relay then distributed to the Axon 200B headstage to realize a multi-channel has been acquired by a high resolution headstage. The Orbit mini (is also a multi-bilayer platform with smaller throughput, four membranes simultaneously; moreover, it allows all four membranes to be monitored at a high sampling rate. Both sets of instrumentation use the same array structure for bilayer formation whose structure is micro fabricated and indicated in an earlier research work of ref. 25.[25] The acquisition unit of the array chip is powered by an eONE amplifier (Elements s.r.l.). Such amplifier is a series of miniaturized devices based on the custom designed chips for low current measurements. It enables the integration of a nanopore measurement circuit in a palm-sized space.[24] At the same time, the basic version of this device is powered by a USB without a voltage boost circuit, so the applied voltage range is only ±380 mV with a resolution of 1 mV. For high voltage applications, the type of eONE-XV functionalizes the voltage ranging from negative 1.65 V to positive 1.65 V. One of the advantages of using the Orbit mini is the temperature control add-on, which allows −10 to 80 °C measurements.

6.4 Integrated Chips for Nanopore Electrochemical Instrumentation

To further lower the noise at high bandwidth, it is required to integrate the current measurement system into a single chip. A low-power composition chip for DNA readout with high accuracy based on nanopores for multichannel applications was developed.[20] By adopting a novel offset cancellation block, the unwanted input offset was reduced, increasing the precision of the DNA readout (Figure 6.11a). The advantage of the circuit structure is the usage of non-inverting input for amplifying the ionic current, which has a high input impedance. This chip was fabricated in a 0.35 μm CMOS process (Figure 6.11b). The feedback resistor of the chip is 498 MΩ, with input-referred noise 3.46 pA_{RMS} at the bandwidth of 10 kHz. It also has a low power consumption of 213 μW.

A novel CMOS-integrated nanopore platform (CNP) based on a micro-chip design was developed (Figure 6.12). The chip is integrated with alow-noise current preamplifier and a high-performance inorganic nanopore.[5] As shown in Figure 6.12a, the SiN_x nanopore was mounted in the chamber above the amplifier to highly reduce the parasitic capacitance. The fabrication of the nanopore is indicated in Figure 6.12b and the layout of the

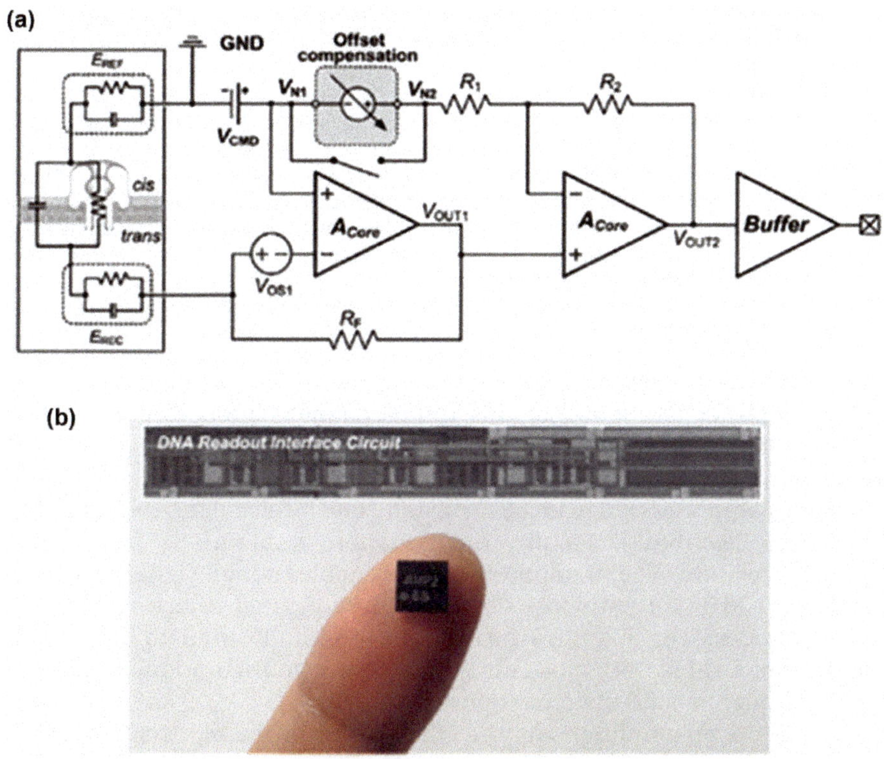

Figure 6.11 (a) The DNA readout chip architecture, which consists of a TIA and a non-inverting amplifier with a high-input impedance. (b) The graph of the chip in a die area of 1.1 mm×0.155 mm (top) and the size of the chip package is much smaller than a finger (bottom).
Reproduced from ref. 20 with permission from Elsevier, Copyright 2016.

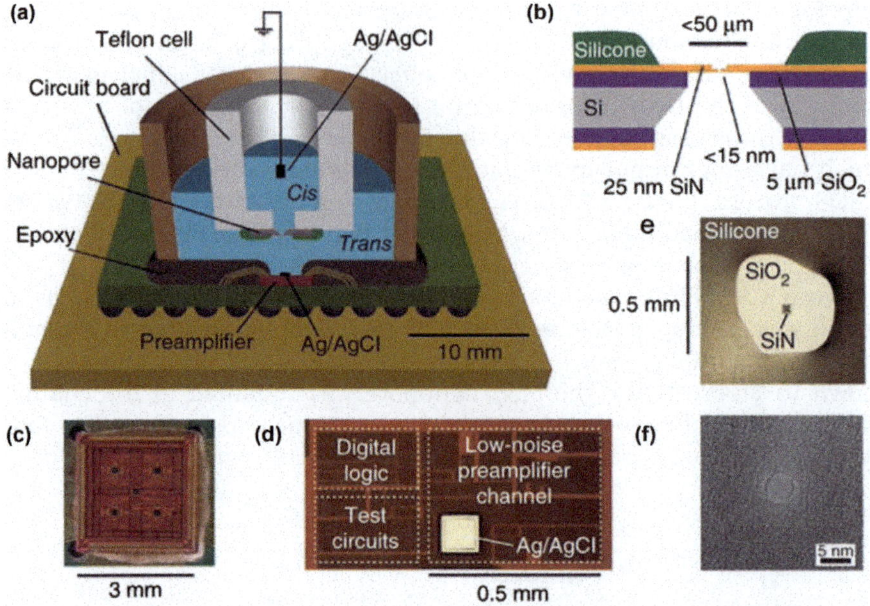

Figure 6.12 (a) Setup of the CNP platform. (b) Cross-section of the low-capacitance thin membrane with a SiN_x nanopore. (c) Chip layout of the eight-channel CMOS current preamplifier for nanopore sensing. (d) The diagram of a single channel of the preamplifier. (e) The fabricated solid-state silicon nitride membrane chip mounted in the chamber. (f) Transmission electron microscope image of the SiN_x nanopore with a diameter of 4 nm.
Reproduced from ref. 5 with permission from Springer Nature, Copyright 2012.

nanopore sensing is given in Figure 6.12c and d. It is noteworthy that the CMOS amplifier is applied to this multi-channel nanopore setup for the purpose of low noise data acquisition. Consequently, it has lower high-frequency noise than in traditional platforms based on external electrophysiology amplifiers such as the Axopatch 200B (Molecular Devices) or EPC-10 (HEKA Electronik). Finally, the nanopore structure is indicated in Figure 6.12e and f. The nanopore platform enables a high signal bandwidth exceeding 1 MHz for nanopore sensing with integrated detection chambers and electrodes. The platform has been successfully used to observe the changing of a short DNA molecule with different configurations as it passes across through a solid-state nanopore.

As a CMOS preamplifier was directly applied in the electrochemical environment, the signal bandwidth of the nanopore was increased by at least an order of magnitude.[5] The size of the chip is only 3×3 mm in dimension with a well-designed integrated circuit. There are eight low-noise preamplifiers assembled and then integrated into a 0.2 mm^2 chip (Figure 6.13). The high bandwidth voltammetry response at such a single molecule interface allows much broader application of nanopore electrochemistry.

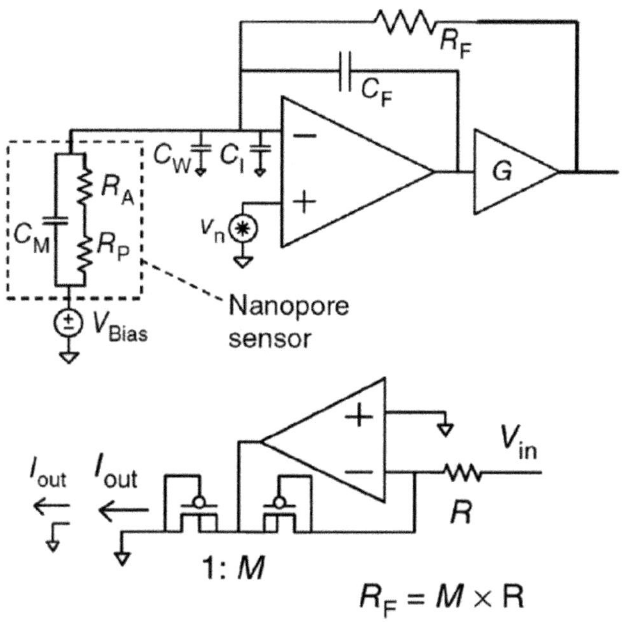

Figure 6.13 The circuit design of the voltage-clamp current preamplifier based on the CMOS process.
Reproduced from ref. 5 with permission from Springer Nature, Copyright 2012.

A custom-designed electronics for high-bandwidth detection of dsDNA with high temporal resolution in nanopipettes was developed (see Figure 6.14).[26] This design of circuit features an integrated CMOS current amplifier with a self-compensating integrator–differentiator based on matched MOSFETs (metal oxide semiconductor field effect transistor) for low-capacitance nanopore applications. The amplifier has a wide bandwidth of 3.5 MHz to record transit signals in the nanopore. In addition, the noise of the circuit is only 6 pA_{RMS} at 100 kHz and 50 pA_{RMS} at 1 MHz acquisition rate (input capacitance $\approx$ 3.5 pF). To measure large DC signals, the amplifier also integrates an additional feedback network with $\pm$100 nA range. This amplifier in this case is applied to the nanopipette for low noise acquisition.

6.5 Data Analysis in Nanopore Electrochemistry

Nanopore studies are mainly based on the detection and analysis of current fluctuations in the current trace and then extracting the amplitude, dwell time and occurrence frequency.[10,27] This information has been used to demonstrate the unknown fundamental chemical and biological question and signal molecule information of analytes such as DNA, RNA, proteins, peptides, polymers, small organic molecules, noble nanoparticles and heavy-metal ions.[28] However, the noisy and fluctuant current signals pose great

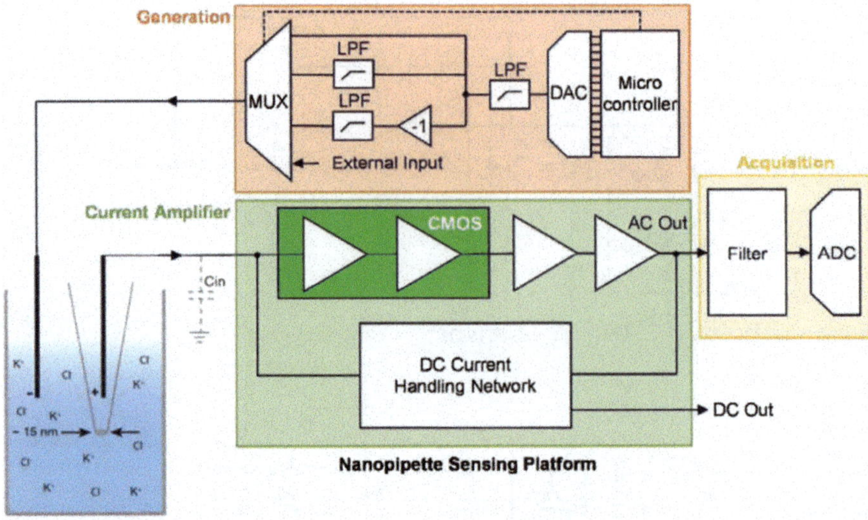

Figure 6.14 Detailed amplifier platform for nanopipette sensing with ultra-low noise current reading and voltage applied.
Reproduced from ref. 26 with permission from the Royal Society of Chemistry.

difficulties for researchers to carry out nanopore data processes and extract the analyte's information. Therefore, appropriate processing algorithms are crucial and indispensable for fast and accurate identification of all the current events and extraction of their information including current amplitude, dwell time and frequency. The single entities induce either an increase or a decrease fluctuation in the ionic current due to the different sensing mechanisms. In this section, we take blockades as the example current response to discuss the nanopore data processing methods.

The conventional method for automatic detecting pulse-like blockades was based on evaluating the mean value of the whole current trace as an open-pore baseline. Once the current pulse deviates from the baseline more than the pre-set threshold, it will be regarded as a blockade.[29] The detection level is usually defined by multiplying a co-factor and the standard deviation of the current baseline (σ), as shown in Figure 6.15; typically the single threshold method is applied. The detection level is chosen appropriately to increase the signal-to-noise ratio, more importantly, so not to lose too much single molecule information.[30]

The feasible single-level approach has commonly been applied to conventional nanopore electrochemistry data analysis. However, it can be hardly used in current data with high noise level. As an improvement, a two-threshold method is used to minimize data loss.[31] Specifically, this approach defines the first threshold (th1) by 2σ from the baseline to avoid around 95% of the noise fluctuations. Subsequently, a higher threshold (th2) defined by 3σ from the baseline was used to distinguish blockade events from the noise.

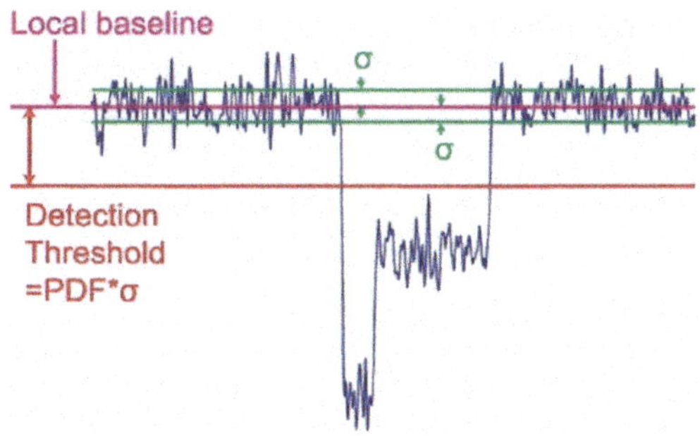

Figure 6.15 The scheme of threshold detection by evaluating the local baseline and noise standard deviation (σ). A constant threshold is introduced for the detection of events by multiplying the noise standard deviation with a coefficient. Events are recognized when points leave the local baseline and drop lower than the threshold. The whole event can be localized by tracking back to the first crossing point and forward to the first points where the current returns to the baseline.
Reproduced from ref. 29, https://doi.org/10.1088/0957-4484/26/8/084003, with permission from IOP Publishing.

Both the single-threshold method and the two-threshold method require the accurate calculation of the baseline of the current trace to serve as a criterion to extract blockade events and analyse the blockade information. Generally, the baseline is based on the Gaussian distribution of the stochastic events.[32] However, during the nanopore experiments, changes in concentration, temperature or a modification of the nanopore interface can induce an increase in the low frequency disturbance part in the Gaussian distribution.[33] Therefore, it is not appropriate to use a constant baseline or threshold for searching blockades, and calculating the dwell time and current amplitude of each current spike precisely due to the inaccurate localization of the start and end points of the blockades.[31] To overcome the deficiencies of a constant baseline, a moving window function was designed to dynamically fetch an updated value baseline level. This method has already been employed in most nanopore data studies, which is typically useful in the case of recordings with obvious current drifts and moderate low-frequency noise by averaging over a limited number of data points (typically 100) to acquire a local baseline.[10]

Threshold-based event detection is just a basis and initial step to detecting and localizing blockade events in the measured current. Afterwards, further data process algorithms will be employed to extract event information and carry out statistical analysis. There are already many different techniques proposed for the analysis of nanopore current traces for event localization and recognizing the multi-levels in events. As for the challenge mentioned above, commercial software, such as pClamp (Axon, Molecular Device), cannot meet the demands of data processing for nanopore experiments. It has become an important and significant target to develop a fast, accurate and automated signal-processing tool for nanopore experimental data. To this end, there are many labs around the world dedicated to developing such a signal-processing tool. The existing nanopore data analysis software such, as Open Nanopore, Pyth-Ion, Nanopore Analysis, Transalyzer and MOSAIC, is discussed below.

After locating and extracting the blockade events, the traditional data analysis for nanopore experiments was based on statistics analysis of whole blockade events. However, these data analysis methods cannot provide detailed information on multiple current levels inside each blockade event and has bad performance when the signal-to-noise ratio (SNR) is very low. To overcome this dilemma, a CUSUM-based (cumulative sum chart) software package named OpenNanopore (shown in Figure 6.16a) has been developed to perform nanopore data analysis in more detail.[33] The CUSUM is an abrupt detection algorithm requiring little prior knowledge of the input data. It has been used to identify different piecewise constant values inside each event and subsequently extracted blockade information of each level of these events (Figure 6.16b). For processing experimental datasets of small biopolymers with multiple current levels, it ensures accurate and fast extraction of dwell time and current amplitude from each current level.

In this program, nanopore data analysis consists of three steps. Firstly, the program detects blockade events using dynamic detection level encoded with moving window to overcome low-frequency variations in the baseline. Then, the CUSUM algorithm is used to fit the levels from the current fluctuations and automatically extracts their dwell time and amplitude information. Finally, the single molecule information of each analyte is stored in a readable database. This algorithm enables distributing the detected events into a histogram and scattering for the further statistical analysis of the multi-level events. To demonstrate the efficiency and validity of the program, a data analysis of a λ-DNA translocation experiment was demonstrated.[33] Making use of the CUSUM algorithm, comparing to the traditional histogram approach, the software achieved a much wider frequency spectrum to extract single molecule information from low SNR cases. Pyth-Ion is a Python-based program that was designed for analysis of nanopore-based ionic current data. The software has a primitive baseline detection. However, for accurate analysis, a user setting baseline is required. After this simple event detection, a database file is generated in the same file folder, which is used to store the event analysis information, including current blockade, fractional current blockade, dwell time, inter-event time, and standard deviation of current values.

In the traditional moving window method, the average of the current baseline in the moving window is heavily influenced by the previous data points. Although the transient in the baseline is eliminated, the typical fluctuation that represents nanopore electrochemical behaviours is also removed. To solve this problem, the software eliminates the effect of the detected events to the baseline evaluation by removing the events from the raw data.[29] As shown in Figure 6.17a, the flowchart enables the detection of each single molecule event. Basically, once a single molecule case is detected, the

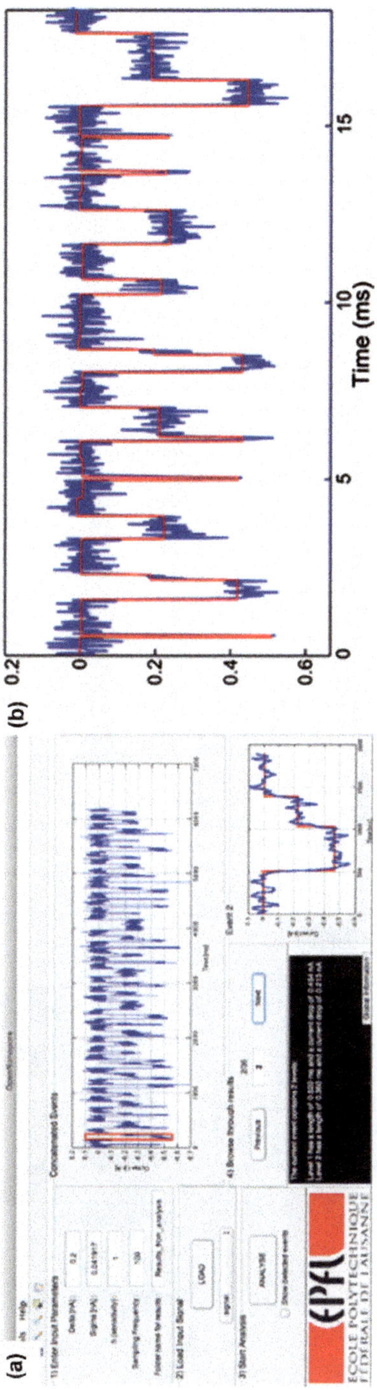

Figure 6.16 (a) GUI of CUSUM-based software OpenNanopore (https://www.epfl.ch/labs/lben/page-79460-en.html/). (b) A segment of raw current signal (blue line) is analysed by the CUSUM method for recognizing the events and labelling their dwell time and amplitude (red line).
Reproduced from ref. 33 with permission from the Royal Society of Chemistry.

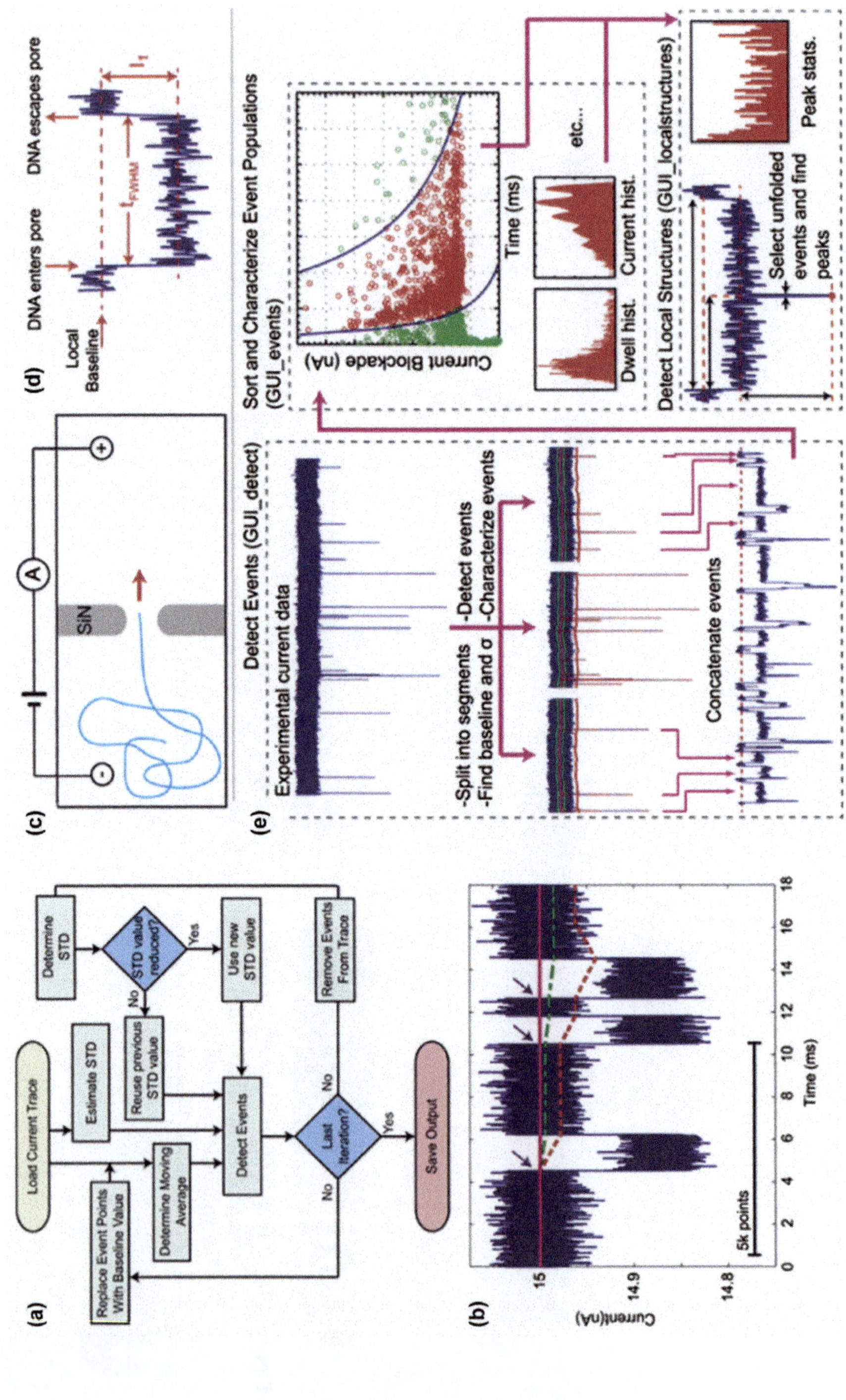
(a)
Load Current Trace
Estimate STD
Determine STD
Replace Event Points With Baseline Value
Determine Moving Average
Reuse previous STD value
STD value reduced?
Use new STD value
Detect Events
Last iteration?
Remove Events From Trace
Save Output
Yes
No
(b)
Current(nA)
Time (ms)
5k points
(c)
SiN
A
(d)
Local Baseline
DNA enters pore
DNA escapes pore
(e)
Detect Events (GUI_detect)
Experimental current data
-Split into segments
-Find baseline and σ
-Detect events
-Characterize events
Concatenate events
Sort and Characterize Event Populations (GUI_events)
Current Blockade (nA)
Time (ms)
Dwell hist.
Current hist.
etc...
Detect Local Structures (GUI_localstructures)
Select unfolded events and find peaks
Peak stats.

baseline information is fetched and returns into the next iteration as a new baseline parameter to gain a more accurate baseline level for the next single molecule case. In the same way, the noise level is also taken into consideration. Figure 6.17b indicates a single molecule current trace using such algorithm (solid line) is not affected due to the current fluctuation compared to the tradition approaches (dashed line). In the case of a SiN_x nanopore detecting a polymer (Figure 6.17c), once the single molecule descriptors from Figure 6.17d are determined, the typical single molecule analysis procedure proceeds as indicated in Figure 6.17e. The algorithm obtaining the original experimental current data then splits it into three segments for determining the baseline, noise level and the single molecule events to sum up a concatenates of all found events. Then it will try to sort and characterize each event for stochastic analysis and, finally, reanalyse the events for the presence of local structures and to generate relevant statistics.

To improve the accuracy and throughput of nanopore-based analysis experiments, a data analysis tool named the modular single-molecule analysis interface (MOSAIC) is used, which was integrated with two pivotal algorithms: ADEPT and CUSUM+. Adaptive time-series analysis (ADEPT, Figure 6.18a) is used to analyse short-lived blockade events that do not reach their steady-state by characterizing the nanopore system with an equivalent circuit model.[34] Cumulative sum analysis (CUSUM+) is an improved cumulative sum statistical method, and is used to acquire more detailed translation information by detecting step-like current curves. As a result, Figure 6.18b shows the comparison of the stochastic analysis from ADEPT (grey) and CUSUM+ (light blue) when analysing the α-hemolysin detecting of PEG molecules. As is expected from the algorithm concept, PEG molecules whose size is larger than 17-mers could be well fitted by both approaches, while for shorter PEG molecules ADEPT performs much better. Herein when adapting the CUSUM+ approach, only single molecule events with residence time larger 5τ are considered, which might be a hinderance for the performance of CUSUM+.

In the ADEPT algorithm, the bilayer with a nanopore inserted is then modelled and equivalent to the electrical elements so that the model system

Figure 6.17 (a) The logic flow of the Transalyzer software for event detection, which measures the value of the local baseline and RMS noise level. (b) Three simulated events with high noise were detected by using the Transalyzer software. (c) Illustration of a typical nanopore setup. (d) Current signal produced by a translocating DNA molecule. (e) Schematic of the typical analysis procedure of a nanopore current trace, which is divided into three parts. The first part splits a current file, detects the events in each segment, characterizes each event, and concatenates all found events. The second part sorts and characterizes the event populations. The final part can be used to sort and reanalyse events for the presence of local structures and to generate relevant statistics.
Reproduced from ref. 29, https://doi.org/10.1088/0957-4484/26/8/084003, with permission from IOP Publishing.

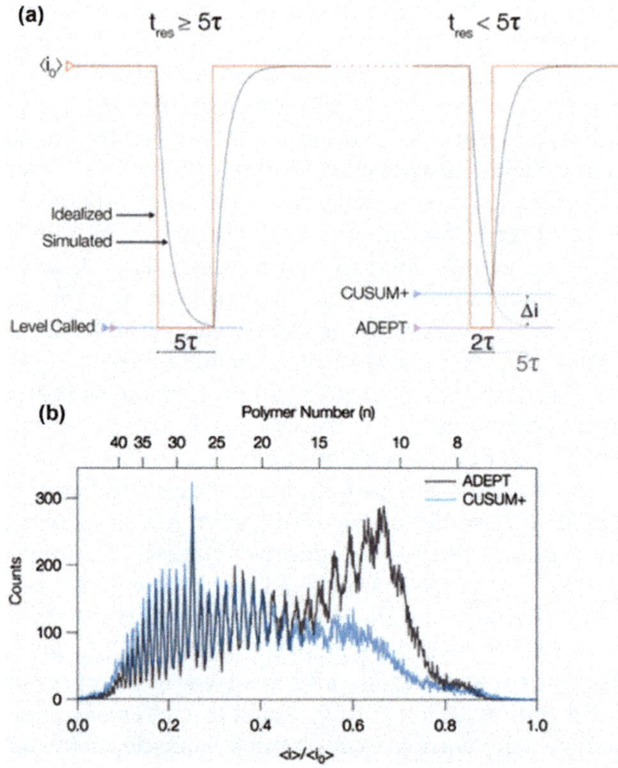

Figure 6.18 (a) ADEPT and CUSUM+ analysis applied to a simulated nanopore measurement (grey). Two events with the same current blockade (red) but different residence times (t_{res}), with respect to the system characteristic relaxation time (τ) are shown. (Left) For a long event ($t_{res} \geq 5\tau$), the ionic current converges close to its steady-state value, and the current levels estimated by ADEPT and CUSUM+ are equivalent. (Right) For short events (*e.g.*, $t_{res} \approx 2\tau$), the current does not reach the steady-state value of the idealized pulse (red). In this case, CUSUM+ and other algorithms used in nanopore analysis systematically underestimate the steady-state current (blue) by an amount Δi (grey; dashed). In contrast, the physical model underlying ADEPT allows the algorithm to accurately estimate an event's steady-state current. (b) Comparison of ADEPT (grey) and CUSUM+ (blue) analysis of a polydisperse polyethylene glycol (PEG) solution measured with an αHL nanopore. A blockade depth histogram of events recovered by each algorithm shows that the number of events recovered by CUSUM+ decreases sharply for small polymers ($n < 20$) that exhibit fast residence times ($<5\tau$) in the pore. Reproduced from ref. 34 with permission from American Chemical Society, Copyright 2016.

could be described into electrical system with the access response (Figure 6.19a). More specifically, the resistance resulting from the electrolyte solution between the two electrodes is modelled as R_{cis} and R_{trans}, with the subscripts denoting each side of the nanopore system. The nanopore inserted in the bilayer is equivalent to a resistor, R_p. Finally, the nanopore support of the lipid bilayer is simplified as a single capacitance (C_m), which is connected in parallel with the aforementioned equivalent circuit. Adapted from this model, the ADEPT algorithm is proposed. This algorithm is validly based on the model that a single entity sojourning inside the nanopore induces the instantaneous increase of the nanopore resistance (R_p) leading to an alternation of ΔR, while the system capacitance can only cause the ionic current over a finite time. As an equivalent model of a nanopore (Figure 6.19a), the correspondence of the electrical circuit is applied, when applying the calibration model to the nanopore model (Figure 6.19b), the impedance is measured after the removal of the artifacts from the instrumentation including the unwanted capacitor and resistance. The electric response of the nanopore model over the frequency domain is plotted in Figure 6.19c. Finally, by applying this technique, a sub-picoampere resolution is achieved from a known sequence of ssDNA indicating a better noise elimination approach.

As to the fast events, which are beyond the bandwidth of the measurement system, they are hardly to be extracted properly from the baseline current due to insufficient amperometric resolution. However, the ADEPT model proposed can simulate the most reality shape of the single molecule event with short dwell time. To evaluate the accuracy and reliability of the algorithm proposed, ADEPT was applied to analyse the interaction between the synthesized polymer with the α-hemolysin. As expected, the output from the newly proposed algorithm figured out 18-fold more single molecule events at higher bandwidth (100 kHz). Regarding the performance at lower bandwidth (10 kHz), the reduced constraints from the on-data acquisition was also achieved with lower noise level. Similarly, an algorithm MOSAIC was adopted to analyse the translocation of a dsDNA strand through an SiN_x nanopore with a *ca.* 2.4 nm opening. The conventional approach could not identify the bouncing between the dsDNA and pore mouth, which should be uncovered by this novel algorithm. In addition, the MOSAIC analysis of the oligonucleotide signals could easily identify the short interaction between dA_{20} and the α-hemolysin nanopore.

An improved MOSAIC (CUSUM+) was developed by integrating with the CUSUM algorithm.[34] This was achieved by firstly calculating the relax time between two neighbouring events. Then, specifying the minimum value of this relax time. This is used to exclude the unwanted transients from the multi-state detection and the size of the events. Short-lived events with distorted blockade depth are calibrated by the CUSUM+ algorithm in MOSAIC. The blockade amplitude is evaluated by the ADEPT algorithm based on fitting the data to an electrical circuit model of the nanopore for reducing the influence of bandwidth limitation. The results when analysing different size PEG molecules using a α-hemolysin nanopore applying ADEPT and CUSUM+ is demonstrated. When the PEG molecules are larger than 17-mer, both algorithms are suitable. Regarding to smaller size molecules, ADEPT shows much better sensitivity; one possibility could be that the CUSUM+ considers events larger than 5τ.

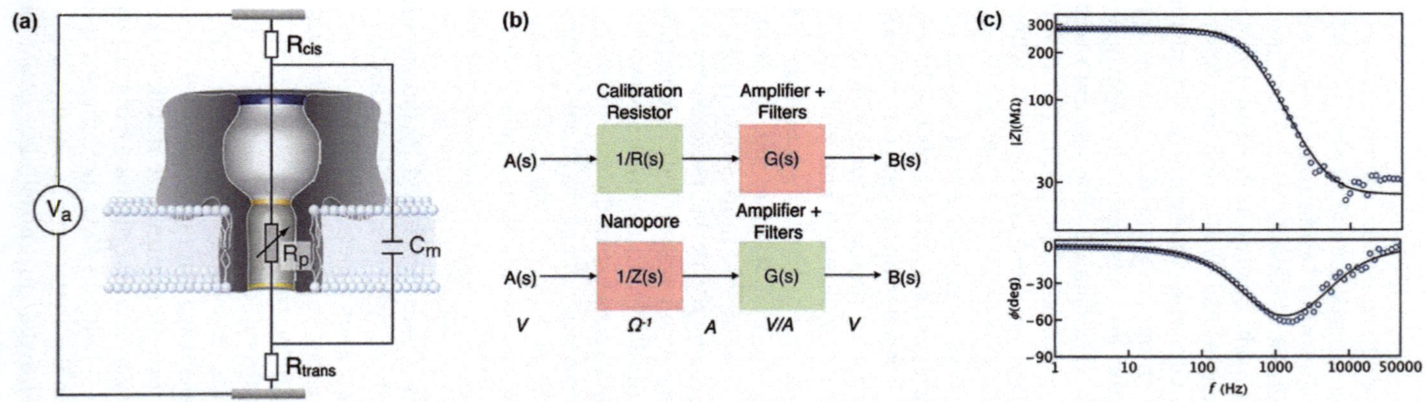

Figure 6.19 (a) Equivalent circuit model of the nanopore system. The nanopore and membrane are modelled as a resistor (R_p) and capacitor (C_m) in parallel, together with the total solution and access resistances on either side of the membrane (R_{cis} and R_{trans}). (b) The nanopore impedance is measured after removing artifacts from the measurement apparatus. First, the transfer function of the measurement apparatus is determined from the response ($B(s)$) of a 300 MΩ calibration resistor to a swept sinusoidal voltage ($A(s)$) stimulus (top). Assuming the resistor frequency response is flat, $G(s) = (B(s)/A(s))R(s)$, where $G(s)$ and $R(s)$ are the transfer functions of the measurement apparatus and the calibration resistor respectively. The corrected nanopore impedance was then determined in an identical manner (bottom) to yield $Z(s) = G(s)/(B(s)/A(s))$. (c) The equivalent circuit model was validated by measuring the impedance of the nanopore–membrane complex as a function of frequency (blue markers). Reproduced from ref. 35 with permission from American Chemical Society, Copyright 2014.

Besides the model fitting method for recovering the events as used in MOSAIC, a novel second-order-differential-based calibration method (DBC) (Figure 6.20a) and a current integration method (Figure 6.20b) were used to accurately extract the amplitude and dwell time of short blockade events.[10] At first, an improved threshold detection method was employed to localize the rough region of the blockade event. Then, the adapted DBC approach was applied to optimize the region of blockades and accuracy of the dwell time. Based on the DBC approach, the precision of determining the start and stop points of a certain blockade was optimized. On the other hand, the current magnitude was investigated by an integration function that was optimized for better blockage recovery. The data analysis procedure was implemented in the software of Nanopore Analysis based on the Matlab platform (http://hysz.nju.edu.cn/ytlong/Data/View/267).

Since the data processing requires massive time and manual resources, the development of an online analysis tool with automatic analysis capability in nanopore techniques is in demand. According to such demand, an online event recognition and analysis package was designed (Figure 6.21). The algorithm provides real-time data storage and online event detection and analysis in real nanopore recording experiments with high accuracy and speed. Firstly, a double buffer data structure is employed to handle the massive amount of data, which is inspired by the Ping-pong buffer principle.

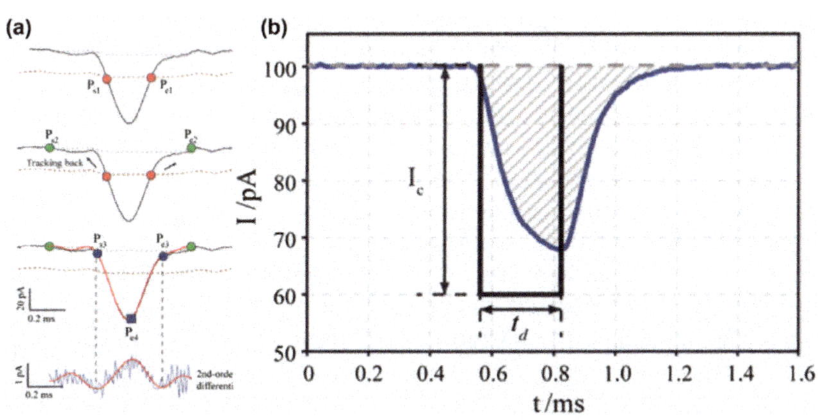

Figure 6.20 (a) Region of blockage recognized by a local threshold. Grey and brown dashed lines represent the local baseline and local threshold, respectively. The recognized region of the blockage is marked between Ps1 and Pe1 (red circles). (b) Example of the measurement of the current amplitude of an individual blockage by an integration method. The blockage was generated with a t_{nom} of 0.25 ms and an I_{nom} of 40 pA. The recorded blockage and its ideal wave pulse are plotted as blue and black lines, respectively. The grey hatched area represents the integrated area of the recorded blockage. The baseline is the grey dashed line. The cutoff frequency of the filter used to record the blockage was 3 kHz. Reproduced from ref. 10 with permission from American Chemical Society, Copyright 2014.

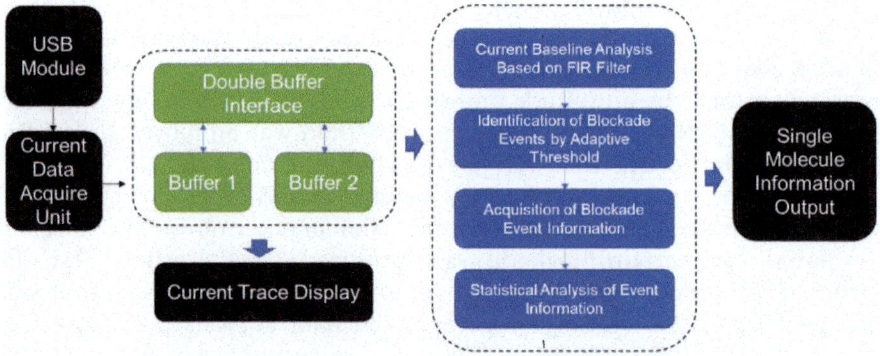

Figure 6.21 Diagram of data process module for nanopore analysis. Adapted from ref. 23 with permission from Elsevier, Copyright 2018 Changchun Institute of Applied Chemistry, Chinese Academy of Sciences.

Buffer 1 and buffer 2 will receive the single molecule acquisition data directly from the USB module; at the same time, the remain steady buffer should process the single molecule data to analysis the single molecule events. According to this novel buffer region, the processing data speed will be optimized, which could reduce the usage of limited RAM space. Consequently, the CPU analysis consumes much less data space for single molecule analysis ensuring the stability of the analysis tool.[23]

Regarding to such double buffer interface, the single molecule cases are analysed with a suitable algorithm. Based on the single molecule data characterization, single molecule information such the residence time should be determined accurately for better analysis. In the later software for post-analysis, the commonly used threshold approach is used. However, the baseline of the current trace needs to be firstly accurately determined as a promising parameter for the threshold approach. In this algorithm, an optimization is introduced by using an FIR filter module. This filter is designed to filter the noise and is based on the previous online data analysis from the double buffer interface; the baseline level could achieve self-assessment.

6.6 Intelligent Big Data for Nanopore Electrochemistry

Nanopore single molecule analysis has become more sophisticated when it shows the capability for sub-molecular information. By reading each single molecule ionic current responses, the inter- and intra-molecular information is unrevealed. Mining data from the stochastic recordings, it is difficult to examine the single-molecule information one-by-one. Thus, a case-to-case intelligent data analysis is required. The stochastic nanopore events contains abundant single molecule information as well as ambient noise, making use of the hidden Markov model (HMM), it is noteworthy to pre-set the starting parameters to only focus on the data categories of interest. To achieve this automatically, a fuzzy c-means (FCM) based on the HMM algorithm could eliminate unwanted short-time elapsed cases then output higher accuracy HMM outputs.[36] Regarding the case that the single molecule event generates multiple current stages, the data procedure could be combined with a modified density-based spatial clustering of applications with noise (DBSCAN) algorithm and a Viterbi training algorithm to retrieve different states of each single molecule event accurately.[37] The stochastic analysis of the massive amount of the single molecule events with various current states drives the analysis to the self-fitted artificial intelligence (AI) approach as well as big data handling. The application of these approaches and data handling strategies is typically useful in the healthcare area. The conventional AI algorithm, the CNN (convolutional neural network) empowered QuipuNet was applied to extract the single molecule characteristics from the current trace.[38] Once the QuipuNet loads all the nanopore electrochemistry data sets, the neural network studies the input with high self-discipline and outputs the final results with predicted multi-state levels as pre-set. This approach is an easy process with high accuracy, and it is potentially possible to extract hidden intermolecular information.

One of the data mining approaches in recognizing the nanopore-based single molecule ionic current is developed on the AdaBoost model.[39] Due to the fact that the interaction between the analyte and the nanopore inducing typical ionic fluctuation, the current signature contains molecular information such as the structure, conformation, chemical groups and charges. From the example work, oligonucleotide 5′-AAAA-3′ and 5′-GAAA-3′ were mixed and applied to the aerolysin nanopore, which produce the same ionic current trace. Usually, the 5′-AAAA-3′ and 5′-GAAA-3′ were hardly identified by the traditional distinguishing of ionic current or duration time. By applying a series of featured vectors, which were calculated from the hidden Markov model to the AdaBoost model for training, the accuracy of the identification outcome from such machine learning models hits 0.991. By further optimizing the training parameters, higher accuracy could be even achieved.

Another machine learning approach based on learning shapelets[40] can also been adopted and applied to the nanopore single molecule data (Figure 6.22). In this work, a shapelet-based machine learning approach is utilized to discriminate the current signal from 5′-AAAA-3′ and 5′-GAAA-3′. Akin to the aforementioned issue, in the conventional analysis approach it is difficult to discriminate the current due to the similar charge density and conformation under experimental condition. To train the machine learning, the time-series dataset of current fluctuation was studied by using time-series shapelets (LTS) algorithm. Then a logistic classifier was applied to identify the two types of substrates to fit the shapelet-transformed descriptors. Lastly, the evaluation was carried out and turned out an accurate F1 score of 0.933.

We anticipate that further introduction of more intelligent algorithms such as deep learning, unsupervised learning and transfer learning promotes the mining of deep information from the blockades instead of conventional evaluations on the dwell time and amplitude. Also, more blockades characteristics should be directly revealed from the ionic current trace, such as current frequency (discussed in Chapter 1),[41] which could be

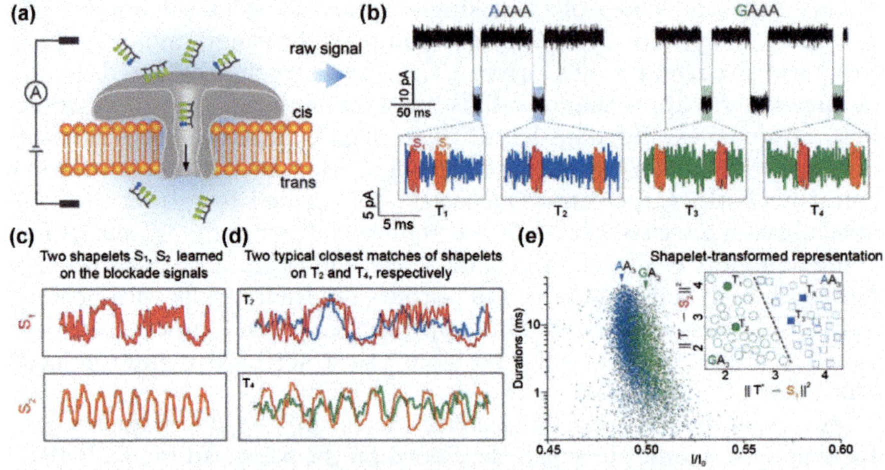

Figure 6.22 Shapelet learning on nanopore data. The blockade current signals are extracted from a limited range of data point numbers, and the most similar segments extracted from the open current are subtracted from them. Then, the LTS algorithm is used to mine shapelets that transform the raw blockade current signal into a new representation of distance or similarity. (a) Illustration of aerolysin nanopore in analysing the mixture of 5′-AAAA-3′ and 5′-GAAA-3′. (b) The raw current traces for 5′-AAAA-3′ (left) and 5′-GAAA-3′ (right). The current trace in each signal is numbered as Tm. The representative shapelets S1 and S2 overlap with the closest match among the subsegments in each blockade. (c) Two typical patterns of shapelets S1 and S2 learned from the blockade current signals. The LTS algorithm can determine a certain number of shapelets. For a clear illustration, this figure shows only two shapelets, S1 and S2, from the blockade current signals. (d) Illustration of the closest matches of two shapelets, S1 and S2, to T2 of 5′-AAAA-3′ (blue) and T4 of 5′-GAAA-3′ (green), respectively. (e) The scatter plots of duration *versus* residual current of 5′-AAAA-3′ and 5′-GAAA-3′. Insert: shapelet-transformed representation in two-dimensional space. The traditional scatter plots of duration *versus* residual current exhibit a serious overlap. Given shapelets S1 and S2, two analytes show a clear distribution in the scatter plots of minimum Euclidean distances $(\| T^* - S_n \| 2)$ between the optimal subsegment in each blockade and S1 and S2. All data were acquired at 22 ± 2 °C with an applied voltage of +100 mV in 1 M KCl, 10 mM Tris and 1 mM EDTA buffer at pH 8.0 in the presence of 5 μM oligonucleotide. Current traces were acquired at a sampling frequency of 250 kHz and filtered by a 5 kHz low-pass filter. Reproduced from ref. 40 with permission from American Chemical Society, Copyright 2019.

of benefit to accurately identify the sub-group of single molecules for potentially achieving wider applications including protein sequencing.

Acknowledgements

Zhen Gu, Cheng-Yu Yang and Cheng-Bing Zhong are especially acknowledged for their contributions in organizing the materials and pictures.

References

1. P. Davies and A. Fletcher, *P. I. Mech. Eng. C-J. Mech.*, 1995, **209**, 409–418.
2. Y.-L. Ying, C. Cao, Y.-X. Hu and Y.-T. Long, *Natl. Sci. Rev.*, 2018, **5**, 450–452.
3. J. D. Uram, K. Ke and M. Mayer, *ACS Nano*, 2008, **2**, 857–872.
4. R. Gao, Y.-L. Ying, B.-Y. Yan and Y.-T. Long, *Chin. Sci. Bull.*, 2014, **59**, 4968–4973.
5. J. K. Rosenstein, M. Wanunu, C. A. Merchant, M. Drndic and K. L. Shepard, *Nat. Methods*, 2012, **9**, 487.
6. D. S. Talaga and J. Li, *J. Am. Chem. Soc.*, 2009, **131**, 9287–9297.
7. A. J. Storm, C. Storm, J. Chen, H. Zandbergen, J.-F. Joanny and C. Dekker, *Nano Lett.*, 2005, **5**, 1193–1197.
8. M. Wanunu, W. Morrison, Y. Rabin, A. Y. Grosberg and A. Meller, *Nat. Nanotechnol.*, 2010, **5**, 160–165.
9. E. A. Manrao, I. M. Derrington, A. H. Laszlo, K. W. Langford, M. K. Hopper, N. Gillgren, M. Pavlenok, M. Niederweis and J. H. Gundlach, *Nat. Biotechnol.*, 2012, **30**, 349–353.
10. Z. Gu, Y. L. Ying, C. Cao, P. He and Y. T. Long, *Anal. Chem.*, 2015, **87**, 907–913.
11. J. J. Kasianowicz, E. Brandin, D. Branton and D. W. Deamer, *Proc. Natl. Acad. Sci. U. S. A.*, 1996, **93**, 13770–13773.
12. Z. Gu, H. Wang, Y.-L. Ying and Y.-T. Long, *Sci. Bull.*, 2017, **62**, 1245–1250.
13. R. Sherman-Gold, Inc. Foster City, CA, 1993, Axon Instruments, Foster City, CA, https://www.moleculardevices.com/en/assets/ebook/dd/cns/axon-guide-to-electrophysiology-and-biophysics-laboratory-techniques.
14. B. Sakmann and E. Neher, *Annu. Rev. Physiol.*, 1984, **46**, 455–472.
15. Z. Gu, Y.-L. Ying and Y.-T. Long, *Sci. China: Chem.*, 2018, **61**, 1483–1485.
16. J. B. Edel and T. Albrecht, *Engineered Nanopores for Bioanalytical Applications*, William Andrew, 2013.
17. R. M. M. Smeets, U. F. Keyser, D. Krapf, M. Y. Wu, N. H. Dekker and C. Dekker, *Nano Lett.*, 2006, **6**, 89–95.
18. D. P. Hoogerheide, S. Garaj and J. A. Golovchenko, *Phys. Rev. Lett.*, 2009, **102**, 256804.
19. H. Landau, *Acta Math.*, 1967, **117**, 37–52.
20. J. Kim and W. B. Dunbar, *Sens. Actuators, B*, 2016, **234**, 273–277.
21. V. Tabard-Cossa, *Instrumentation for Low-noise High-bandwidth Nanopore Recording*, William Andrew, ch. 3, 2003.

22. G. Zhen, Y. Yi-Lun and L. Yitao, *Sci. China: Chem.*, 2018, **61**, 1483–1485.
23. W. Hui-Feng, F. Huang, G. Zhen, H. Zheng-Li, Y. Yi-Lun, Y. Bing-Yong and L. Yi-Tao, *Chin. J. Anal. Chem.*, 2018, **46**, 843–850.
24. C. Weichbrodt, H. Bajaj, G. Baaken, J. Wang, S. Guinot, M. Kreir, J. C. Behrends, M. Winterhalter and N. Fertig, *Analyst*, 2015, **140**, 4874.
25. G. Baaken, M. Sondermann, C. Schlemmer, J. Ruhe and J. C. Behrends, *Lab Chip*, 2008, **8**, 938–944.
26. R. L. Fraccari, M. Carminati, G. Piantanida, T. Leontidou, G. Ferrari and T. Albrecht, *Faraday Discuss.*, 2016, **193**, 459–470.
27. N. Arjmandi, W. Van Roy, L. Lagae and G. Borghs, *Anal. Chem.*, 2012, **84**, 8490–8496.
28. C. Dekker, *Nat. Nanotechnol.*, 2007, **2**, 209–215.
29. C. Plesa and C. Dekker, *Nanotechnology*, 2015, **26**, 084003.
30. C. Cao, Y. L. Ying, Z. Gu and Y. T. Long, *Anal. Chem.*, 2014, **86**, 11946–11950.
31. A. Oukhaled, L. Bacri, M. Pastoriza-Gallego, J. M. Betton and J. Pelta, *ACS Chem. Biol.*, 2012, **7**, 1935–1949.
32. D. Pedone, M. Firnkes and U. Rant, *Anal. Chem.*, 2009, **81**, 9689–9694.
33. C. Raillon, P. Granjon, M. Graf, L. Steinbock and A. Radenovic, *Nanoscale*, 2012, **4**, 4916–4924.
34. J. H. Forstater, K. Briggs, J. W. Robertson, J. Ettedgui, O. Marie-Rose, C. Vaz, J. J. Kasianowicz, V. Tabard-Cossa and A. Balijepalli, *Anal. Chem.*, 2016, **88**, 11900–11907.
35. A. Balijepalli, J. Ettedgui, A. T. Cornio, J. W. F. Robertson, K. P. Cheung, J. J. Kasianowicz and C. Vaz, *ACS Nano*, 2014, **8**, 1547–1553.
36. S. Winters-Hilt, W. Vercoutere, V. S. DeGuzman, D. Deamer, M. Akeson and D. Haussler, *Biophys. J.*, 2003, **84**, 967–976.
37. J. Zhang, X. Liu, Y. L. Ying, Z. Gu, F. N. Meng and Y. T. Long, *Nanoscale*, 2017, **9**, 3458–3465.
38. K. Misiunas, N. Ermann and U. F. Keyser, *Nano Lett.*, 2018, **18**, 4040–4045.
39. X.-J. Sui, M.-Y. Li, Y.-L. Ying, B.-Y. Yan, H.-F. Wang, J.-L. Zhou, Z. Gu and Y.-T. Long, *J. Anal. Test.*, 2019, **3**, 134–139.
40. Z. X. Wei, Y. L. Ying, M. Y. Li, J. Yang, J. L. Zhou, H. F. Wang, B. Y. Yan and Y. T. Long, *Anal. Chem.*, 2019, **91**, 10033–10039.
41. S.-C. Liu, M.-X. Li, M.-Y. Li, Y.-Q. Wang, Y.-L. Ying, Y.-J. Wan and Y.-T. Long, *Faraday Discuss.*, 2018, **210**, 87–99.
42. G. Zhen, East China University of Science and Technology (Shanghai), Doctorate Thesis, 2018.

CHAPTER 7

Summary and Outlook

YI-LUN YING,[a] SI-MIN LU,[a] JIAJUN WANG*[a,b] AND YI-TAO LONG[a]

[a] Nanjing University, Nanjing, China; [b] East China University of Science and Technology, Shanghai, China
*Email: jiajunwang@nju.edu.cn

Over the last few decades, inspired by transmembrane ionic channels, various nanopores, including biological nanopores and synthetic nanopores, have been designed and engineered to fabricate electrochemically confined spaces. Beyond DNA sequencing, nanopore electrochemistry has a huge impact upon the measurement of single entities, including single molecules, single cells, or single particles.[1] As a result, nanopore confinement is at the forefront of single entity electrochemistry (SEE) that facilitates the studies of stochastics and heterogeneities in 'single entity' behaviors. The pore-based structure makes it possible to unveil the intrinsic characteristics of individual sub-populations at a single entity level, such as nucleic acids, amino acids, peptides, proteins, and even in a complete biological-system. The emergence of nanopore-based electrodes also benefits the control and monitor of molecular flux and electron-transfer processes during electrochemical reactions, contributing to revealing the structure-and-function relationships of individual entities. Moreover, nanopores can be integrated into optical sensing platforms to focus electromagnetic waves within a nano-scale confined space for the enhancement of optical read-outs. Overall, nanopore-based electrochemical confinement has opened up a new avenue for the better understanding of biological processes and chemical reactions.

Detection Science Series No. 17
Confining Electrochemistry to Nanopores
Edited by Yi-Tao Long

Published by the Royal Society of Chemistry, www.rsc.org

The first attempt of using the nanopore for single bio-molecule measurement is the achievement in the analysis of single nucleic acids in 1996.[2] Since then, the pore-based sensing tool has stepped forward into the single molecule level. Driven by the demand of advanced DNA sequencing, a variety of biological nanopores with diameter size from 0.7 nm to 2.8 nm were successfully explored to achieve better sensitivity for single base discrimination over conventional methodologies. However, DNA sequencing is not the end of nanopore sensing system, as it presents many potential applications in the identification of a range of nucleic acids, proteins, metabolomic biomarkers, and even single molecule dynamics. The performance of biological nanopores for single molecule sensing is associated with the structure and the chemical properties of the nanopore. To further improve the performance (*i.e.* sensitivity and selectivity) of the nanopore-based single biomolecule interface for analytes of interest, massive efforts and progress have been made with nanopore design and fabrication. For instance, it has been demonstrated that the structure and the composition of the nanopore lumen dominates the sensing capability. One of the most prominent examples, aerolysin, from *Aeromonas hydrophila*, can spontaneously assemble into a heptameric transmembrane barrel with a diameter of approximately 1.0 − 1.4 nm. Early in 2006, it was for the first time applied to peptide investigation.[3] ssDNA was demonstrated to translocate through the aerolysin nanopore. The narrow constriction of aerolysin nanopore's sensing region along with the unique non-covalent interactions between ssDNA and a single biomolecule interface can significantly enhance the spatial resolution to ~1 nm. When the ssDNA transits through the WT aerolysin nanopore at *ca.* 2 ms/base, the relatively strong non-covalent interactions enable the ultra-sensitivity identification of DNA molecules and the real-time monitoring of Exonuclease I at the single molecule level.[4] The translocation of short oligonucleotides with only three and four bases through the nanopore confinement could also be differentiated respectively, which further demonstrates the superior spatial and temporal resolution of such biomolecule sensing interface.[5] Furthermore, such a sensing interface that can be controlled at the single atomic precision is further extended to the determination of DNA methylation in clinical serums at a single molecule level.[6] It is worth mentioning that the maturity of nanotechnology spurs the emergence of synthetic nanopores (*e.g.* inorganic porous material, nanopipette, and nanochannel). The mechanical robustness and the ease of modification of the synthetic nanopore also permits the possibility of nucleic acid detection.[6,7] In general, typical velocities of nucleic acids for transit through the solid-state nanopore is fast, to typically 10–1000 ns per base, which is a critical limitation to the progress of solid-state nanopore in single molecule sensing. By virtue of the ease of modification, a nanofiber mesh was integrated into a SiN_x nanopore to regulate the translocation speed of individual double strand DNA (dsDNA) molecules. In the experiment, either the *cis* or *trans* side of the nanopore was treated with a tunable and highly porous mesh, and the effect of nanofiber mesh on the translocation speed of dsDNA was studied by employing a set of meshes with various mesh composition. The narrower space of mesh provided a more effective confinement to hamper the mobility of dsDNA. The research indicated that compared with the translocation speed of dsDNA in the bare nanopore system, the passage speed of dsDNA can be effectively reduced even 100-fold after the modification by the nanofiber mesh, which enhanced the resolution of the nanopore sensor.[8] Similarly, the strategy based on the chemical modification was further applied in the nanopipette sensing system. After modification by hydrogel, the electrochemical confined space in the nanopipette is able to dynamically capture the mobility of short nucleic acid fragments as small as 100 base pairs, showing tremendous potential for molecular diagnosis.[9]

As the major building blocks of life, proteins dictate most biological functions among all the living species; more importantly, proteins are directly related to the phenotype of a cell. Consequently, to study the protein tertiary structures, active sites, *etc.* is a major concern in protein science. Although proteomic study is critical, unlike genomic study the development of peptide and protein science are hindered mainly due to the shortage of characterization approaches. In addition to single nucleic acid molecules, nanopore-based confined space was also applied in the investigation of peptides/proteins.[2] Compared with the structure of DNA, the complex structure of a tertiary structure and charge distribution of a protein and peptide make it a challenge for nanopore sensing. In order to simplify the complexity of the structure, model molecules were studied initially. Peptides of different repeats of [-CSA-(Gly-Pro-Pro)*n*] ($n = 1$, 2 and 3 denoted as P1, P2 and P3) were measured by an α-hemolysin nanopore. In recent years, some important characterizations of proteins such as the stability, conformation change, folding state of protein and the enzymatic kinetics in solution have been further studied using the nanopore technique. For example, to further understand the interaction of peptides and nanopores, peptides with different conformation such as α-helix or β-sheet were discriminated. According to the statistical signals, both nanopore properties (*e.g.* shape and surface charge of the nanopore cap) and peptide properties (*e.g.* sequence, length, charge distribution, conformation, *etc.*) dominate the pore–analyte interactions. In addition, nanopores have been applied in the direct detection of disease proteins such as amyloid protein and α-synuclein in the presence of Congo red (CR) and cyclodextrin (CD). In 2011, α-hemolysin was used to study the aggregation of Aβ1-42, which is related to Parkinson's disease, and the protein generated different signals while interacting with the drug molecules.[10] This method could be beneficial for the evaluation of effects on the structure of peptides and hence drug screening. Furthermore, Aβ25-35 and Aβ35-25 could be directly discriminated by the α-hemolysin nanopore. Through the statistical analysis of the events frequency *versus* time, the dynamic process of the aggregation of the proteins could be acquired. Additionally, α-synuclein of which the fibril formation is critical in the pathogenesis of Parkinson's disease, was also studied by the nanopore.[10]

According to the outcomes, the real-time monitoring of α-synuclein and voltage-induced conformational change was achieved. Based on the interactions of biomolecules, nanopores could also be applied to the sensing of complex of protein–ligand (*e.g.* biotin–streptavidin) and antigen–antibody (*e.g.* IgC–protein G). Owing to the adjustable dimension of the nanopipette, the nano-sized volume of the pipette rendered the monitoring of the dynamic aggregation of Aβ1-42. By precisely controlling the fluid and electric field in the nanopipette confinement, the single Aβ1-42 molecules in various aggregation states were driven toward the inner channel of the pipette. The variation of the aggregation state resulted in the differences in ion current modulation, and the dynamic oligomerization of Aβ1-42 molecules was thus successfully revealed by the electric read-out by a nanopipette with diameter of 30 nm.[11]

DNA sequence tools have been developed in a robust manner over the few past decades. However, the sensing tools for peptide and protein sequencing seeking higher sensitivity and higher throughput properties, face the following three main bottleneck problems: (i) proteins are built from 20 distinctive amino acids while DNA is comprised of only four different bases; (ii) amino acid charges are distributed widely; (iii) disruption of the secondary and tertiary structures. Moreover, DNA samples can be amplified using polymerases, while protein sequencing platforms cannot benefit from such an approach. Nanopores have shown a great potential for label-free protein/peptide characterization.

Thus, beyond DNA sequencing, they also hold the possibility for peptides and proteins sequence. Nanopore electrochemistry is very sensitive owing to its nano-sized confined space compared to peptides or proteins. The probing of the tertiary structure or active sites of a protein could be achieved by altering the physical parameters of the aqueous solution; downsizing the nanopore geometry; controlled tuning of the balance between the electrostatic and electroosmotic forces; coating nanopores with a fluid lipid bilayer; employing a pressure-voltage biased pore: one needs to boost the current resolution with a controllable translocation manner by all means. Early in 2004, an example was shown to discriminate the peptide conformational change through a well-studied biological nanopore, α-hemolysin.[3] Moreover, thioredoxin with unphosphorylated, monophosphorylated and dephosphorylated residues had also been studied through such nanopores by attaching the protein with a primer composed of oligonucleotide.[12–16] In addition, ClpX, a motor enzyme, could unfold and pull the polypeptide chain in a controlled manner through α-hemolysin with a reasonable sequencing speed.

By concerning the tertiary structure and non-uniform charge distribution, chemical treatments such as sodium dodecyl sulfate (SDS) or guanidine hydrochloride (GdnHCl), and physical conditions such as heating or high voltage, were used to unfold proteins in both biological and solid-state nanopores.[38,39] Moreover, the coating of SDS could also make the charge distribution of a peptide uniform. The biological nanopore could commonly withstand denaturing conditions (4 M urea and 1.5 M GdnHCl) and the solid-state nanopore can suffer extreme buffer conditions (8 M urea, 6 M GdnHCl, 1% SDS).[14] Last but not least, a series of facile combinations between the nanopore electrochemistry and advanced sensing technologies were explored for the protein sequencing. For instance, a solid-state nanopore was employed to innovate a mass spectrometer, in which the electrospray ionization was activated from a solid-state nanopore. Upon the single proteins being fragmented in the nanopore, each individual amino acid was subsequently ionized and analyzed.[17] It is possible to achieve protein sequencing by integrating the plasmonic system into the nanopore electrochemistry. Recently, a plasmonic nanopore was developed for SERS discrimination of single biomolecules. The electromagnetic field was confined within the narrow volume of the nanopore, which resulted in the balance between the optical force and electric field force.[18] The translocation speeds of individual molecules were subsequently regulated, and a hot spot generated by the nanopore confinement contributed to the SERS read-out. This method is expected to be extended to label-free protein sequencing.

Beyond single molecule sensing, the term 'nanopore electrochemistry' has been successfully extended to the ultrasensitive characterization of single particles and individual cells based on nanopore confinement. The nanopore structure provides a confined volume for the formation of nanoelectrode or nanopore electrode that has been widely applied for single particle electrochemistry. For instance, the confined effect of nanoelectrodes and nanopore electrodes improves not only the collision frequencies of single nanoparticles at the electrode surface, but also enhances the signal-to-noise ratio (SNR) by reducing the background current associated with the double layer charging process.[19,20] Moreover, the confined space enables the electrochemical measurement in the nanosecond scale by decreasing the time constant, R_sC_{dl}, where R_s is defined as the solution resistance and C_{dl} represents the double layer capacitance.[21,22] By single particle electrochemistry, the kinetic information of nanoparticles can be determined *via* nanoelectrodes, which contributes to the preparation of powerful electrocatalysts.[23] The single particle electrochemistry also makes it possible to visualize the extracellular release processes when vesicles collide with the electrode surface.[24] As we mentioned in Chapter 5, a wireless nanopore electrode (WNE) based on the nanopore confined space was recently developed for single particle discrimination.[25] Taking advantage of the electrochemically confined effect, bipolar electrochemistry was effectively activated in the nanopipette, which facilitates the visualization of the electron transfer process within a single living cell.[26] A bare nanopipette, a kind of special nanosyringe, successfully performed the one-step single-cell injection. A confined nanopipette with 100 nm diameter was used to manipulate the movement of fluorescent probes, the fluorescent molecules were precisely delivered into the single MCF-7 cells under the effect of electroosmotic flow. Taking the advantage of simultaneous and precise injection of multi-probe molecules based on the nanopore sensing tools, the evaluation of the pesticide effect was achieved at the subcellular scale. In addition to the single cell analysis, the nanopore is also regarded as a powerful tool to achieve the manipulation of individual living cells.[27] A nanopore-based nano-tweezer was recently developed for single-cell biopsies, which was composed of two individual nanopore electrodes separated by an insulating septum of ~10–20 nm. The dielectrophoresis (DEP) force resulting from a non-uniform electric field in the nanopore confined space was exerted on a single living cell to trap individual molecules of interest with sub-cellular precision.[28] The application of the nanopore-based nano-tweezer filled the gap between single cell manipulation and cell biology. Moreover, researchers took advantage of both the confined nanopore and atomic force microscope (AFM) to develop a unique force-controlled scanning nanopore microscope (SNM).[29] The clever combination permits the possibility of achieving the simultaneous measurement of ion-channel activity and secreted molecules in a single living cell. The mechanical confinement between the substrate and the SNM tip controlled the speed of single molecule movement, ensuring the single cell manipulation at a sub-cellular level.

Perspective

Despite the mentioned advantages only a few applications have been suggested. For example, nanopores have been previously used for biomolecular detection at the single molecule level, such as probing the secondary structure of peptides or monitoring protein binding. Several teams have developed similar approaches using both biological nanopores or solid-state nanopores for detecting enzymes and characterizing single proteins. Aerolysin acting as a promising single biomolecule interface became very attractive as a recently published structure that shows two sensing regions with many possible mutation sites. These mutation sites can enhance the sensing performance for single molecule detection. One could achieve real-time monitoring of the Exonuclease I step-wise cleavage of nucleotides by using aerolysin nanopore.[4] Whereas the aerolysin nanopore shows a promising application in single molecule sensing, there are several challenges in the sensing system. A lower volumetric consumption, reasonably good throughput, longer lifetime span, increased amperometric sensitivity and fast data analysis are the key demands required to further step into the commercialized nanopore products. In addition, solid-state nanopores including SiN_x membrane and quartz nanopipettes have been widely used for the determination and quantification of a wide range of single molecules. The remarkable electrochemical confinement effects enable nanopore techniques as powerful tools for obtaining a wealth of valuable information buried in every individual biomolecule. With the emerging modification methodologies and techniques for solid-state nanopores, the size, geometry and chemical properties can be well tuned, which significantly enhances their sensing ability for a variety of applications. Integration of optical detection into the solid-state nanopore offers the ability to acquire complementary information of a single molecule, increase the analytical throughput and measure positional and structural information with spatial accuracy. We anticipate the new detection mechanism to be incorporated into solid-state nanopore sensing, which will continuously push the development of new types of nanopores and lead to a series of vivid and exciting discoveries in single molecule sensing.

Whereas nanopore electrochemistry has been playing a huge role in single entity sensing, the transient nature of the signal of the nanopore presents a tough barrier for nanopore electrochemistry. The movements of analytes of interest are too fast to transit through the nanopore at the scale of a few microseconds, which is far from the sampling rate of most commercialized electrochemical instruments (<25 kHz). In addition, a higher signal-to-noise ratio is desired to unveil more intrinsic signals of individual entities previously hidden in the high background noise. Driven by the nanopore technique, relative instruments and software have also been developing rapidly. However, the typical nanopore experiments are carried out with bandwidths below 100 kHz to restrict the background noise. Therefore, continued efforts are needed for the development of an electrochemical instrument to overcome these limitations. For example, a low noise system created by integrating complementary metal-oxide semiconductor (CMOS) pre-amplifiers and solid nanopores was developed in thin silicon nitride films, which could acquire high bandwidths of up to 5 at 1 MHz.[30,31] The application of CMOS-based amplifier provided nanopore sensing with sub-microsecond temporal resolution and high signal-to-noise resolution, contributing to the study of single molecule dynamics. The CMOS technique has been applied to nanopore analysis due to low input current noise and high frequency response. Such optimization caused an improvement in the temporal and current resolution reaching 10 μs and 0.5 pA, respectively. Another example, a micro-current electrochemical detection system, was specifically developed for nanopore electrochemical analysis, which uses an integrated circuit composed of three parts, including current amplifier, digital–analog converter and a software module.

Another challenge in nanopore technology is how to process complex data in a high-quality and precise manner. For instance, in order to effectively analyze the data in a nanopore experiment, comprehensive software called 'Nanopore Analysis' has been developed *via* MATLAB software.[32] Such a system can be used to evaluate the blockage current signals, automatically extract and recover the current signals in low signal-to-noise ratio and further analyze the extracted current signals. The stochastic process in nanopore electrochemistry generates massive amounts of data points in the experimental process with high reproducibility; however, aligning, categorizing and automatic clustering of the data are very challenging. To optimize the data processing requires the aid of multidisciplinary teams to resolve and align the unique single molecule information to the predicted model to the heterogeneous signature current. In recent years, it has been promising to apply machine learning to nanopore electrochemistry. For example, although measurement instrumentation has been developed to achieve ultra-low current measuring (pA) at ultra-high temporal resolution (μs), there is still a great amount of hidden information, including more information about the structure adjustment, charge dynamic redistribution and single molecule interactions, in the recorded ionic current. To uncover the intrinsic information of single molecule interactions, a unique algorithm termed as the Hilbert–Huang transform (HHT) was explored to extract the frequency information from the current recording, which can reflect the dynamic non-covalent interactions between analytes and pores (as shown in Figure 1.3).[33] The energy–frequency–time spectrum obtained from current recording by HHT could not only provide the interaction details between the pore and analyte based on the frequency and related energy, but also suggest the structural details of the nanopore according to the change in interaction characteristics during the entire translocation process. It provides us with a new strategy to identify and quantify the non-covalent interaction with high throughput and label-free merits. Furthermore, by clustering the frequency–energy eigenvalues of the dynamic interactions between biomolecules in different systems, combined with big-data analysis and machine learning, one can gradually establish a non-covalent interactions library with different types of dynamic interaction to understand the relationship between interaction and function and achieve accurate analysis of biomolecules. In addition to the fruitful intrinsic information of individual molecules hidden in the background noise, an efficient algorithm to analyze the interferent signals for a complex system is also urgently desired. For instance, there are almost identical signals in an aerolysin sensing system for the translocations of 5′-AAAA-3′(AA_3) and 5′-GAAA-3′(GA_3). In order to eliminate the overlap in the ion-current signal distribution, shapelets, a kind of primitive machine learning methodology, were recently developed to achieve the accurate discrimination of the complex sample containing AA_3 and GA_3. The shapelets used here are a kind of machine learning concept that relies on short time-series segments to predict the categories of targets. As a result, AA_3 and GA_3 in the mixture can be effectively identified by the time-series current that also facilitates the in-depth understanding of dynamic interactions between single biomolecule interface and analytes.[34]

It is foreseeable that every segment of the medical field, from diagnosis, monitoring and treatment to drug delivery, will fully open up to an era of intelligence: precision medicine. To meet the urgent demand for precision medicine, we can also expect a clever marriage between nanopore electrochemistry and artificial intelligence to achieve multiplex target sensing for precision medicine, including DNA sequencing and protein sequencing.[35,36] By using integrated life-group data to conduct risk assessments and preventive treatment of familial genetic diseases, accurately diagnose complex diseases caused by multiple factors and using the most effective drugs for treatment, the cure rate and survival rate for patients can be enhanced. The prognosis and quality of life will improve the health of people. In addition, the advent of protein sequencing *via* nanopore technology may facilitate the

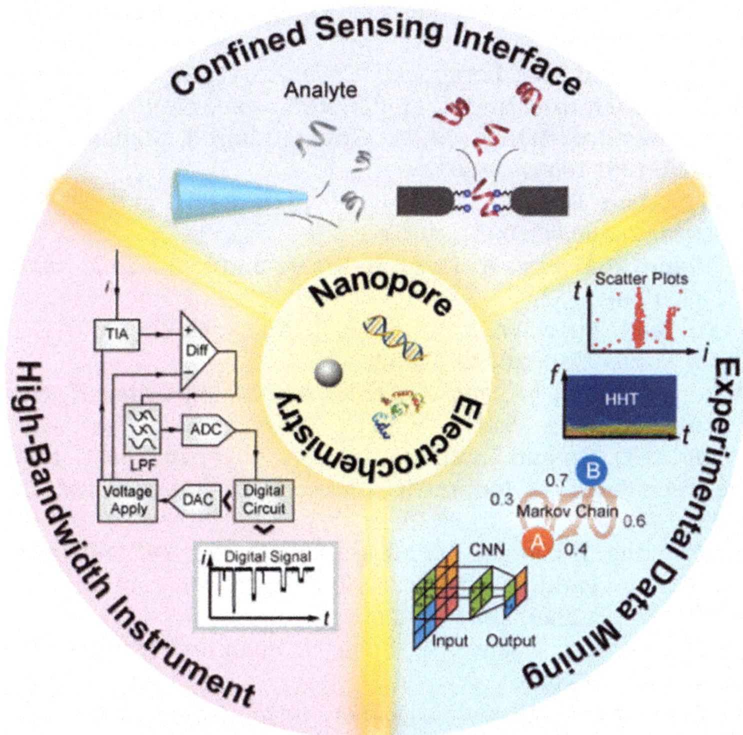

Figure 7.1 Nanopore electrochemistry for single entity sensing. Reprinted from ref. 37 with permission from American Chemical Society, Copyright 2020.

achievement of protein storage one day. Undoubtedly, we are at the point of booming electrochemically confined nanopore sensing and of collaboration in multi-disciplinary teams, including chemists, biologists, physicists and informationists, and this will open up new avenues for the next generation nanopore-based single entity measurement (Figure 7.1).

References

1. Y.-L. Ying, J. Wang, A. R. Leach, Y. Jiang, R. Gao, C. Xu, M. A. Edwards, A. D. Pendergast, H. Ren, C. K. T. Weatherly, W. Wang, P. Actis, L. Mao, H. S. White and Y.-T. Long, *Sci. China: Chem.*, 2020, **63**, 589–618.
2. J. J. Kasianowicz, E. Brandin, D. Branton and D. W. Deamer, *Proc. Natl. Acad. Sci. U. S. A.*, 1996, **93**, 13770–13773.
3. R. Stefureac, Y.-T. Long, H.-B. Kraatz, P. Howard and J. S. Lee, *Biochemistry*, 2006, **45**, 9172–9179.
4. C. Cao, Y. L. Ying, Z. L. Hu, D. F. Liao, H. Tian and Y. T. Long, *Nat. Nanotechnol.*, 2016, **11**, 713–718.

5. Y.-L. Ying, C. Cao, Y.-X. Hu and Y.-T. Long, *Natl. Sci. Rev.*, 2018, **5**, 450–452.
6. J. Yu, C. Cao and Y.-T. Long, *Anal. Chem.*, 2017, **89**, 11685–11689.
7. A. Zrehen, D. Huttner and A. Meller, *ACS Nano*, 2019, **13**, 14388–14398.
8. A. H. Squires, J. S. Hersey, M. W. Grinstaff and A. Meller, *J. Am. Chem. Soc.*, 2013, **135**, 16304–16307.
9. D. Al Sulaiman, P. Cadinu, A. P. Ivanov, J. B. Edel and S. Ladame, *Nano Lett.*, 2018, **18**, 6084–6093.
10. H. Y. Wang, Y. L. Ying, Y. Li, H. B. Kraatz and Y. T. Long, *Anal. Chem.*, 2011, **83**, 1746–1752.
11. R.-J. Yu, S.-M. Lu, S.-W. Xu, Y.-J. Li, Q. Xu, Y.-L. Ying and Y.-T. Long, *Chem. Sci.*, 2019, **10**, 10728–10732.
12. N. Varongchayakul, J. Song, A. Meller and M. W. Grinstaff, *Chem. Soc. Rev.*, 2018, **47**, 8512–8524.
13. Y. Wang, L. Q. Gu and K. Tian, *Nanoscale*, 2018, **10**, 13857–13866.
14. L. Restrepo-Perez, C. Joo and C. Dekker, *Nat. Nanotechnol.*, 2018, **13**, 786–796.
15. J. Houghtaling, J. List and M. Mayer, *Small*, 2018, **14**, e1802412.
16. T. Luchian, Y. Park, A. Asandei, I. Schiopu, L. Mereuta and A. Apetrei, *Acc. Chem. Res.*, 2019, **52**, 267–276.
17. W. Maulbetsch, B. Wiener, W. Poole, J. Bush and D. Stein, *Phys. Rev. Appl.*, 2016, **6**, 054006.
18. D. Verschueren, X. Shi and C. Dekker, *Small Methods*, 2019, **3**, 1800465.
19. K. Fu and P. W. Bohn, *ACS Cent. Sci.*, 2018, **4**, 20–29.
20. M. Zhou, Y. Yu, K. Hu, H. L. Xin and M. V. Mirkin, *Anal. Chem.*, 2017, **89**, 2880–2885.
21. S. Chen, Y. Liu and J. Chen, *Chem. Soc. Rev.*, 2014, **43**, 5372–5386.
22. Y. L. Ying, Z. Ding, D. Zhan and Y. T. Long, *Chem. Sci.*, 2017, **8**, 3338–3348.
23. X. Shan, I. Diez-Perez, L. Wang, P. Wiktor, Y. Gu, L. Zhang, W. Wang, J. Lu, S. Wang, Q. Gong, J. Li and N. Tao, *Nat. Nanotechnol.*, 2012, **7**, 668–672.
24. N. T. N. Phan, X. Li and A. G. Ewing, *Nat. Rev. Chem.*, 2017, **1**, 0048.
25. R. Gao, Y. L. Ying, Y. X. Hu, Y. J. Li and Y. T. Long, *Anal. Chem.*, 2017, **89**, 7382–7387.
26. Y. X. Hu, Y. L. Ying, R. Gao, R. J. Yu and Y. T. Long, *Anal. Chem.*, 2018, **90**, 12352–12355.
27. R.-C. Qian, J. Lv and Y.-T. Long, *Anal. Chem.*, 2018, **90**, 13744–13750.
28. B. P. Nadappuram, P. Cadinu, A. Barik, A. J. Ainscough, M. J. Devine, M. Kang, J. Gonzalez-Garcia, J. T. Kittler, K. R. Willison, R. Vilar, P. Actis, B. Wojciak-Stothard, S.-H. Oh, A. P. Ivanov and J. B. Edel, *Nat. Nanotechnol.*, 2019, **14**, 80–88.
29. M. Aramesh, C. Forró, L. Dorwling-Carter, I. Lüchtefeld, T. Schlotter, S. J. Ihle, I. Shorubalko, V. Hosseini, D. Momotenko, T. Zambelli, E. Klotzsch and J. Vörös, *Nat. Nanotechnol.*, 2019, **14**, 791–798.
30. H. Li, X. Liu, L. Li, X. Mu, R. Genov and A. J. Mason, *Sensors*, 2017, **17**, 74.

31. S. Fouladi and R. R. Mansour, *IEEE Trans. Microwave Theory Tech.*, 2010, **58**, 478–486.
32. Z. Gu, Y. L. Ying, C. Cao, P. G. He and Y. T. Long, *Anal. Chem.*, 2015, **87**, 907–913.
33. S. C. Liu, M. X. Li, M. Y. Li, Y. Q. Wang, Y. L. Ying, Y. J. Wan and Y. T. Long, *Faraday Discuss.*, 2018, **210**, 87–99.
34. Z.-X. Wei, Y.-L. Ying, M.-Y. Li, J. Yang, J.-L. Zhou, H.-F. Wang, B. Yan and Y.-T. Long, *Anal. Chem.*, 2019, **91**, 10033–10039.
35. K. Misiunas, N. Ermann and U. F. Keyser, *Nano Lett.*, 2018, **18**, 4040–4045.
36. K. K. Chen, J. B. Zhu, F. Bošković and U. F. Keyser, *Nano Lett.*, 2020, **20**, 3754–3760.
37. S. M. Lu, Y. Y. Peng, Y. L. Ying and Y. T. Long, *Anal. Chem.*, 2020, **92**, 5621–5644.
38. A. Oukhale, M. Pastoriza-Gallego, L. Bacri, J. Mathé, L. Auvray and J. Pelta, *Protein Pept. Lett.*, 2014, **21**(3), 266–274.
39. K. J. Freedman, S. R. Haq, J. B. Edel, P. Jemth and M. J. Kim, *Sci. Rep.*, 2013, **3**, 1638.

Subject Index